Modeling and Computation in Environmental Sciences

Edited by
Rainer Helmig
Willi Jäger
Wolfgang Kinzelbach
Peter Knabner
and Gabriel Wittum

Notes on Numerical Fluid Mechanics (NNFM) Volume 59

Volume 58 ECARP – European Computational Aerodynamics Research Project: Validation of CFD Codes and Assessment of Turbulence Models (W. Haase / E. Chaput / E. Elsholz / M. A. Leschziner / U. R. Müller, Eds.)

Volume 57 Euler and Navier-Stokes Solvers Using Multi-Dimensional Upwind Schemes and Multigrid Acceleration. Results of the BRITE/EURAM Projects AERO-CT89-0003 and AER2-CT92-00040, 1989–1995 (H. Deconinck / B. Koren, Eds.)

Volume 55 EUROPT – A European Initiative on Optimum Design Methods in Aerodynamics. Proceedings of the Brite/Euram Project Workshop „Optimum Design in Aerodynamics", Barcelona, 1992 (J. Periaux / G. Bugeda / P. K. Chaviaropoulos / T. Labrujere / B. Stoufflet, Eds.)

Volume 54 Boundary Elements: Implementation and Analysis of Advanced Algorithms. Proceedings of the Twelfth GAMM-Seminar, Kiel, January 19–21, 1996 (W. Hackbusch / G. Wittum, Eds.)

Volume 53 Computation of Three-Dimensional Complex Flows. Proceedings of the IMACS-COST Conference on Computational Fluid Dynamics, Lausanne, September 13–15, 1995 (M. Deville / S. Gavrilakis / I. L. Ryhming, Eds.)

Volume 52 Flow Simulation with High-Performance Computers II. DFG Priority Research Programme Results 1993–1995 (E. H. Hirschel, Ed.)

Volumes 1 to 51 are out of print.
The addresses of the Editors are listed at the end of the book.

Modeling and Computation in Environmental Sciences

Proceedings of the First GAMM-Seminar at ICA Stuttgart, October 12–13, 1995

Edited by
Rainer Helmig
Willi Jäger
Wolfgang Kinzelbach
Peter Knabner
and Gabriel Wittum

vieweg

Softcover reprint of the hardcover 1st edition 1997

Vieweg ist a subsidiary company of Bertelsmann Professional Information.

http://www.vieweg.de

Produced by Langelüddecke, Braunschweig
Printed on acid-free paper

ISSN 0179-9614
ISBN 978-3-322-89567-7 ISBN 978-3-322-89565-3 (eBook)
DOI 10.1007/978-3-322-89565-3

Preface

The GAMM Committee for Scientific Computing organizes workshops on subjects in Scientific Computing. These workshops are intended to bring researchers from engineering application and mathematical theory together, to provide a platform for discussion and scientific exchange and to further new research fields.

The series of such workshops was continued in 1995, October 12-13, with the 1st GAMM-Seminar at ICA Stuttgart on the special topic

"Modelling and Computation
in Environmental Sciences"

at the Institute for Computer Applications, Universitiy of Stuttgart. The seminar was attended by 75 scientists from 5 countries and 26 lectures were given.

The list of topics contained lectures on ground water and soil water flow and transport, heterogeneity, density driven ground water flow, homogenization and multi-scale modelling, special discretization schemes, adaptivity, multi-grid methods, and parameter identification.

Special thanks are due to Prudence Lawday, Oktavia Klassen and Gerd Griesheimer, who carefully compiled the contributions to this volume.

July 1996

Rainer Helmig
Willi Jäger
Wolfgang Kinzelbach
Peter Knabner
Gabriel Wittum

Contents

Page

Page

Contents (continued)

THE APPLICATION OF MIXED METHODS TO SUBSURFACE SIMULATION

Todd Arbogast, Clint N. Dawson, Philip T. Keenan,
Mary F. Wheeler, and Ivan Yotov

Texas Institute for Computational and Applied Mathematics
Center for Subsurface Modeling, Taylor Hall 2.400
The University of Texas at Austin, Austin, Texas 78712, U.S.A.

SUMMARY

We consider the application of mixed finite element and finite difference methods to groundwater flow and transport problems. We are concerned with accurate approximation and efficient implementation, especially when the porous medium may have geometric irregularities, heterogeneities, and either a tensor hydraulic conductivity or a tensor dispersion. For single-phase flow, we develop an expanded mixed finite element method defined on a logically rectangular, curvilinear grid. Special quadrature rules are introduced to transform the method into a simple cell-centered finite difference method. The approximation is locally conservative and highly accurate. We also show that the highly nonlinear two-phase flow problem is well approximated by mixed methods. The main difficulty is that the true solution is typically lacking in regularity.

INTRODUCTION

Our primary goal is to develop discretization methods that accurately and efficiently approximate the equations governing subsurface multi-phase flow and transport. We can judge the accuracy of an approximation by many criteria. Asymptotic convergence results tell us that we have an accurate solution when the mesh spacing h is *small enough.* Often, we cannot use as fine a mesh resolution as we would like, because of the computational effort needed to solve the equations. An equally important criterion to consider is the ability of the numerical scheme to preserve important qualitative properties of the governing equations so that physically meaningful results are obtained on a relatively coarse discretization scale. The most important qualitative property in subsurface simulation is conservation of mass. Mass should be conserved locally, that is, element-by-element.

Several additional physical phenomena need to be addressed by our numerical schemes. They should handle tensor permeabilities and dispersivities. Dispersivities are naturally tensors, and tensor permeabilities can arise from the use of homogenization or scale-up techniques. Subsurface aquifers are irregularly shaped and contain layers with differing material properties. Nonlinear effects are also prevalent especially in multi-phase flow.

We present here some of our work on mixed finite element and finite difference methods [9, 6, 10, 5, 8, 4, 7]. These methods are "mixed" in that they approximate directly *both* pressure and velocity (in the flow problem), and they are asymptotically accurate and conserve mass locally. The standard mixed finite element method was developed by Raviart and Thomas [19, 21, 12], and we restrict our attention to their lowest-order method. It was first used to solve subsurface problems by Douglas, Ewing, and Wheeler [14], although Russell and Wheeler [20] pointed out that the often used cell-centered finite difference method on rectangular grids [18] for problems with diagonal permeabilities is the lowest order Raviart-Thomas mixed finite element method approximated by applying appropriate quadrature rules to some of the integrals.

A problem with mixed methods that we address is that they can be difficult to implement directly, especially if the aquifer domain is not rectangular or the permeability is a tensor. There has also been very little theoretical basis for concluding that the approximation of highly nonlinear multi-phase problems is accurate.

Our discretization schemes are based on an expanded mixed finite element method that we define below. An approximation to this expanded mixed method reduces it to cell-centered finite differences; thus, it is easy to implement and has only one unknown per element. The elements can be deformed rectangles or bricks, although many of our results extend to triangles and tetrahedra [6].

FINITE ELEMENT APPROXIMATION OF SINGLE-PHASE FLOW

To illustrate the numerical schemes, we begin by considering incompressible, single phase subsurface flow on the aquifer domain $\Omega \subset \mathrm{R}^d$, $d = 2$ or 3. We solve for the pressure p and the velocity $\mathbf{u}$ satisfying

$$\mathbf{u} = -\mathbf{K}\nabla p, \quad \mathbf{x} \in \Omega, \tag{1}$$

$$\nabla \cdot \mathbf{u} = q, \quad \mathbf{x} \in \Omega, \tag{2}$$

$$p = p_0, \quad \mathbf{x} \in \partial\Omega_D, \tag{3}$$

$$\mathbf{u} \cdot \nu = g, \quad \mathbf{x} \in \partial\Omega_N, \tag{4}$$

where $\mathbf{K}$ is the hydraulic conductivity tensor, q is a source term, ν is the outer unit normal vector to $\partial\Omega$, p_0 gives a Dirichlet boundary condition on $\partial\Omega_D$, and g gives a Neumann condition. This is a second order elliptic equation.

Lowest order Raviart-Thomas spaces.

Let $L^2(\Omega)$ denote the space of square integrable functions, and let $H(\Omega; \mathrm{div})$ denote the space of vector functions that have a divergence; that is,

$$L^2(\Omega) = \left\{ w(\mathbf{x}) : \int_\Omega |w|^2 \, dx < \infty \right\},$$

$$H(\Omega; \mathrm{div}) = \{ \mathbf{v}(\mathbf{x}) : \mathbf{v} \in (L^2(\Omega))^d \text{ and } \nabla \cdot \mathbf{v} \in L^2(\Omega) \}.$$

We suppose that the domain Ω is partitioned into a finite number of non-overlapping elements or cells E of maximal diameter h. In the lowest order Raviart-Thomas mixed spaces [19, 16, 11], pressures can be approximated over elements or on element faces (or edges in 2-D). Element pressures are approximated in

$$W_h = \{ w : w \text{ is constant on each element} \} \subset L^2(\Omega),$$

and, on the Neumann part of the exterior domain boundary, element face "Lagrange multiplier" pressures [11] are approximated in

$$\Lambda_h^N = \{ \mu : \mu \text{ is constant on each element face of } \partial\Omega_N \} \subset L^2(\partial\Omega_N).$$

The nodal degrees of freedom can be considered as the function values at the centers of the elements or faces.

The velocity $\mathbf{u}$ is approximated in a space of vector valued functions $\mathbf{V}_h$ such that

$$\mathbf{V}_h \subset \{\mathbf{v} \in (L^2(\Omega))^d : \mathbf{v} \cdot \nu \text{ is constant on each element face and continuous across elements}\} \subset H(\Omega; \text{div}).$$

On a 2-D (or 3-D) rectangle E, this space of functions is

$$\mathbf{V}_h|_E = \{\mathbf{v} : \mathbf{v}_i = a_i + b_i x_i \text{ for some constants } a_i \text{ and } b_i,\ i = 1, 2(,3)\},$$

where we use the standard Cartesian decomposition of the vectors $\mathbf{x} = (x_1, x_2(, x_3))$ and $\mathbf{v} = (v_1, v_2(, v_3))$; that is, the ith component of $\mathbf{v}$ is linear in the ith coordinate direction and constant in the other direction(s). The important fact is that $\mathbf{v} \cdot \nu$ is a constant; therefore, the nodal degrees of freedom can be considered as the values of $\mathbf{v} \cdot \nu$ at the centers of the element faces.

For a relatively general shaped element E, assume that there is a map $F : \hat{E} \to E$ from a rectangle or brick $\hat{E}$ to E. Following Thomas [21], we use the Piola transform to define $\mathbf{V}_h|_E$ from $\hat{\mathbf{V}}_h|_{\hat{E}}$; this transform preserves normal fluxes in an average sense (i.e., it is locally mass conservative). Let the Jacobian matrix be $DF = (\partial F_i / \partial x_j)$. Then

$$\mathbf{v}(\mathbf{x}) = \frac{1}{J} DF\, \hat{\mathbf{v}}(\hat{\mathbf{x}}), \tag{5}$$

where $J = |\det(DF)|$.

The expanded mixed method.

Unlike the standard mixed method, we introduce a symmetric and positive definite tensor $\mathbf{G}$ and define the "adjusted" pressure gradient $\tilde{\mathbf{u}}$ by

$$\mathbf{G}\tilde{\mathbf{u}} = -\nabla p. \tag{6}$$

Then the system of equations is

$$\mathbf{K}\mathbf{G}\tilde{\mathbf{u}} - \mathbf{u} = 0, \tag{7}$$

$$\mathbf{G}\tilde{\mathbf{u}} + \nabla p = 0, \tag{8}$$

$$\nabla \cdot \mathbf{u} = q. \tag{9}$$

Denote inner-products over a set S by

$$(\varphi, \psi)_S = \int_S \varphi(\mathbf{x})\, \psi(\mathbf{x})\, d\mathbf{x} \quad \left(\text{or } \int_S \varphi(\mathbf{x}) \cdot \psi(\mathbf{x})\, d\mathbf{x}\right),$$

and inner-products over a boundary set ∂S by

$$\langle \varphi, \psi \rangle_S = \int_{\partial S} \varphi(\mathbf{x})\, \psi(\mathbf{x})\, da(\mathbf{x}),$$

where S is omitted if $S = \Omega$. The expanded mixed finite element method is then: Find $\mathbf{u} \in \mathbf{V}_h$, $\tilde{\mathbf{u}} \in \mathbf{V}_h$, $p \in W_h$, and $\lambda \in \Lambda_h^N$ such that

$$(\mathbf{G}\mathbf{K}\mathbf{G}\tilde{\mathbf{u}}, \mathbf{v}) - (\mathbf{G}\mathbf{u}, \mathbf{v}) = 0 \quad \text{for all } \mathbf{v} \in \mathbf{V}_h, \tag{10}$$

$$(\mathbf{G}\tilde{\mathbf{u}}, \mathbf{v}) - (p, \nabla \cdot \mathbf{v}) = -\langle p_0, \mathbf{v} \cdot \nu \rangle_{\partial\Omega_D} - \langle \lambda, \mathbf{v} \cdot \nu \rangle_{\partial\Omega_N} \quad \text{for all } \mathbf{v} \in \mathbf{V}_h, \tag{11}$$

$$(\nabla \cdot \mathbf{u}, w) = (q, w) \quad \text{for all } w \in W_h, \tag{12}$$

$$\langle \mathbf{u} \cdot \nu, \mu \rangle_{\partial\Omega_N} = \langle g, \mu \rangle_{\partial\Omega_N} \quad \text{for all } \mu \in \Lambda_h. \tag{13}$$

We remark that if $\mathbf{G} = \mathbf{K}^{-1}$, then $\dot{\mathbf{u}} = \mathbf{u}$ and we recover the standard mixed method [19, 21, 12]. If $\mathbf{G} = \mathbf{I}$, we recover the expanded mixed method considered in [24, 15, 13, 8, 9]. Later we will make a special choice of $\mathbf{G}$.

The algebraic system of equations that results is a symmetric saddle point linear system of the form

$$\begin{pmatrix} M & -A & 0 & 0 \\ -A^T & 0 & B & -L \\ 0 & B^T & 0 & 0 \\ 0 & -L^T & 0 & 0 \end{pmatrix} \begin{pmatrix} \dot{U} \\ U \\ P \\ \Lambda \end{pmatrix} = \begin{pmatrix} 0 \\ P_0 \\ Q \\ -G \end{pmatrix}, \tag{14}$$

where we represent $\mathbf{u}$ by U, $\dot{\mathbf{u}}$ by $\dot{U}$, p by P, and λ by Λ in the nodal bases $\{\mathbf{v}_i\}$ for $\mathbf{V}_h$, $\{w_i\}$ for W_h, and $\{\mu_i\}$ for Λ_h. In particular,

$$M_{ij} = (\mathbf{GKGv}_i, \mathbf{v}_j) \quad \text{and} \quad A_{ij} = (\mathbf{Gv}_i, \mathbf{v}_j) \tag{15}$$

are symmetric and positive definite. To reduce the size of the linear system, we can solve for the Shur complement by eliminating

$$U = A^{-1}M\dot{U} \quad \text{and} \quad \dot{U} = A^{-1}(BP - L\Lambda - P_0) \tag{16}$$

to obtain

$$(\mathcal{B}A^{-1}MA^{-1}\mathcal{B}^T)\begin{pmatrix} P \\ \Lambda \end{pmatrix} = \begin{pmatrix} Q \\ -G \end{pmatrix} + (\mathcal{B}A^{-1}MA^{-1})P_0, \tag{17}$$

where $\mathcal{B} = \begin{pmatrix} B \\ L \end{pmatrix}$. This system is symmetric, positive definite, and relatively small (one unknown per element plus a few boundary nodes). Unfortunately, although A is sparse, A^{-1} is in general full. Iterative solution will require the following steps for the application of the matrix: a matrix vector multiply $x = \mathcal{B}^T \begin{pmatrix} P \\ \Lambda \end{pmatrix}$; the solution of the system $Ay = x$; and another matrix vector multiply $\mathcal{B}y$. Thus, we need inner iterations within our overall iterative solution, which can become somewhat expensive.

CELL-CENTERED FINITE DIFFERENCE APPROXIMATION

We now use approximate integration to reduce A in (17) to an easily inverted diagonal matrix.

The rectangular cell-centered finite difference approximation.

Assume in this subsection that the grid is rectangular. Take $\mathbf{G} = I$ and use the trapezoidal quadrature rule to approximate the first three integrals (i.e., those involving a vector-vector product) in (10)–(13). This diagonalizes the matrix A and the Shur complement system becomes sparse [20, 23], even when $\mathbf{K}$ is a tensor [8, 9].

It is easy to unravel the procedure in terms of the nodal degrees of freedom of $\mathbf{u}$, $\tilde{\mathbf{u}}$, p, and λ. Consider an element E (not adjacent to the outer boundary). Equation (12) requires that the divergence of $\mathbf{u}$ be set equal to the source term q. This involves differences of the normal velocities that live on the four edges or six faces of the element. Equation (10) relates the velocities to the gradients of pressure. The velocity $\mathbf{u}$ on a given edge or face is related to the gradient $\tilde{\mathbf{u}}$ that lives on the given edge or face and to those that live on the adjacent but perpendicular four edges or eight faces (if $\mathbf{K}$ is not diagonal). Finally, (11) relates $\tilde{\mathbf{u}}$ living on an edge or face to the difference of the adjacent pressures. Combining this together, we get a 9 point stencil for the pressure on E if $d = 2$, and 19 points if $d = 3$. More details are given later and also in [9].

The Geometry Mapping.

To handle irregular geometry, we assume that there is a smooth mapping F of a rectangular, computational domain $\hat{\Omega}$ onto the domain Ω. Given a rectangular grid on $\hat{\Omega}$, F defines a smooth, logically rectangular, curvilinear grid on Ω. (In practice, there are grid generation codes available for creating F at the grid points. We use finite differences to approximate DF.)

The Transformed, Computational Problem.

In the expanded mixed method, take

$$\mathbf{G} = J(DF^{-1})^T DF^{-1}. \tag{18}$$

Transform (10)–(13) to the computational domain $\hat{\Omega}$. Vector and scalar basis functions transform by the Piola and natural transforms, respectively; that is,

$$\begin{aligned} \mathbf{v}(\mathbf{x}) &= \frac{1}{J(\hat{\mathbf{x}})} DF(\hat{\mathbf{x}})\hat{\mathbf{v}}(\hat{\mathbf{x}}) \qquad &&\text{(velocity)},\\ w(\mathbf{x}) &= \hat{w}(\hat{\mathbf{x}}) &&\text{(pressure)}. \end{aligned}$$

Thus we need to find $\hat{\mathbf{u}} \in \hat{\mathbf{V}}_h$, $\hat{\tilde{\mathbf{u}}} \in \hat{\mathbf{V}}_h$, $\hat{p} \in \hat{W}_h$, and $\hat{\lambda} \in \hat{\Lambda}_h$ such that

$$(\mathcal{K}\hat{\tilde{\mathbf{u}}}, \hat{\mathbf{v}}) = (\hat{\mathbf{u}}, \hat{\mathbf{v}}) \quad \text{for all } \hat{\mathbf{v}} \in \hat{\mathbf{V}}_h, \tag{19}$$

$$(\hat{\tilde{\mathbf{u}}}, \mathbf{v}) - (\hat{p}, \hat{\nabla} \cdot \hat{\mathbf{v}}) = -\langle \hat{p}_0, \hat{\mathbf{v}} \cdot \hat{\nu} \rangle_{\partial\hat{\Omega}_D} - \langle \hat{\lambda}, \hat{\mathbf{v}} \cdot \hat{\nu} \rangle_{\partial\hat{\Omega}_N} \quad \text{for all } \hat{\mathbf{v}} \in \hat{\mathbf{V}}_h, \tag{20}$$

$$(\hat{\nabla} \cdot \hat{\mathbf{u}}, \hat{w}) = (\hat{q}J, \hat{w}) \quad \text{for all } \hat{w} \in \hat{W}_h, \tag{21}$$

$$\langle \hat{\mathbf{u}} \cdot \hat{\nu}, \hat{\mu} \rangle_{\partial\hat{\Omega}_N} = \langle \hat{g}J_{\hat{\nu}}, \mu \rangle_{\partial\Omega_N} \quad \text{for all } \hat{\mu} \in \hat{\Lambda}_h, \tag{22}$$

where $\mathcal{K} = JDF^{-1}\mathbf{K}(DF^{-1})^T$. This is the discrete problem in $\hat{\Omega}$:

$$\hat{\mathbf{u}} = -\mathcal{K}\hat{\nabla}p, \quad \hat{\mathbf{x}} \in \hat{\Omega}, \tag{23}$$

$$\hat{\nabla} \cdot \hat{\mathbf{u}} = \hat{q}J, \quad \hat{\mathbf{x}} \in \hat{\Omega}, \tag{24}$$

$$\hat{p} = \hat{p}_0, \quad \hat{\mathbf{x}} \in \partial\hat{\Omega}_D, \tag{25}$$

$$\hat{\mathbf{u}} \cdot \hat{\nu} = \hat{g}J_{\hat{\nu}}, \quad \hat{\mathbf{x}} \in \partial\hat{\Omega}_N. \tag{26}$$

All computations are performed on the rectangular grid of $\hat{\Omega}$ after preprocessing the coefficients: $\mathbf{K}$ becomes $\mathcal{K}$, q becomes $\hat{q}J$, and g becomes $\hat{g}J_{\hat{\nu}}$.

The logically rectangular Cell-Centered Finite Difference Approximation.

To problem (19)–(22), we use the trapezoidal quadrature rule for approximating the three integrals involving a vector-vector product to obtain our cell-centered finite difference method on the logically rectangular mesh. As an illustration, consider a 2-D, uniform grid, with a constant $\mathcal{K}$. Denote the grid points and cell centers by

$$(\hat{x}_{i+1/2}, \hat{y}_{j+1/2}) \quad \text{and} \quad (\hat{x}_i, \hat{y}_j).$$

Then $\hat{\mathbf{u}} = -\hat{\nabla}\hat{p}$ is

$$\hat{\tilde{u}}^{\hat{x}}_{i+1/2,j} = -\frac{\hat{p}_{i+1,j} - \hat{p}_{i,j}}{\hat{h}}, \qquad \hat{\tilde{u}}^{\hat{y}}_{i,j+1/2} = -\frac{\hat{p}_{i,j+1} - \hat{p}_{i,j}}{\hat{h}}, \tag{27}$$

and $\hat{\mathbf{u}} = \mathcal{K}\hat{\tilde{\mathbf{u}}}$ is

$$\hat{u}^{\hat{x}}_{i+1/2,j} = \mathcal{K}_{11}\hat{\tilde{u}}^{\hat{x}}_{i+1/2,j} + \frac{\mathcal{K}_{12}}{4}\{\hat{\tilde{u}}^{\hat{y}}_{i+1,j-1/2} + \hat{\tilde{u}}^{\hat{y}}_{i+1,j+1/2} + \hat{\tilde{u}}^{\hat{y}}_{i,j-1/2} + \hat{\tilde{u}}^{\hat{y}}_{i,j+1/2}\}, \tag{28}$$

with a similar expression for $\hat{u}^{\hat{x}}_{i,j+1/2}$. Finally, for each element E_{ij}, $\hat{\nabla} \cdot \hat{\mathbf{u}} = \hat{q}J$ is

$$\left[\frac{\hat{u}^{\hat{x}}_{i+1/2,j} - \hat{u}^{\hat{x}}_{i-1/2,j}}{\hat{h}} + \frac{\hat{u}^{\hat{y}}_{i,j+1/2} - \hat{u}^{\hat{y}}_{i,j-1/2}}{\hat{h}}\right]\hat{h}^2 = \int_{\hat{E}_{ij}} \hat{q}J\, d\hat{x}. \tag{29}$$

The solution $\mathbf{u}$ on Ω is obtained from $p = \hat{p}$ and $\lambda = \hat{\lambda}$ using (16).

CONVERGENCE RESULTS

Let $\|\cdot\|$ denote the L^2-norm, $\|\varphi\|^2 = \int_\Omega |\varphi(\mathbf{x})|^2\, dx$, and let $|||\cdot|||_M$ denote the L^2-norm approximated by the midpoint quadrature rule. Before stating our result that the scheme is optimally convergent, we need the following definition.

Definition: An asymptotic family of grids is said to be generated by a C^2 map if each grid is an image by a fixed map of a grid that is uniform in each coordinate direction. Each component of the map must be strictly monotone and in $C^2(\bar{\Omega})$.

Theorem 1: There exists a constant C depending on the smoothness of F, $\mathbf{K}$, and the solution, but independent of the maximum grid spacing h, such that the cell-centered finite difference approximation satisfies

$$\begin{aligned}
&\|p_{\mathrm{true}} - p_{\mathrm{approx}}\| \le Ch,\\
&\|\mathbf{u}_{\mathrm{true}} - \mathbf{u}_{\mathrm{approx}}\| + \|\tilde{\mathbf{u}}_{\mathrm{true}} - \tilde{\mathbf{u}}_{\mathrm{approx}}\| \le Ch,\\
&\|\nabla \cdot (\mathbf{u}_{\mathrm{true}} - \mathbf{u}_{\mathrm{approx}})\| \le Ch.
\end{aligned}$$

Moreover,

$$\begin{aligned}
&|||p_{\mathrm{true}} - p_{\mathrm{approx}}|||_M \le Ch^2,\\
&|||\mathbf{u}_{\mathrm{true}} - \mathbf{u}_{\mathrm{approx}}|||_M + |||\tilde{\mathbf{u}}_{\mathrm{true}} - \tilde{\mathbf{u}}_{\mathrm{approx}}|||_M \le Ch^r,
\end{aligned}$$

where

$$r = \begin{cases} 2 & \text{if } K \text{ is diagonal and } \partial\Omega_N = \partial\Omega,\\ 3/2 & \text{if } K \text{ is diagonal or the grids are generated by a } C^2 \text{ map},\\ 1 & \text{otherwise.} \end{cases}$$

When K is diagonal or the grids are generated by a C^2 map, a half power of h is lost in the super-convergence result for the velocity. This is due strictly to effects near the boundary of the domain.

Theorem 2 (Interior estimates): Let Ω' be compactly contained in Ω, and suppose that either K is diagonal or the grids are generated by a C^2 map. For any $\epsilon > 0$, there exists a constant C_ϵ depending on the smoothness of F, $\mathbf{K}$, and the solution, but independent of the maximum grid spacing h, such that

$$|||\mathbf{u}_{\text{true}} - \mathbf{u}_{\text{approx}}|||_{M,\Omega'} \leq C_\epsilon h^{2-\epsilon}.$$

The proofs of these results can be found in [6] (see also [23, 9]). These results are sharp in the sense that they are seen computationally in practical settings [9, 6].

DISCONTINUOUS MATERIAL PROPERTIES AND THE ENHANCED CELL-CENTERED FINITE DIFFERENCE METHOD

We show by example that our cell-centered scheme has difficulties approximating the solution when the material properties are discontinuous. On the unit square, let the true solution p and $\mathbf{K}$ be

$$p(x,y) = \begin{cases} xy & \text{for } x \leq 1/2, \\ xy + (x-1/2)(y+1/2) & \text{for } x > 1/2, \end{cases} \tag{30}$$

$$\mathbf{K}(x,y) = \begin{cases} \begin{pmatrix} 2 & 1 \\ 1 & 2 \end{pmatrix} & \text{for } x < 1/2, \\ \begin{pmatrix} 1 & 0 \\ 0 & 1 \end{pmatrix} & \text{for } x > 1/2. \end{cases} \tag{31}$$

Note that the eigenvectors for $\mathbf{K}$ are at 45 degrees to the grid for $x < 1/2$. Computationally, we see the following convergence rates:

$$|||p_{\text{true}} - p_{\text{approx}}|||_M \leq 0.06\, h^{0.96},$$
$$|||\mathbf{u}_{\text{true}} - \mathbf{u}_{\text{approx}}|||_M \leq 0.23\, h^{0.50}.$$

These are much worse that predicted by Theorem 1. The error is concentrated along the line $x = 1/2$.

If $\mathbf{K}$ or the map F is not smooth along an interface, then $\mathbf{u} = \mathbf{K}\mathbf{G}\tilde{\mathbf{u}}$ but not $\tilde{\mathbf{u}}$ is continuous in the normal direction. However, we have approximated $\tilde{\mathbf{u}}$ in the same space as $\mathbf{u}$, i.e., $\mathbf{V}_h$, which has continuous normal components. We must relax this continuity along any interface where the material properties change discontinuously. To do so for the approximation of $\tilde{\mathbf{u}}$ but not $\mathbf{u}$ would make the matrix A in (14) non-square and therefore not fully invertible. We therefore need to relax the continuity of $\mathbf{V}_h$ for both $\tilde{\mathbf{u}}$ and $\mathbf{u}$.

This idea originates in the hybrid form of the mixed method of Arnold and Brezzi [11]. Introduce Lagrange multiplier pressures living on the element edges or faces of the discontinuity interface Γ (we already have them on $\partial\Omega_N$), and add a condition that specifies that $\mathbf{u}$ alone is continuous across Γ. That is, replace (11) in (10)–(13) and add (33):

$$(\mathbf{G}\tilde{\mathbf{u}}, \mathbf{v}) - \sum_E (p, \nabla\cdot\mathbf{v})_E = -\langle p_0, \mathbf{v}\cdot\nu\rangle_{\partial\Omega_D} - \sum_E \langle\lambda, \mathbf{v}\cdot\nu\rangle_{\partial E\cap(\partial\Omega_N\cup\Gamma)} \quad \text{for all } \mathbf{v}\in\mathbf{V}_h, \tag{32}$$

$$\sum_E \langle\mathbf{u}\cdot\nu, \mu\rangle_{\partial E\cap\Gamma} = 0 \quad \text{for all } \mu\in\Lambda_h. \tag{33}$$

We call this scheme the enhanced cell-centered finite difference scheme. It can be formulated without explicit reference to Lagrange pressures by taking a mesh with zero width cells on Γ. The infinitely thin cell's pressure is the Lagrange pressure [22].

Using the enhanced method, the same computational example shows the following convergence rates:

$$|||p_{\text{true}} - p_{\text{approx}}|||_M \leq 0.18\, h^{2.02},$$
$$|||\mathbf{u}_{\text{true}} - \mathbf{u}_{\text{approx}}|||_M \leq 0.10\, h^{1.49}.$$

The Hybrid Form of the Mixed Method.

It should be noted that the hybrid form of the mixed method uses Lagrange multiplier pressures along all faces. By eliminating all but the Lagrange pressures leads to a Shur complement system that is sparse, symmetric, and positive definite. Unfortunately, there are more unknowns than the number of elements. In 2-D, if deformed rectangles are used, there are two times as many unknowns. In 3-D, deformed brick elements need three times as many unknowns.

TWO-PHASE FLOW

We consider the accuracy of mixed finite element methods for approximating the highly nonlinear problem of two-phase flow of incompressible water and air (or oil or a NAPL). This represents work of two of the authors and Nai-Ying Zhang [10]. The governing equations are

$$\phi\frac{\partial s}{\partial t} - \nabla \cdot [K\lambda_w(s)\nabla p_w] = q_w(s), \tag{34}$$

$$-\phi\frac{\partial s}{\partial t} - \nabla \cdot [K\lambda_a(s)\nabla p_a] = q_a(s), \tag{35}$$

$$p_c(s) = p_a - p_w, \tag{36}$$

where $0 \leq s(\mathbf{x},t) \leq 1$ is the (normalized) wetting fluid saturation, ϕ is the porosity, $\lambda_\alpha = k_{r\alpha}(s)/\mu_\alpha$ is the relative mobility of phase $\alpha = w, a$, and p_c is the capillary pressure.

Define total velocity

$$\mathbf{v} = -K\lambda_w\nabla p_w - K\lambda_a\nabla p_a. \tag{37}$$

By rearranging (34)–(35) and using (36), we can obtain the pressure equation

$$\nabla \cdot \mathbf{v} = -\nabla \cdot (K(\lambda_w + \lambda_a)\nabla p_w + K\lambda_a\nabla p_c) = q_w + q_a \tag{38}$$

and the saturation equation

$$\phi\frac{\partial s}{\partial t} + \nabla \cdot \left(K\frac{\lambda_w\lambda_a}{\lambda_w + \lambda_a}\nabla p_c + \frac{\lambda_w}{\lambda_w + \lambda_a}\mathbf{v}\right) = q_w. \tag{39}$$

The pressure equation is a well behaved elliptic equation; however, the saturation equation is degenerate parabolic, so concentrate on it.

Kirchhoff Transformation.

Define the Kirchhoff Transformation

$$P(s) = -\int_0^s \left(\frac{\lambda_w \lambda_a}{\lambda_w + \lambda_a} p_c'\right)(\sigma)\, d\sigma. \tag{40}$$

Then $\nabla P(s) = -\dfrac{\lambda_w \lambda_a}{\lambda_w + \lambda_a} \nabla p_c$, and the saturation equation takes the form

$$\phi \frac{\partial s}{\partial t} - \nabla \cdot [\alpha \nabla P(s) + \beta(P(s))] = \gamma(P(s)). \tag{41}$$

Regularity (Smoothness) of the Solution.

We assume that $P(s)$ is strictly monotone increasing, that $P'(s)$ may be zero (degenerate parabolic), but there is a constant $C_0 > 0$ such that

$$\|P(\varphi_1) - P(\varphi_2)\|^2 \le C_0 (P(\varphi_1) - P(\varphi_2), \varphi_1 - \varphi_2).$$

We also assume that β and γ are Lipschitz continuous. These assumptions can be justified on physical grounds [10, 3].

Introduce a new variable

$$\psi = -\alpha \nabla P(s) - \beta(P(s)). \tag{42}$$

It is known [2, 1, 3] that s is continuous and

$$s \in L^\infty(L^\infty), \quad \frac{\partial s}{\partial t} \in L^2(H^{-1}), \quad \psi \in L^2((L^2)^d), \quad \gamma(P(s)) \in L^2(L^2),$$

where H^{-1} is the dual of H_0^1 and the outer function space refers to the time interval of interest and the inner one to Ω. The low regularity is due to the degenerate diffusion. As we pass from a strictly two-phase region to a region with a single phase, the solution is not particularly smooth; thus, it is very difficult to approximate the solution accurately.

Many authors have considered the approximation properties of the continuous, piecewise linear finite element Galerkin method. However, cell-centered finite difference methods are commonly used to solve this problem [18]. Since the mixed finite element method is strongly related to cell-centered finite difference methods, we present a mixed method analysis. Such an analysis has not previously appeared. Our approach is to consider the problem from the point of view of optimal approximation; that is, can we approximate the solution as well as possible, given the finite element approximating space. Since typically the interfaces between the single and two-phase regions occupy a fairly small part of the domain, we can then expect to have very good approximation of the solution.

A NONLINEAR MIXED METHOD

In a standard mixed variational form for (41),

$$\left(\phi \frac{\partial s}{\partial t}, w\right) + (\nabla \cdot \psi, w) = (\gamma(P(s)), w) \quad \text{for all } w \in H_0^1. \tag{43}$$

Since we can only expect in general that $\partial s/\partial t \in L^2(H^{-1})$, the trial functions must belong to $H_0^1(\Omega)$. To avoid this, following Nochetto [17], we integrate (41) in time to obtain

$$s(\mathbf{x},t) + \nabla\cdot\int_0^t \psi\, d\tau = \int_0^t \gamma(P(s))\, d\tau + s_0(\mathbf{x}),$$

where s_0 is the initial saturation. Note that

$$\int_0^t \psi\, d\tau \in H^1((L^2)^d) \cap L^2(H(\text{div})),$$

so we can formulate a mixed variational form as

$$(s(\cdot,t),w) + \Big(\nabla\cdot\int_0^t \psi\, d\tau, w\Big) = \Big(\int_0^t \gamma(P(s))\, d\tau, w\Big) + (s_0,w) \quad \text{for all } w \in L^2, \tag{44}$$

$$(\alpha^{-1}[\psi + \beta(P(s))],\mathbf{v}) - (P(s),\nabla\cdot\mathbf{v}) = -\langle P(s_D),\mathbf{v}\cdot\nu\rangle \quad \text{for all } \mathbf{v} \in H(\text{div}), \tag{45}$$

where s_D is, say, a Dirichlet boundary condition.

Now let Ω be partitioned into a conforming finite element mesh with maximal element diameter h. Let $W_h \times \mathbf{V}_h \subset L^2 \times H(\text{div})$ be any standard mixed finite element space. Let $\Delta t > 0$, $t_n = n\Delta t$, and $\varphi^n = \varphi(t_n)$.

Our nonlinear mixed method is to find $s^n \approx S^n \in W_h$ and $\psi^n \approx \Psi^n \in \mathbf{V}_h$ satisfying

$$(S^n,w) + \Big(\nabla\cdot\sum_{j=1}^n \Psi^j\,\Delta t, w\Big) = \Big(\sum_{j=1}^n \gamma(P(S^j))\,\Delta t, w\Big) + (s_0,w) \quad \text{for all } w \in W_h, \tag{46}$$

$$(\alpha^{-1}[\Psi^n + \beta(P(S^n))],\mathbf{v}) - (P(S^n),\nabla\cdot\mathbf{v}) = -\langle P(s_D^n),\mathbf{v}\cdot\nu\rangle \quad \text{for all } \mathbf{v} \in \mathbf{V}_h. \tag{47}$$

The first equation is equivalent to the usual backward Euler form

$$\Big(\frac{S^n - S^{n-1}}{\Delta t}, w\Big) + (\nabla\cdot\Psi^n, w) = (\gamma(P(S^n)), w) \quad \text{for all } w \in W_h, \tag{48}$$

$$(S^0,w) = (s_0,w) \quad \text{for all } w \in W_h. \tag{49}$$

Let $\mathcal{P}_{W_h}$ and $\mathcal{P}_{V_h}$ denote L^2-projection into W_h and $\mathbf{V}_h$, respectively. Let Π denote the usual element based flux-preserving projection operator for mixed methods [19, 12]. Such projections are optimal in their approximation properties.

Theorem 3: For the nonlinear mixed finite element approximation,

$$\begin{aligned}
&\sum_{j=1}^n (S^j - s^j, P(S^j) - P(s^j))\,\Delta t + \Big\|\sum_{j=1}^n \Psi^j\,\Delta t - \mathcal{P}_{V_h}\int_0^{t_n}\psi\, d\tau\Big\|^2 \\
&\quad \le C\sum_{j=1}^n \Big\{\|\mathcal{P}_{W_h}s^j - s^j\|^2 + \Big\|(\mathcal{P}_{V_h} - I)\int_0^{t_j}\psi\, d\tau\Big\|^2 + \Big\|\nabla\cdot(\Pi - \mathcal{P}_{V_h})\int_0^{t_j}\psi\, d\tau\Big\|^2 \\
&\qquad + \Big\|\sum_{i=1}^j \gamma(P(s^i))\,\Delta t - \int_0^{t_j}\gamma(P(s))\, d\tau\Big\|^2 + \Big\|\frac{1}{\Delta t}\int_{t_{j-1}}^{t_j}\psi\, d\tau - \psi^j\Big\|^2\Big\}\Delta t,
\end{aligned}$$

and

$$\begin{aligned}
\|S^n - s^n\|_{H^{-1}} &\le C\Big\{h\|\mathcal{P}_{W_h}s^n - s^n\| + \Big\|\sum_{j=1}^n \Psi^j\,\Delta t - \Pi\int_0^{t_n}\psi\, d\tau\Big\| \\
&\quad + \sum_{j=1}^n \|P(S^j) - P(s^j)\|\,\Delta t + \Big\|\sum_{j=1}^n \gamma(P(s^j))\,\Delta t - \int_0^{t_n}\gamma(P(s))\, d\tau\Big\|^2\Big\}.
\end{aligned}$$

The form $\left\{\int_0^t (S-s, P(S)-P(s))\,d\tau\right\}^{1/2}$ bounds the size of $S-s$; for example, it bounds the norm $\|P(S)-P(s)\|$. It is *not*, however, a norm itself. It may even fail to be a metric. Also, $\|\cdot\|_{H^{-1}} = \sup_{\varphi\in H_0^1} (\cdot,\varphi)/\|\varphi\|_{H^1}$. The theorem says that the saturation, in a weak H^{-1}-norm sense, is bounded by optimal approximation terms, time truncation terms, and a term involving the divergence that is essentially optimal in character.

CONCLUSIONS

We have developed a cell-centered finite difference mixed method as an approximation to an expanded mixed finite element method. It is suitable for groundwater flow and transport problems on general geometry with tensor permeabilities or dispersivities. It is both locally (element-by-element) mass conservative and highly accurate.

General geometry can be handled by a mapping between the computational and physical domains. The coefficients are transformed in a simple way before commencing the computation. On a logically rectangular mesh, special quadrature rules are applied to the expanded finite element method to transform it into a cell-centered finite difference method with a stencil of 9 points in 2-D and 19 points in 3-D, allowing easy and efficient implementation. Similar results hold for triangular meshes [4, 5, 6].

The solution of the logically rectangular cell-centered finite difference scheme converges to the true solution at the optimal order as the mesh is refined; moreover, superconvergence is attained by the velocity at certain discrete points of the domain on C^2-smooth grids away from the boundary. We saw that grid geometry strongly affects the approximation error, so it is necessary to define and refine grids in a C^2-smooth manner.

Lagrange multiplier pressures or infinitely thin cells need to be introduced along faces where the tensor or geometry changes discontinuously. This allows the adjusted pressure gradient $\tilde{\mathbf{u}}$ to be discontinuous, so that the coefficient times $\tilde{\mathbf{u}}$ approximates well the continuous velocity $\mathbf{u}$.

Mixed methods accurately approximate two-phase flow. The Kirchoff transformation may be useful in computations. The interface between single and two-phase regions is captured in an approximate (H^{-1}-norm) sense.

ACKNOWLEDGMENTS

This work was supported in part by the U.S. Department of Energy, the U.S. National Science Foundation, and the State of Texas Governor's Energy Office.

REFERENCES

[1] H. W. Alt and E. DiBenedetto, *Nonsteady flow of water and oil through inhomogeneous porous media*, Ann. Scuola Norm. Sup. Pisa Cl. Sci., **12** (1985), 335–392.

[2] H. W. Alt and S. Luckhaus, *Quasilinear elliptic-parabolic differential equations*, Math. Z., **183** (1983), 311–341.

[3] T. Arbogast, *The existence of weak solutions to single porosity and simple dual-porosity models of two-phase incompressible flow*, Nonlinear Analysis, Theory, Methods and Applications, **19** (1992), 1009–1031.

[4] T. Arbogast, *Mixed Methods for Flow and Transport Problems on General Geometry*, in Finite Element Modeling of Environmental Problems, G. F. Carey, ed., Wiley, 1995, pp. 275–286.

[5] T. Arbogast, C. N. Dawson, and P. T. Keenan, *Efficient mixed methods for groundwater flow on triangular or tetrahedral meshes*, in Computational Methods in Water Resources X, Vol. 1, A. Peters, *et al.*, eds., Kluwer, 1994, pp. 3–10.

[6] T. Arbogast, C. N. Dawson, P. T. Keenan, M. F. Wheeler, and I. Yotov, *Cell-centered finite difference methods for elliptic equations on general geometry*, submitted to SIAM J. Sci. Stat. Comput.

[7] T. Arbogast, P. T. Keenan, M. F. Wheeler, and I. Yotov, *Logically rectangular mixed methods for Darcy flow on general geometry*, SPE 29099, in Proceedings of the 13th SPE Symposium on Reservoir Simulation, Society of Petroleum Engineers, 1995, pp. 51–59.

[8] T. Arbogast, M. F. Wheeler, and I. Yotov, *Logically rectangular mixed methods for groundwater flow and transport on general geometry*, in Computational Methods in Water Resources X, Vol. 1, A. Peters *et al.*, eds., Kluwer, 1994, pp. 149–156.

[9] T. Arbogast, M. F. Wheeler, and I. Yotov, *Mixed Finite Elements for Elliptic Problems with Tensor Coefficients as Finite Differences*, SIAM J. Numer. Anal., to appear.

[10] T. Arbogast, M. F. Wheeler, and Nai-Ying Zhang, *A nonlinear mixed finite element method for a degenerate parabolic equation arising in flow in porous media*, SIAM J. Numer. Anal., **33** (1996), to appear.

[11] D. N. Arnold, and F. Brezzi, *Mixed and nonconforming finite element methods: implementation, postprocessing and error estimates*, R.A.I.R.O. Modél. Math. Anal. Numér., **19** (1985), 7–32.

[12] F. Brezzi and M. Fortin, *Mixed and hybrid finite elements*, Springer Series in Computational Mathematics, Vol. 15, Springer-Verlag, Berlin, 1991.

[13] Zhangxin Chen, *BDM mixed methods for a nonlinear elliptic problem*, J. Comp. Appl. Math., **53** (1994), 207–223.

[14] J. Douglas, Jr., R. E. Ewing, and M. F. Wheeler, *Approximation of the pressure by a mixed method in the simulation of miscible displacement*, R.A.I.R.O. Modél. Math. Anal. Numér., **17** (1983), 17–33.

[15] J. Koebbe, *A computationally efficient modification of mixed finite element methods for flow problems with full transmissivity tensors*, Numer. Meth. for PDE's, **9** (1993), 339–355.

[16] J. C. Nedelec, *Mixed finite elements in* Re^3, Numer. Math., **35** (1980), 315–341.

[17] R. H. Nochetto, *Error estimates for two-phase Stefan problems in several space variables, I: linear boundary conditions*, Calcolo, **22** (1985), 457–499.

[18] D. W. Peaceman, *Fundamentals of numerical reservoir simulation*, Elsevier, 1977.

[19] P. A. Raviart and J. M. Thomas, *A mixed finite element method for 2nd order elliptic problems*, in Mathematical Aspects of the Finite Element Method, Lecture Notes in Math., Springer-Verlag, Berlin, 1977.

[20] T. F. Russell and M. F. Wheeler, *Finite element and finite difference methods for continuous flows in porous media*, in The Mathematics of Reservoir Simulation, R. E. Ewing, ed., SIAM, Philadelphia, 1983.

[21] J. M. Thomas, *These de Doctorat d'etat à l'Universite Pierre et Marie Curie*, 1977.

[22] A. Weiser, personal communication, 1991.

[23] A. Weiser and M. F. Wheeler, *On convergence of block-centered finite differences for elliptic problems*, SIAM J. Numer. Anal., **25** (1988), 351–375.

[24] M. F. Wheeler, K. R. Roberson, and A. Chilakapati, *Three-dimensional bioremediation modeling in heterogeneous porous media*, in Computational Methods in Water Resources IX, Vol. 2, T. F. Russell *et al.*, eds., Computational Mechanics Publications, Southampton, U. K., 1992, pp. 299–315.

NUMERICAL SIMULATIONS BY HOMOGENIZATION OF TWO-PHASE FLOW THROUGH RANDOMLY HETEROGENEOUS POROUS MEDIA

A. Badea and A. Bourgeat

Equipe d'Analyse Numérique, URA 740,
Faculté des Sciences, 23, rue du Dr. P. Michelon,
42023 Saint-Etienne Cedex 2, France

SUMMARY

We consider the behavior of incompressible two-phase flow in heterogeneous reservoirs with randomly placed heterogeneities; that is, a porous medium with permeability A and porosity Φ which are statistically homogeneous random fields oscillating at the dimensionless scale ε. Using the tools of stochastic homogenization we get the nonlinear effective equations which govern the flow behavior in a homogeneous medium being equivalent, in the sense of homogenization theory, to the original one. The computation of the effective permeability tensor A^{hom} is done by solving auxiliary stochastic problems, similar to the ones for the linear one-phase flow case. Under ergodicity assumption, and using the primal and dual formulation of these auxiliary problems, we design a numerical algorithm computing both the effective parameters and the minimal volume on which these effective properties are valid. The validity of our algorithm is tested on a two-phase groundwater flow with injection of one-phase from a well.

INTRODUCTION

Groundwater two-phase flow could describe for instance the pollution (by salt, nytrogen, ...) of a pure water storage zone. It could be modelized by the system of equations for the incompressible two-phase flow in a porous medium i.e. by the conservation equation in each i-phase:

$$\Phi\frac{\partial(S_i\rho_i)}{\partial t}+\operatorname{div}(\rho_i q_i)=0 \qquad i=\alpha,\beta \tag{1}$$

with q_i the Darcy velocity, given by

$$q_i=-\frac{k_i(S_i)}{\mu_i}A(x)(\nabla p_i-\rho_i g) \qquad i=\alpha,\beta, \tag{2}$$

where p_i, ρ_i and μ_i are the intrinsic phase averages of the presure, density and viscosity for each i-phase respectively; $k_i(S_i)$ is the relative permeability of phase i and function of S_i, the saturation of phase i; A is the rock permeability tensor, Φ is the porosity and g is the gravity acceleration vector. The saturations verify, by definition

$$S_\alpha+S_\beta=1 \tag{3}$$

and the difference of pressures p_i is given by the capillary pressure curve

$$p_c=p_\alpha-p_\beta. \tag{4}$$

Natural reservoirs are made of a heterogeneous porous medium and the rock characteristics are highly oscillating at the scale of the geological data ($\approx$ 10 cm). But, on an other hand, simulation of two-phase flow through such a medium requires using meshes whose scale is of order 10 meters. Homogenization is the mathematical technique used then to derive the behavior (equations) at the higher scale (the reservoir scale) from the small scale (geological data) behavior. This technique is based on finding the mathematical limit of the small scale equation as ε, the characteristic scales ratio ($\varepsilon = \frac{l}{L}$, $l \equiv$ heterogeneities scale, $L \equiv$ reservoir scale) tends to zero. The need for explicit global solution disconnected from the local behavior leads to assume for instance periodicity of heterogeneities. It should be noticed that this assumption is not only a tool to effectively compute the global behavior, but it is also the simplest example of an ergodic dynamical system which is the general context in which randomly distributed heterogeneities could be mathematically studied. Most of the results already obtained under the periodicity assumption could be extended to the random case by using the stochastic two-scale convergence concept as defined in [13].

Several situations occur, according to how and what rock properties are oscillating and to what is the leading phenomenon in the flow (diffusion due to the capillary pressure, or convection).

By mean of homogenization, under the periodicity assumption and assuming moreover capillary pressure driven diffusion, the following situations have already been investigated.

The first studied case was when the porosity $\Phi(x)$ and the rock permeability $A(x)$ are the only oscillating parameters describing the heterogeneities, see for instance [8], [7] or [3].

An other situation occurs when all the rock properties, i.e. not only the porosity $\Phi(x)$ and the rock permeability $A(x)$, but also the relative permeability curves $k_i(S,x)$, $(i = \alpha, \beta)$ and the capillary pressure curve $p_c(S,x)$, are allowed to oscillate at the heterogeneities scale. In that case, see [9] and [18], even periodicity assumption does not disconnect the equation at the small scale from the equation at the big scale. Then, other physical assumptions are necessary in order to have the big scale equations totally disconnected from the small scale, see [3], [9], [18].

The double porosity model has been obtained by homogenization in [11]. In this model, the porous medium is made of a system of high-conductivity porous fissures together with a large number of porous matrix blocks containing most of the β-phase (the non-wetting phase). This situation leads to consider not only the two $\mathcal{O}(\varepsilon)$ and $\mathcal{O}(1)$ spatial scales but also two different time scaling. If the characteristic time for the rescaled flow, by $y = x/\varepsilon$, in a single porous matrix block is of order $\mathcal{O}(1)$, then the characteristic time for the flow through the entire system of all the fractures is then of order $\mathcal{O}(\varepsilon^{-2})$. Interaction of these two kinds of scaling, in time and in space, leads in the final homogenized equation to the addition of a non linear memory like term.

When the capillary pressure effects are neglected, the equations become of hyperbolic type, and we lose uniqueness of the solutions. In order to keep only the physically meaningfull solutions it is then necessary to consider only the entropic solutions, defined for instance as the limit of the solution with capillary effects when capillary forces tend to zero. Making the capillary forces tend to zero leads to a singular perturbation problem which could be mathematically handled through the theory of vanishing viscosity as in [15], developed only for a scalar equation. In the case of a two-phase flow through a porous core, which is a 1-D displacement in a porous medium, equations (1) and (2) lead to the scalar Buckley-Leverett equation. Assuming that either the porosity $\Phi(x)$ and the rock permeability $A(x)$ or the given initial condition $q_T(t = 0)$ are oscillating at the lower scale, the homogenization coupled with the singular perturbation technique in the 1-D case have been considered in [12].

Theoretical results have been obtained discarding the periodicity assumption but assuming randomly ergodic distribution of the heterogeneities for the first situation, i.e. with the porosity $\Phi(x)$ and the rock permeability $A(x)$ being the only randomly oscillating parameters, see [10]. It has been proved in [10], by stochastic homogenization, that there exist upscaled parameters A^{hom} and Φ^{hom} corresponding to an effective (deterministic) behavior of the flow at a sufficiently large scale.

Now we are herein this paper giving a way to really compute these effective parameters A^{hom} and Φ^{hom}. Some examples of computations comparing the heterogeneous simulations to the global homogenized ones are shown in the last part.

DEFINITION OF THE PROBLEM

Starting from equations (1)-(4), assuming incompressibility, leads to

$$\Phi\frac{\partial S}{\partial t} - \text{div}\ (a(S)A(x)\nabla S - b_G(S)\nabla P_G - b(S)q_T) = 0 \tag{5}$$

$$\text{div}\ q_T = 0 \tag{6}$$

where $S = S_\alpha$ is the saturation of the wetting fluid, $q_T = q_\alpha + q_\beta$ is the total superficial velocity, $d(S) = \Lambda_\alpha(S) + \Lambda_\beta(S)$ is the total mobility, with $\Lambda_i(S_i) = \dfrac{k_i(S_i)}{\mu_i}$ being the mobility of phase i and $a(S) = \dfrac{\Lambda_\alpha(S)\Lambda_\beta(S)}{d(S)}\dfrac{dp_c(S)}{dS}$; $b(S) = \dfrac{\Lambda_\alpha(S)}{d(S)}$ the α-fractional flow, $b_G(S) = \dfrac{2\Lambda_\alpha(S)\Lambda_\beta(S)}{d(S)}\dfrac{\rho_\alpha - \rho_\beta}{\rho_\alpha + \rho_\beta}$ and $P_G = \dfrac{1}{2}(\rho_\alpha + \rho_\beta)\, gz$ the gravity potential.

Using the reduced pressure P as defined in [2], [4] or in [14], we get

$$q_T = -A(x)d(S)(\nabla P + \gamma_2(S)\nabla P_G) \tag{7}$$

with $\gamma_2(S) = \dfrac{2(\rho_\alpha\Lambda_\alpha(S) + \rho_\beta\Lambda_\beta(S))}{d(S)(\rho_\alpha + \rho_\beta)}$ and $P = \dfrac{1}{2}(p_\alpha + p_\beta) + \displaystyle\int_{S_c}^{S}\left[b(\xi) - \frac{1}{2}\right]\frac{dp_c}{d\xi}d\xi.$

The rock permeability tensor $A(x)$ and the rock porosity $\Phi(x)$ are random functions of position, changing rapidly as x varies over lenghts comparable to the size of the region G in $\mathbb{R}^n$ occupied by the reservoir. The parameter $\varepsilon > 0$ is defined as the ratio of a typical length scale associated with the heterogeneities size, i.e. with the variation in permeability and porosity, to the size of the region G. Capillary pressure and relative permeability curves are assumed to be unique.

The permeability tensor A and the porosity Φ are random functions of the form $A\left(\dfrac{x}{\varepsilon},.\right)$ and $\Phi\left(\dfrac{x}{\varepsilon},.\right)$ where $A(x,\omega)$ and $\Phi(x,\omega)$, $\omega \in \Omega$ (the sample space), are realizations of given statistically homogeneous random functions. The saturation S and the pressure P in the system of equations are now random variables and depending on the size of the heterogeneities. They are then denoted by $S_\varepsilon^\omega, p_\varepsilon^\omega$.

In this work, our main concern is the large-scale characterization of the flow field within naturally heterogeneous formation as well as the computation of the "effective" properties of these porous media, and for simplicity, we will neglect the effect of gravity.

The system (5)-(7) reduces then to the stochastic partial differential equations

$$\begin{cases} q_\varepsilon^\omega = -d(S_\varepsilon^\omega)A(x/\varepsilon,\omega)\nabla p_\varepsilon^\omega \\ \text{div } q_\varepsilon^\omega = 0 \\ \Phi(x/\varepsilon,\omega)\dfrac{\partial S_\varepsilon^\omega}{\partial t} - \text{ div } (A(x/\varepsilon,\omega)a(S_\varepsilon^\omega)\nabla S_\varepsilon^\omega - b(S_\varepsilon^\omega)q_\varepsilon^\omega) = 0 \\ (t,x,\omega) \in]0,\tau[\times G \times \Omega \end{cases} \tag{8}$$

with, for instance, the boundary and initial conditions

$$\begin{cases} q_\varepsilon^\omega.\nu\,|_{\Sigma_1} = g_1; \quad q_\varepsilon^\omega.\nu\,|_{\Sigma_3} = 0; \quad p_\varepsilon^\omega\,|_{\Sigma_2} = p_p; \\ S_\varepsilon^\omega\,|_{\Sigma_1} = 1; \quad S_\varepsilon^\omega\,|_{\Sigma_2} = 0; \quad a(S_\varepsilon^\omega)A(x/\varepsilon,\omega)\nabla S_\varepsilon^\omega.\nu\,|_{\Sigma_3} = 0; \\ S_\varepsilon^\omega(x,0) = s_0(x) \quad \text{in } G; \end{cases} \tag{9}$$

ν being the outer normal unit vector and g_1 and p_p some constants. The reservoir G is a bounded domain in $I\!R^n$ whose boundary Γ splits up into three parts such that $\Gamma = \overline{\Gamma_1} \cup \overline{\Gamma_2} \cup \overline{\Gamma_3}$ and $\Gamma_k \cap \Gamma_j = \emptyset$ for $k \neq j$; Γ_1 is the part of the boundary where the wetting fluid is injected, Γ_2 is the producing part of the boundary and Γ_3 is the impervious part of the boundary and $\Sigma_j = \Gamma_j \times]0,\tau[$. The system (8)-(9) is valid before the breakthrough, i.e. before the α-phase reaches the producing part of the boundary.

The main results of [10] are the following:

- *There exists an effective purely deterministic behavior of the flow at large scale, determined by a constant porosity Φ^{hom} and a constant rock permeability tensor A^{hom}. This effective behavior is obtained as the almost sure (a.e. $\omega \in \Omega$) limit of the system of equations* (8)-(9) *in [10].*

- *These global parameters Φ^{hom} and A^{hom} are obtained exactly like if it was a one-phase flow, i.e. if the flow was described by the only equation:*

$$- \, div\,(A(x/\varepsilon,\omega)\nabla p_\varepsilon^\omega = f. \tag{10}$$

- *Global parameters are given by:*

$$\begin{aligned} \Phi^{hom} &= \int_\Omega \Phi(y,\omega)d\mu \\ A^{hom}e_k &= \int_\Omega A(y,\omega)(e_k + \nabla_y u_k)d\mu \end{aligned} \tag{11}$$

where e_k, $k = 1,\ldots,n$ form the orthonormal basis of $I\!R^n$ and u_k, $k = 1,\ldots,n$ are defined by the auxiliary problems:

$$\begin{cases} -\, div_y(A(y,\omega)(e_k + \nabla_y u_k)) = 0 \quad in\ I\!R^n, \quad for\ any\ \omega \in \Omega \\ u_k \in H^1_{loc}(I\!R^n)\ ; \ \displaystyle\int_\Omega \nabla_y u_k d\mu = 0. \end{cases} \tag{12}$$

ALGORITHM FOR COMPUTING THE EFFECTIVE PROPERTIES

Finally in [6], we design an algorithm avoiding the time consuming computation of A^{hom} given by (11). This algorithm is based on the solutions of the two following problems, defined on a "representative elementary volume" $V = v \times]0,1[^n$, $v \in I\!R$, for $k = 1,\ldots,n$:

$$\begin{cases} -\text{ div } \sigma_k = 0 \ \text{ in } V \\ \sigma_k = A(x)\nabla u_k \ \text{ in } V \\ u_k|_{\partial V} = e_k \cdot x|_{\partial V} \end{cases} \tag{13}$$

$$\begin{cases} -\operatorname{div} \Sigma_k = 0 \ \text{ in } V \\ \Sigma_k = A(x)\nabla w_k \ \text{ in } V \\ \Sigma_k \cdot \vec{n}|_{\partial V} = e_k \cdot \vec{n}|_{\partial V}. \end{cases} \tag{14}$$

Due to the linearity of the above solutions with respect to e_k, we may define two $n \times n$ tensors $K(V, A(x))$ (from (13)) and $S(V; A(x))$ (from (14)) by:

$$M_V\{\sigma_k\} = K(V, A(x))e_k \tag{15}$$

$$M_V\{\Sigma_k\} = S(V, A(x))e_k \tag{16}$$

($M_V\{\cdot\}$ denotes $\int_V \cdot\, dx$). Under the additional assumption of ergodicity, using the Akcoglu - Krengel theorem [1] and the H-convergence [16] as is the case of elasticity [17], we show in [6] that at the limit, $\lim_{v\to\infty} K(V, A(x)) = A^{hom}$ and $\lim_{v\to\infty} S(V, A(x)) = (A^{hom})^{-1}$, i.e. we mathematically prove that doing the so-called "volume averaging" but like in equations (13) and (14), leads to the right homogenized coefficients.

Finally this algorithm gives both an approximation of A^{hom}, denoted A^0, and the minimal volume on which this approximation should be done.

This algorithm could be summarized in

```
begin
                initialise  v  (volume size)
                V  := v × ]0, 1[^n
                i  := 1
        repeat
                increase  v  (according to a predefined strategy)
                V  := v × V
                i  := i+1
                compute  σ_k  solution of (13) for  k = 1,... n
                compute  ∇w_k from the solution of (14) for  k = 1,... n
                compute  K(V, A(x))  from (15);   K := K(V, A(x))
                compute  S(V, A(x))  from (16);   S := S(V, A(x))
        until               ||K · S - I|| ≤ tol
                V  is then the Representative Elementary Volume
                A^0 := K
end
```

PRESENTATION OF NUMERICAL TESTS

Numerical tests have been made using the above algorithm on ten realizations of a random porous medium, which are constructed as follows.

We assume the reservoir to be two-dimensional and made of a union of isotropic squares, with permeability given by a rescaled log-normal law, i.e.

$$\begin{aligned} A(x,\omega) &= cI, \\ c &= 0.1\exp(c^*) \end{aligned} \tag{17}$$

where c^* is obeying a centered normal law of standard deviation 1.0986, and with porosity given by the empirical law:

$$\Phi(x,\omega) = 0.552 + 0.229 \ln c. \tag{18}$$

First we generate ten realizations of a 20×20 permeability chart and we start to compute $K(V, A(x))$ according to our numerical algorithm with $v = 2, \ldots, 20$ and, according to [5], with a mesh size h equal to the heterogeneities size.

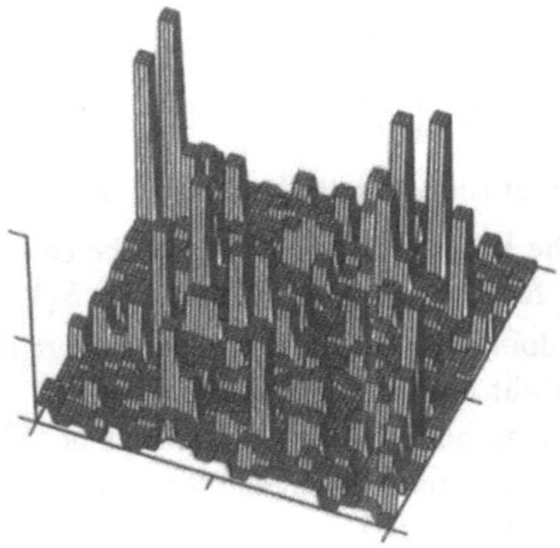

Figure 1: Permeability chart for $v = 20$.

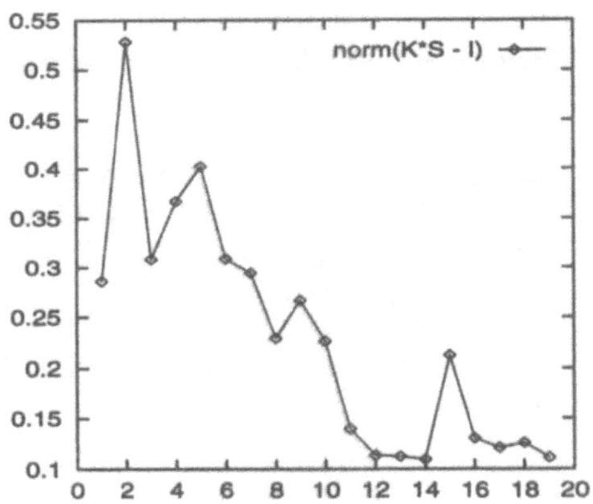

Figure 2: Accuracy of the approximation of A^0 by $K(V, A(x))$.

In Figure 1 we present one realization (realization number 3) of $A(x, \omega)$ on V for $v = 20$ and in Figure 2 we plot the accuracy of the approximation of A^0 by $K(V, A(x))$ (i.e. the norm $\|K \cdot S - I\|$, where $\|M\| = [$ spectral radius of $(M^* M)]^{1/2}$) versus the size of V for the realization number 3.

For the realization number 3, the size $v = 20$ gives:

$$K(V, A(x)) = \begin{pmatrix} 9.59764 \cdot 10^{-2} & 2.40112 \cdot 10^{-4} \\ 2.41018 \cdot 10^{-4} & 9.23277 \cdot 10^{-2} \end{pmatrix}. \tag{19}$$

After considering all the ten realizations, we conclude that, like in Figure 2 for the realization number 3, a control volume of order $V = 12\times]0,1[^2$ will be enough to have a good approximation of the homogenized coefficients.

Now we apply our previous result to the simulation of a two-phase flow through a quarter five spot reservoir. In the reservoir there are five wells, one injector and four producers symmetrically reparted around the injection well. By symmetry, the reservoir is cut in four domains, as in Figure 3.

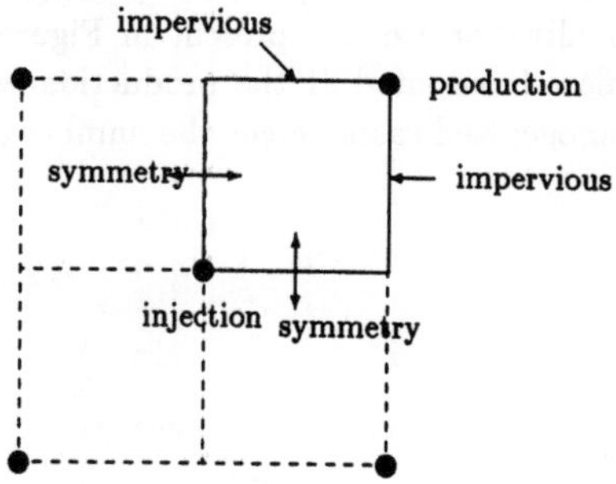

Figure 3: A five spot reservoir.

The permeability and porosity charts for this reservoir are assumed to be given by (17) and (18) respectively.

In order to apply our theoretical results, see [10], [7], [8], we assume the relative permeability curves $k_\alpha(S)$ and $k_\beta(S)$ and the capillary pressure curve $p_c(S)$ not depending on $x \in V$ or $\omega \in \Omega$ (which means that we suppose to have the same type of rock). They are defined in Figure 4.

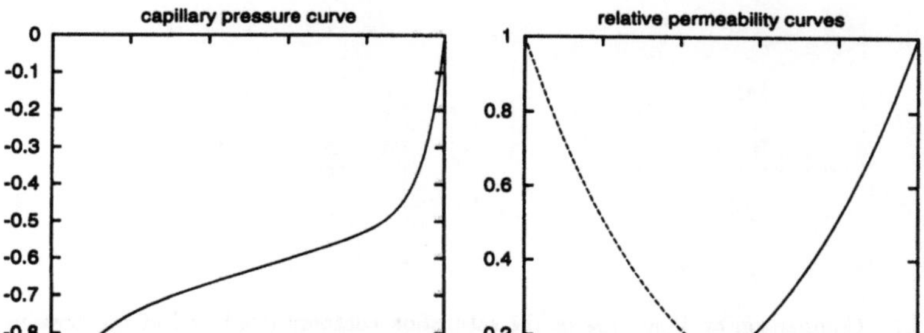

Figure 4: Capillary pressure for the whole reservoir, $p_c(S) = -47\{10000[1-\exp(-20(1-S))]+7000(1-S)+2650\frac{(1-S)^{10}}{(0.1+S)^{0.2}}\}\ 10^{-6}$; wetting and non-wetting permeability curves for the whole reservoir, $k_\alpha(S) = S^3$, $k_\beta(S) = (1-S)^3$.

The viscosities are $\mu_\alpha = 1$ cp and $\mu_\beta = 1.43$ cp and the densities are $\rho_\alpha = 1$ g/cm^3

and $\rho_\beta = 0.75$ g/cm^3. The injection total rate is $q_T = 35$ m^3/day and the pressure at the production well is fixed to $p_r = 100$ bars.

The simulator used is based on a traditional block-centered finite volume formulation, with IMPES discretization in time associated to upwinding of the saturation.

For this quarter type reservoir we first show at a time before the breakthrough, in Figure 5, the saturation and pressure contours for both the heterogeneous case and the homogenized one (with $\Phi^0 = M_V\{\Phi(x)\}$ and $A^0 = \mathrm{diag}(K(V, A(x)))$) for the realization number 3. For this realization we also present in Figure 6 the production curves and the fractional water flow (computed at the production well) respectively, for both the heterogeneous and homogenized cases versus the number of days.

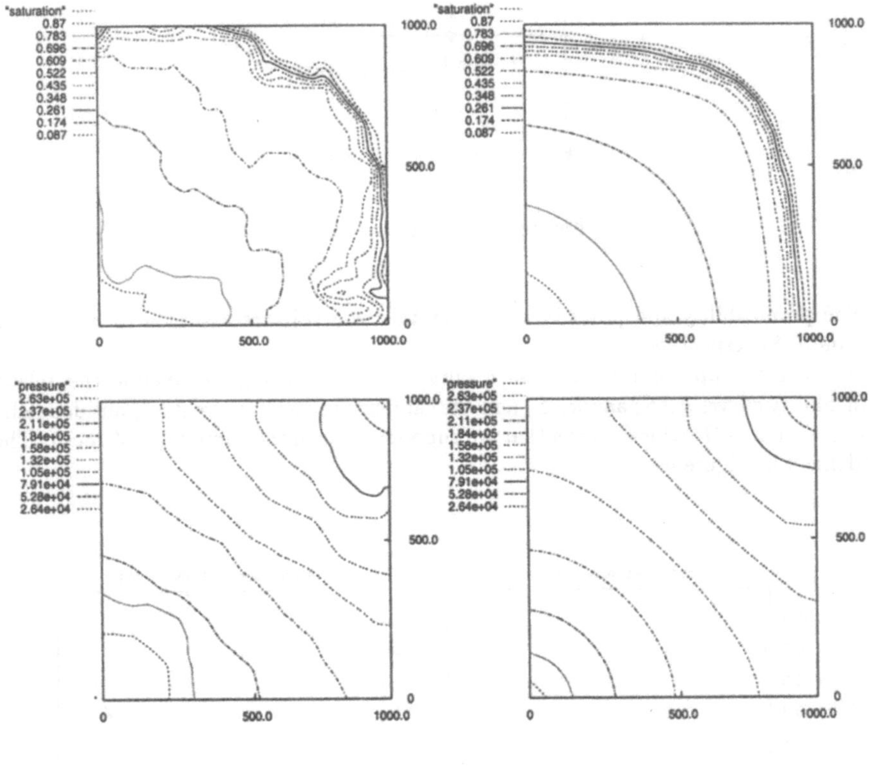

Figure 5: **Comparison at 1800 days of the saturation contours (top) and of the pressure contours (bottom), between the solutions for the randomly heterogeneous reservoir with $A(x,\omega)$ and $\Phi(x,\omega)$ (left) and for the globally homogenized reservoir (right), with $\Phi^0 = 0.120$ and $A^0_{11} = 0.09598$, $A^0_{22} = 0.09233$, $A^0_{12} = A^0_{21} = 0$, computed on $V = 20\times]0,1[^2$.**

In Figure 7 we plot the production curves for all the realizations in the heterogeneous case and in the corresponding homogeneous case versus the number of days.

Finally, we compute the breakthrough for every realization in the heterogeneous and

Table 1: **Breakthrough and recuperation rate at 1500 days and at 2500 days for the heterogeneous and the homogeneous reservoir.**

	Heterogeneous case			Homogeneous case		
realization number	break-through	REC at 1500 days	REC at 2500 days	break-through	REC at 1500 days	REC at 2500 days
1	1820	0.4277	0.6471	2230	0.4228	0.6726
2	1890	0.4514	0.6534	2120	0.4514	0.683
3	2160	0.4366	0.6701	2190	0.4366	0.6768
4	1910	0.4387	0.6544	2180	0.4387	0.6777
5	2030	0.4366	0.6515	2190	0.4366	0.6768
6	2010	0.4534	0.6676	2110	0.4534	0.6838
7	1910	0.4805	0.6807	1990	0.4805	0.6933
8	2090	0.433	0.6619	2210	0.433	0.675
9	2120	0.4116	0.6474	2320	0.4116	0.6639
10	1740	0.476	0.6774	2010	0.476	0.6918
$\bar{x}$	1968	0.4445	0.6612	2155	0.4445	0.679
Var(x)	16716	$4.08 \cdot 10^{-4}$	$1.35 \cdot 10^{-4}$	9045	$4.08 \cdot 10^{-4}$	$7.02 \cdot 10^{-5}$

in the corresponding homogeneous case. We also compute the recuperation rate (denoted REC) for every realization in the heterogeneous and in the corresponding homogeneous case at a time (t_1 = 1500 days) inferior to the breakthrough and at a time (t_2 = 2500 days) superior to the breakthrough. We summarize our results in Table 1.

In Figures 8, 9, 10 we plot the histograms for the breakthrough, for the recuperation rate at 1500 days and for the recuperation rate at 2500 days respectively.

CONCLUSIONS

After having obtained rigorous results for the homogenization of incompressible two-phase flow in randomly porous media, we have presented herein a method for approximating the theoretical homogenized coefficients. The numerical comparisons show first the efficiency of our algorithm, a square of 12×12 meshes gives results close to the theory; and secondly its robustness considering the boundary conditions. Although the theory is valid locally, i.e. for any domain not touching the boundary, we get fairly good results on a quarter five spot problem including the wells vicinity.

It should be moreover noticed that the theory used to design this algorithm gives, on a rigorous mathematical ground, the right way to do "volume averaging" and proves the ability of our algorithm to approximate the homogenized parameters.

ACKNOWLEDGEMENTS

This work was partially supported by the "Région Rhône-Alpes" and by the contract ELF-EAP / UJM no.9813.

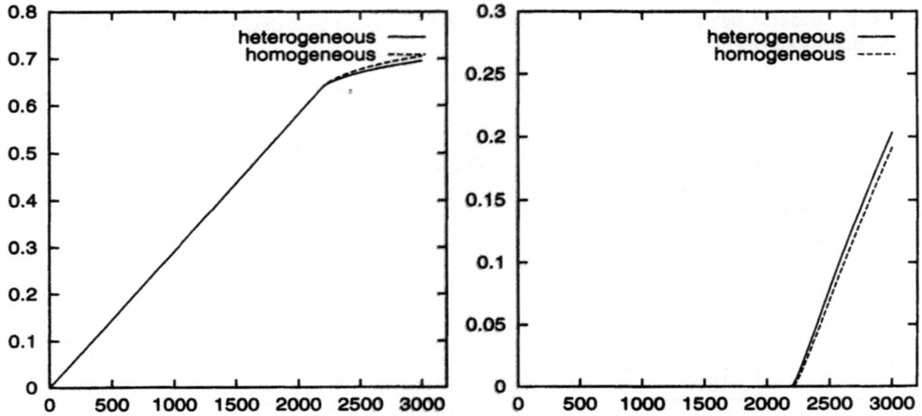

Figure 6: Production curves (left) and fractional water flow at the production well (right) for both heterogeneous and homogeneous reservoir.

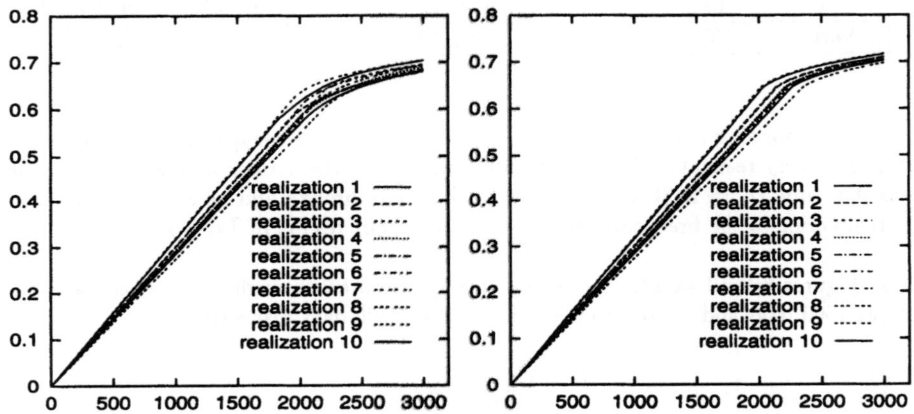

Figure 7: Production curves for all the realizations, heterogeneous case (left) and homogeneous one (right).

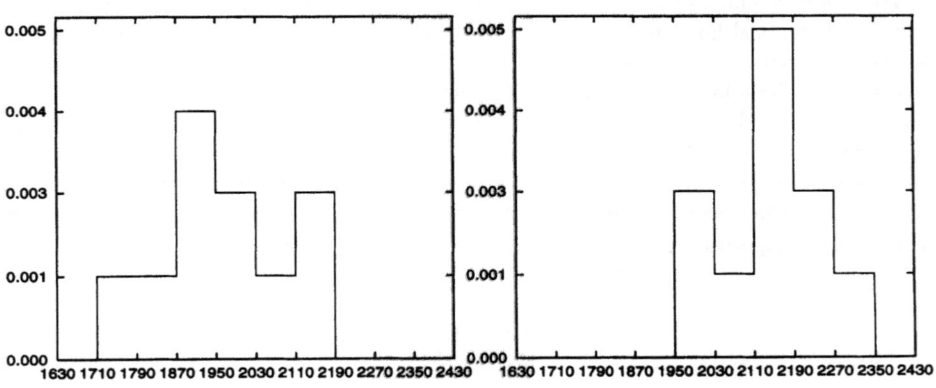

Figure 8: Histograms for the breakthrough, heterogeneous case (left) and homogeneous case (right).

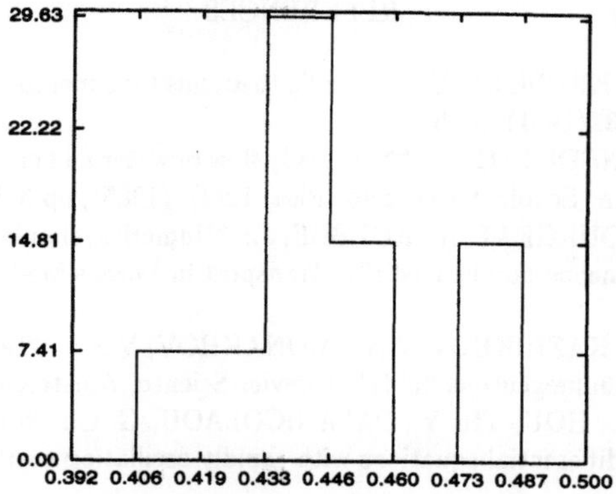

Figure 9: **Histogram for the recuperation rate at 1500 days for the heterogeneous case; in the homogeneous case we obtain exactly the same histogram.**

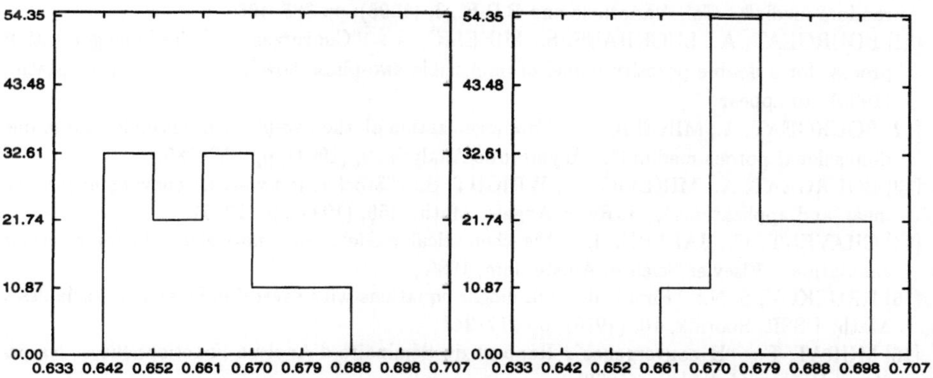

Figure 10: **Histograms for the recuperation rate at 2500 days, heterogeneous case (left) and homogeneous case. (right)**

REFERENCES

[1] AKCOGLU, M. A., KRENGEL, U.: "Ergodic theorems for superadditive processes", J.Reine Angew.Math, 323, (1981), pp 53–67.

[2] ALT, H. W., DI BENEDETTO, E.: "Nonsteady flow of water and oil through inhomogeneous porous media", Ann. Scuola Norm. Sup. Pisa, 12(4), (1985), pp 335–392.

[3] AMAZIANE, B., BOURGEAT, A., KOEBBE, J.: "Numerical simulation by homogenization in periodic heterogeneous porous media", Transport in Porous Media, II, 6, (1991), pp 519–548.

[4] ANTONSEV, S.N., KAZHIKHOV, A.V., MONAKHOV, V.N.: "Boundary value problems in mechanics of nonhomogeneous fluids", Elsevier Science, Amsterdam, 1990.

[5] AVELLANEDA, M., HOU, TH. Y., PAPANICOLAOU, G. C.: "Finite difference approximations for partial differential equations with rapidly oscillating coefficients", M^2AN, 25(6), (1991), pp 693–710.

[6] BADEA, A., BOURGEAT, A.: "Homogenization of two-phase flow through randomly heterogeneous porous media", International Conference on Mathematical Modeling of Flow through Porous Media, Saint-Etienne, 1995. Scientific World Pub., (1995), pp 44–58.

[7] BOURGEAT, A.: "Homogenization method applied to the behavior of a naturally fissured reservoir", In K. J. Gross, editor, Mathematical Method in Energy Research, SIAM, (1984), pp 181–193.

[8] BOURGEAT, A.: "Homogenized behavior of two-phase flows in naturally fractured reservoir with uniform fracture distribution", Comput. Methods in App. Mech. and Eng., 47, (1984), pp 205–217.

[9] BOURGEAT, A., HIDANI, A.: "Effective model of two-phase flow in a porous medium made of different rock types", Applicable Analysis, 58, (1995), pp 1–29.

[10] BOURGEAT, A., KOZLOV, S., MIKELIĆ, A.: "Effective equations of two-phase flow in random media", Cal. Variations and P.D.E., 3, (1995), pp 385–406.

[11] BOURGEAT, A., LUCKHAUS, S., MIKELIĆ, A.: "Convergence of the homogenization process for a double porosity model of immiscible two-phase flow", SIAM J. Appl. Math., (1995). to appear.

[12] BOURGEAT, A., MIKELIĆ, A.: "Homogenization oh the two-phase immiscible flow in one dimensional porous medium", Asymptotic Analysis, 9, (1994), pp 359–380.

[13] BOURGEAT, A., MIKELIĆ, A., WRIGHT, S.: "Stochastic two-scale convergence in the mean and applications", J. Reine Angew. Math., 456, (1994), pp 19–51.

[14] CHAVENT, G., JAFFRÉ, J.: "Mathematical models and finite elements for reservoir simulation", Elsevier Science, Amsterdam, 1986.

[15] KRUZKOV, S. N.: "First order quasilinear equations with several independent variables", Math. USSR, Sbornik, 10, (1970), pp 217–243.

[16] MURAT, F.: "H–convergence", Rapport du Séminaire d'Analyse Fonctionnelle et Numérique, Université d'Alger, (1978).

[17] SAB, K.: "On the homogenization and the simulation of random materials", Eur. J. Mech. A Solids, 11, (1992), pp 585–607.

[18] SAEZ, A. E., OTERO, C. J., RUSINEK, I.: "The effective homogeneous behavior of heterogeneous porous media", Transport in Porous Media, 4, (1989), pp 213–238.

A Dynamic Adaptive Method for the Computation of Highly Advective or Highly Dispersive Transport Processes in Fractured Rock

C.Barlag AND W.Zielke
Institut für Strömungsmechanik
und Elektronisches Rechnen im Bauwesen
Universität Hannover, Appelstraße 9A, 30167 Hannover

SUMMARY

The numerical simulation of flow and transport processes in fractured rock is continuously gaining in importance, because locations in this kind of rock are intended to be used for biological, chemical or radioactive waste deposits.

The description of such complex geological formations is difficult, because the critical regions for simulation, such as sources and sinks as well as transitory regions from fracture zones to the rock matrix, require a much finer resolution than other parts of the domain. An appropriate strategy of mesh generation in the preprocessing stage is presented in this paper.

The main topic of this contribution deals with an adaptive method for the computation of highly advective and highly dispersive transport processes using dynamic data structures, refinement and recoarsement strategies. The model is based on an 'Upwind Finite Element' formulation. The performance of the implemented algorithm (modelling error, cpu time, storage requirement) is discussed, using a sharp front problem as an example.

1 Introduction

The deposit of biological, chemical or radioactive waste requires a quite dense subsurface. Rock is interesting in this context since it appears to be a geological barrier. Rock, however, contains fractures, where flow and transport processes can occur (fig. 1). A numerical model was developed at our institute for simulation of such processes in a subsurface aquifer [8].

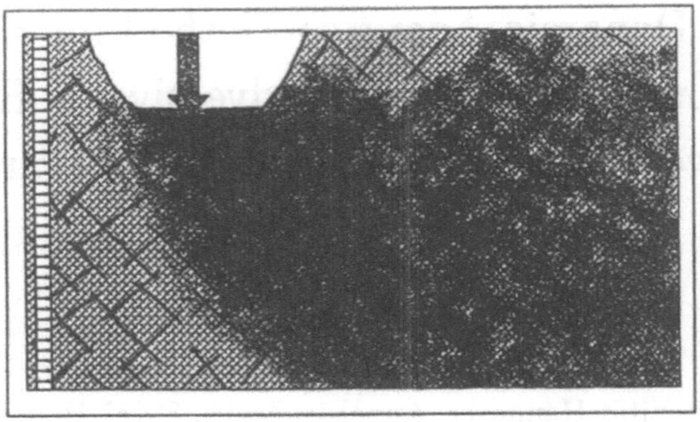

Figure 1: Waste injection in a fracture aquifer [6]

Because of scaling effects which occur when discretizing the rock matrix and the much smaller fractures, the use of varying element types is appropriate. We divide the porous rock matrix into two–dimensional planes, and the fractures into one–dimensional line elements using a conform coupling of both. Further works are going on to discretize the rock matrix into three–dimensional volumes, fractures into two–dimensional planes, and main pathways into one–dimensional line elements. For such complex geometries it is necessary to use a special preprocessor to generate the 1d/2d respectively 1d/2d/3d grids and to visualize them (fig. 2).

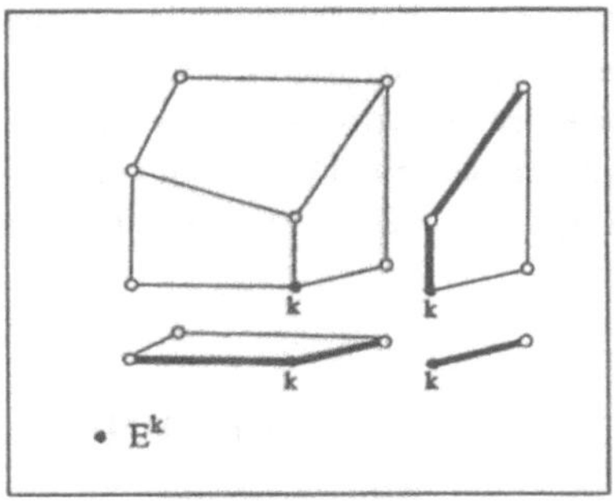

Figure 2: Conform coupling of different element dimensions [5]

The solution of the flow and transport differential equations in fracture/matrix–systems is demanding. It is necessary to calculate problems with injection or extraction points (high velocity gradients), sharp concentration fronts (inside fractures), and interaction zones between fractures and matrix (highly–curved streamlines).

The aim is to obtain reliable forecasts with regard to the possible propagation of tracers in the fractured subsurface with reasonable effort. Therefore, it is necessary to minimize storage and cpu time requirements to get an efficient and accurate model. A possible way is to apply adaptive methods in space and time.

2 Mathematical and Numerical Basics

The partial differential equation of groundwater flow is given by

$$S_0 \frac{\partial h}{\partial t} + \operatorname{div} \mathbf{v}_f - q = 0 \tag{1}$$

including Darcy's law

$$\mathbf{v}_f = -\frac{\varrho g}{\nu} \mathbf{k} \cdot \mathbf{grad}\, h$$

where

S_0	:	specific storage coefficient	ϱ	:	density
h	:	piezometric head	g	:	gravitational acceleration
t	:	time	ν	:	viscosity
$\mathbf{v}_f$	:	seepage velocity vector	$\mathbf{k}$	:	permeability tensor.
q	:	flux			

The transport equation reads as follows

$$\underbrace{n\varrho \frac{\partial c}{\partial t}}_{\text{mass term}} + \underbrace{\varrho \mathbf{v}_f \cdot \mathbf{grad}\, c}_{\text{advection term}} - \underbrace{\operatorname{div}(n \mathbf{D} \cdot [\varrho\, \mathbf{grad}\, c])}_{\text{dispersion term}} + \underbrace{\varrho q c - r(c)}_{\text{well terms}} = 0 \tag{2}$$

where

n	:	porosity	$\mathbf{D}$	:	dispersion and diffusion tensor
c	:	concentration	$r(c)$	:	non–conservative transport processes (e.g. sorption, decay).

With the assumption of an incompressible flow and an ideal tracer ($\varrho = \eta = const.$), the flow and transport equations are decoupled and linear. Velocity field and flux are the connecting parameters between the two models.

The numerical solution is based on the method of weighted residua. A semi–discrete finite element method with linear shape functions is used. The flow equation (1) is treated with the 'Bubnov Galerkin' method in space and the implicit 'Euler' method in time.

The solution of the transport equation (2) is more complex. This type of equation is mixed parabolic/hyperbolic, depending on advection–dispersion ratio defined by the 'Peclet' number

$$Pe = \frac{\mathbf{v}_f \Delta \mathbf{s}}{2\mathbf{D}} \begin{cases} \gg 1 & \text{advection–dominated} \\ < 1 & \text{dispersion–dominated} \end{cases} \tag{3}$$

where

$\Delta \mathbf{s}$: element size.

The weak formulation of equation 2 without the well terms is given by

$$\begin{aligned}
&\left[\frac{1}{\Delta t}\int \Phi\, n\, \Omega\, dV + \frac{1}{2}\int \Phi\, \mathbf{v}_f \cdot \mathbf{grad}\, \Omega\, dV \right. \\
&\qquad \left. - \frac{1}{2}\int \Phi\, \mathrm{div}\,(n\, \mathbf{D} \cdot \mathbf{grad}\, \Omega)\, dV \right] \hat{c}^{i+1} \\
-&\left[\frac{1}{\Delta t}\int \Phi\, n\, \Omega\, dV - \frac{1}{2}\int \Phi\, \mathbf{v}_f \cdot \mathbf{grad}\, \Omega\, dV \right. \\
&\qquad \left. + \frac{1}{2}\int \Phi\, \mathrm{div}\,(n\, \mathbf{D} \cdot \mathbf{grad}\, \Omega)\, dV \right] \hat{c}^{i} = 0
\end{aligned} \tag{4}$$

where

$\Delta t = t^{i+1} - t^i$	:	time step	Ω	:	shape functions
Φ	:	test functions	$\hat{c}$	:	approximated solution.

The 'Streamline Upwind/Petrov Galerkin' method [4] in space smoothes the resulting curve, especially in highly advective cases, by means of a flow oriented modification of the 'Bubnov Galerkin' method (fig. 3):

$$\Phi = \Omega + \alpha\, \mathbf{grad}\, \Omega$$

with

α : upwind parameter .

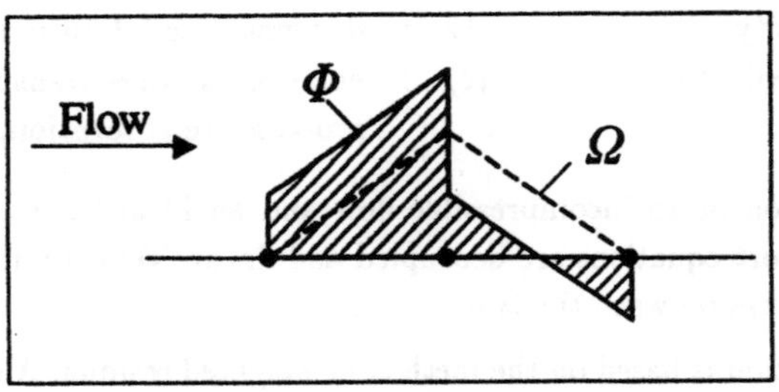

Figure 3: Streamline Upwind/Petrov Galerkin method

3 Grid Generating

For an optimum use of adaptive methods it is necessary to provide a well–discretized start grid. The complex physical relations have to be taken into account, thus the initial generation of grids is important.

In the fracture aquifer, discretization zones vary widely. One of them is the zone surrounding punctiform injection or extraction wells, due to high velocity gradients. For modeling transport processes the velocity field must be suitable, so the discretization has to take into account the discontinuity point. Since the spatial position will not change during a simulation, it is possible to generate an e.g. logarithmically adapted grid around wells or sinks [1].

The same effect occurs in the interaction zone between fracture and matrix. Here, the main problems are the highly curved streamlines between the highly-permeable fractures and the weakly-permeable rock matrix. An improvement is achieved by adaptation of the grid to the different fracture and matrix permeabilities. Figure 4 shows a typical injection/extraction problem in fractured rock. The system is described in figure 4a; at the left and the right of a fracture boreholes exist. The injection well is discretized by a point, the extraction well by modelling of an inner boundary. The fracture is discretized by 1d elements with a much higher permeability than that of the 2d matrix elements. Results of the transport simulation are shown in figure 4b. The isolines of the concentration are smooth, there are no oscillations.

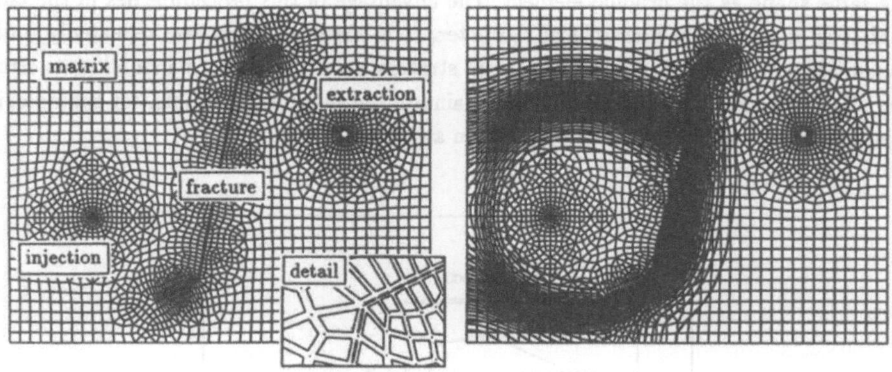

a) Problem adapted FE-mesh

b) Velocity field and concentration isolines

Figure 4: Injection/extraction problem

4 Transient Adaptive Method

Time-dependent problems, as the moving sharp concentration fronts in the fracture, cannot be efficiently treated by stationary criteria, e.g. a priori grid generation. In this case an hr-adaptive method is very suitable in order to refine the mesh size exclusively in the area of the sharp front. Based on dynamic data structures, the implemented algorithm is developed in order to obtain efficiency with minimum storage and cpu time requirements for any desired accuracy of the solution [2]. This problem is solved with a transient refinement and recoarsement strategy for coupled element types.

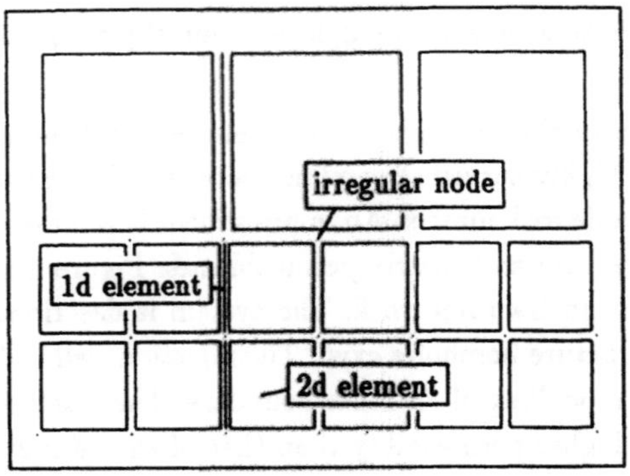

Figure 5: Refinement strategy

Figure 5 shows the refinement in the neighbourhood of different element types. The refinement structure makes use of irregular nodes to establish new elements, which have the same shape as the original element. The advantage of this procedure lies in the fact that it is of a strongly hierarchical character. It is possible to recoarse repeatedly until the initial grid is rebuilt. The recoarsement strategy generally leads to a small mass error (fig. 6), which smoothes the solution. To maintain the sharpness of fronts, for example, it is necessary to implement a mass correction algorithm.

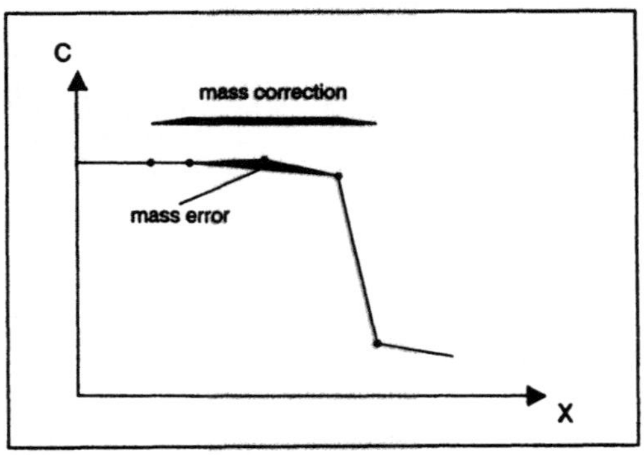

Figure 6: Patchwise mass correction

The relation between the two gray areas in figure 6 is given by

$$\int \Phi\, c^f \, dV \stackrel{!}{=} \int \Phi\, c^g \, dV \qquad (5)$$

with

c^f : results of the fine mesh
c^g : results of the coarsened mesh.

In order to obtain sufficient stability as well as optimum accuracy, the time step has to be adapted to the varying mesh size. The sensitive time step is controlled automatically. For advective processes the 'Courant' number is the decisive stability limit:

$$\Delta t_{Cr} = \min \left| \frac{\Delta s}{v} \right| . \qquad (6)$$

For highly dispersive transport processes, the 'Neumann' criterion is applied:

$$\Delta t_{Ne} = \min \left| \frac{\Delta s^2}{2D} \right| . \qquad (7)$$

5 Error Estimation

In the first place the efficiency of a refinement strategy depends on the quality of the used error indicators. Here, the problem lies in the mixed type of the partial differential equation (2).

For advection–dominated processes (hyperbolic type of equation) in the fracture zone, however, it is difficult to formulate universally valid error indicators. Therefore, heuristic indicators are used. When calculating highly advective transport processes, it is possible to use a difference indicator

$$\eta_j(t) = \Delta c_j = c_j - c_{j-1} \qquad j = 1,2,3,\ldots,n_e. \qquad (8)$$

where

η	:	error indicator
x	:	x–direction
n_e	:	number of elements.

Refinement is limited to zones with large concentration differences, where the finite element solution based on a coarse grid cannot approximate the exact solution satisfactorily. Recoarsement is performed in areas with much lower differences.

Numerical simulation of the rock matrix leads to a parabolic type. There exist some mathematically conclusive error criteria. The adaptive solution of the diffusion problem in section 6.2 uses the residuum indicator of 'Bietermann & Babuška' [3] to indicate the elements to be refined or recoarsed. For a 1d linear parabolic equation, the indicator is given by

$$\eta_j(t) = \left\{ \frac{|x_j - x_{j-1}|^2}{12\,D_x} \int\limits_{x_{j-1}}^{x_j} \left[\frac{\partial \hat{c}}{\partial t} - D_x \frac{\partial^2 \hat{c}}{\partial x^2} \right]^2 dx \right\}^{1/2} \qquad j = 0,1,2,\ldots,n_e. \qquad (9)$$

6 Example

The following example shows a 1d transport problem in a stationary flow field. At the left inflow boundary, the 1d fracture system has a concentration boundary condition. The example is calculated as a highly advective transport problem (sec. 6.1), as well as a diffusion problem (sec. 6.2). The start mesh for adaption contains 100 elements. Every element is being refined up to three times. The solutions are compared with those of a uniform refined mesh of 800 elements (reference solution) and analytical solutions [7]. The table gives the general calculation data:

$$\begin{aligned} \text{system length:} \quad & L &=& \ 100\,m \\ \text{concentration b.c.:} \quad & \bar{c} &=& \ 10\% \\ \text{porosity:} \quad & n &=& \ 1\,. \end{aligned}$$

6.1 Highly Advective Transport Problem

This variant shows the behaviour of the developed algorithm for a highly hyperbolic type of equation. The following parameters are a result of a 'Peclet' number (eqn. 3) of 100 and a 'Courant' number (eqn. 6) of 1:

$$\begin{aligned} \text{seepage velocity:} \quad & v_f &=& \ 10^{-6}\,m/s \\ \text{diffusion coeff.:} \quad & D &=& \ 0.625 \cdot 10^{-9}\,m^2/s \\ \text{initial time step:} \quad & \Delta t_0 &=& \ 10^{+6}\,s\,. \end{aligned}$$

The discretization errors are defined by a difference indicator. The concentration difference of one element should not exceed the refinement limit $\Delta c^f = 1\%$. The recoarsement limit is selected to $\Delta c^g = 0.1\%$. Figure 7 shows the results after $0.25 \cdot 10^7\,s$ (200 time steps).

Figure 7a shows the mesh refinement in the neighbourhood of the sharp concentration front. A difference indicator (eqn. 8) refines such elements where the solution shows very large concentration differences. The positive influence of the SU/PG method (fig. 3) is shown in figure 7b, the wiggles behind the front are smoothened. A comparison of the adaptive solution with a solution of a uniform refined grid is shown in figure 7c. Both curves coincide, especially the gradient of the front is depicted very well. The oscillations behind the peak are smoothened. The difficulties of the finite element method to approximate such a sharp front problem is demonstrated in figure 7d. The adaptive solution just as the reference solution cannot represent the analytical solution.

An advantage of using adaptive methods lies in the minimization of storage and cpu time requirements. In this highly advective example, the maximum number of elements in the adaptive case is 134, whereas the uniform refined solution uses 800 elements. Compared with this reduction of storage requirements, it was not possible to reduce cpu time. Since the calculation effort of the adaptive finite element model is much higher (several adaption steps par time step) than that of the reference system, the cpu time is almost identical ($\approx 100\,s$).

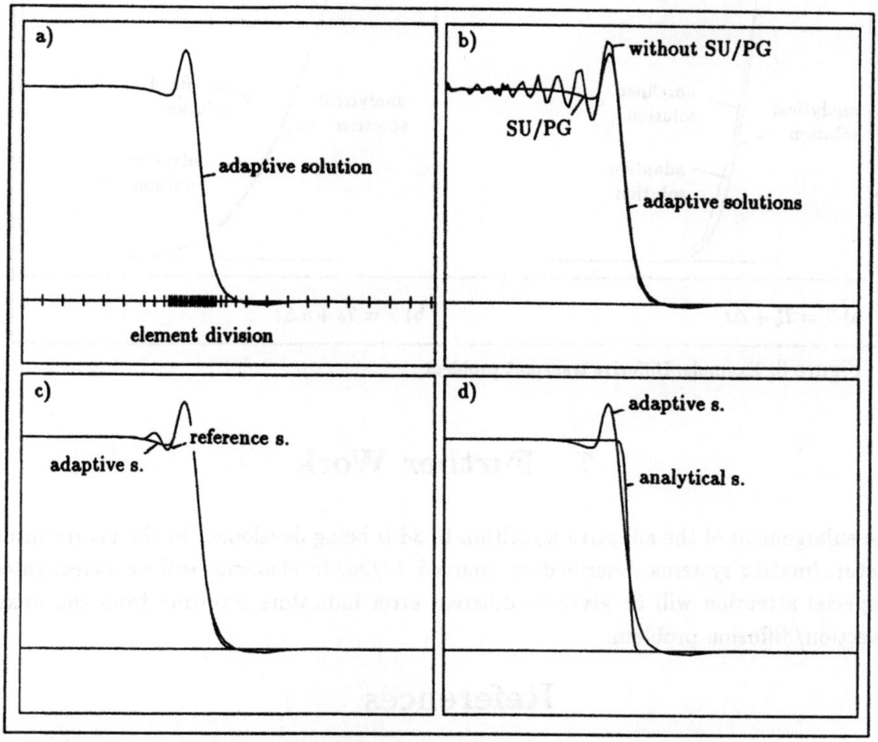

Figure 7: Example: Highly advective transport problem

6.2 Diffusive Transport Problem

Figure 8a shows one analytical, one adaptively refined, and one unrefined solution at the beginning of the simulation. The analytical and the adaptive solutions coincide, whereas the unrefined solution differs strongly. But with increasing time (fig. 8b) the solutions vary less. Because of this smoothening character of parabolic types of equations, adaptive methods for pure diffusion problems are not relevant in this context.

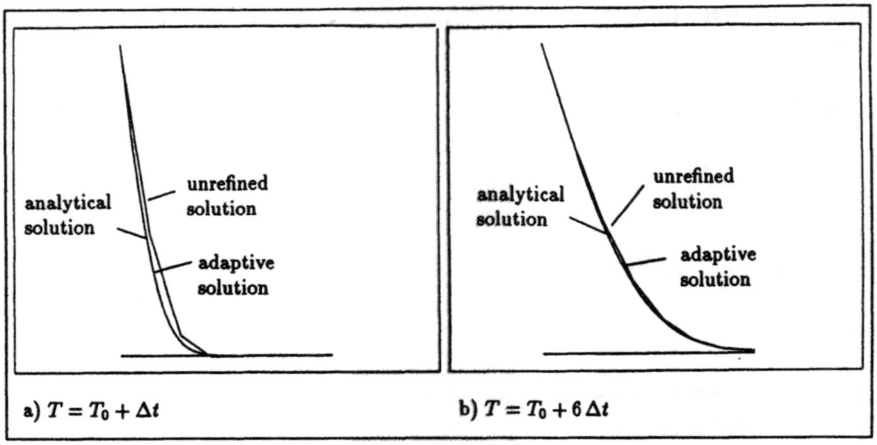

Figure 8: Example: Diffusive transport problem

7 Further Work

The enlargement of the adaptive algorithm to 3d is being developed. In the nearer future fracture/matrix systems described by coupled 1d/2d/3d elements will be investigated. A special attention will be given to different error indicators resulting from the mixed advection/diffusion problem.

References

[1] Barlag,C., Zielke,W. *Untersuchung zur punktweisen Einleitung von Injektionsmengen.* Technischer Bericht, Institut für Strömungsmechanik und Elektron. Rechnen im Bauwesen, Universität Hannover, 1994.

[2] Barlag,C., Zielke,W. *Adaptive Modellierung dominant-advektiver Transportprozesse.* Technischer Bericht, Institut für Strömungsmechanik und Elektron. Rechnen im Bauwesen, Universität Hannover, 1995.

[3] Bietermann,M., Babuška,I. The Finite Element Method for Parabolic Equations. *Num. Math.*, 40:339–406, 1982.

[4] Brooks,A.N., Hughes,T.J.R. Streamline Upwind/Petrov–Galerkin Formulations for Convection Dominated Flows with Particular Emphasis on the Navier–Stokes Equations. *Comp. Meth. Appl. Mech. Eng.*, 32:199–259, 1982.

[5] Helmig,R. *Theorie und Numerik der Mehrphasenströmungen in geklüftet-porösen Medien.* Dissertation, Institut für Strömungsmechanik und Elektron. Rechnen im Bauwesen, Universität Hannover, Bericht Nr.34, 1993.

[6] Kröhn,K.-P. *Simulation von Transportvorgängen im klüftigen Gestein mit der Methode der Finiten Elemente.* Dissertation, Institut für Strömungsmechanik und Elektron. Rechnen im Bauwesen, Universität Hannover, Bericht Nr.29, 1991.

[7] Lege,T., Kolditz,O., Zielke,W. *Strömungs- und Transportmodellierung (Handbuch zur Erkundung des Untergrundes von Deponien und Altlasten; Bd.2).* Springer Verlag, 1996.

[8] ROCKFLOW. *Theorie und Benutzeranleitung zum Programmsystem.* Auszug:
Teil 1: Wollrath,J., Helmig,R. (1992): SM2 — Strömungsmodul für inkompressible Fluide;
Teil 2: Kröhn,K.-P., Lege,T. (1992): TM2 — Transportmodul für inkompressible Fluide (Tracertransport);
Teil 5: Behrendt,F., Kasper,H. (1993): XDALI — Prä- und Postprozessor;
Teil 10a: Lege,T., Taniguchi,T., Barlag,C. (1994): NG2D — Netzgenerator für beliebige 1D- und 2D-Gebiete;
Teil 11: Barlag,C. (n.n.): ATM — Adaptives Transportmodul für inkompressible Fluide.
Technischer Bericht, Institut für Strömungsmechanik und Elektron. Rechnen im Bauwesen, Universität Hannover, 1986–1995.

Lagrange–Galerkin Approximation For Advection–Dominated Contaminant Transport With Nonlinear Equilibrium Or Non–equilibrium Adsorption *

J.W. BARRETT[‡], H. KAPPMEIER[†] and P. KNABNER[†]

(†) University of Erlangen–Nürnberg
Institute of Applied Mathematics
Martensstr. 3, D 91058 Erlangen
Germany

(‡) Imperial College
Department of Mathematics
London SW7 2BZ
United Kingdom

Summary

An extension of the Lagrange–Galerkin approach is developed for advection-dominated problems with nonlinear adsorption, either being in equilibrium or in non–equilibrium, possibly with isotherms of the Freundlich type. The scheme should be feasable also for the hyperbolic limit case.

The basic problems to deal with are the possibility of non–unique characteristics due to the nonlinear isotherms and the incorporation of the time–dependent non–equilibrium adsorption kinetics, avoiding a strong restriction on the CFL–number.

The one–dimensional scheme derived is able to handle shock solutions.

In the non–equilibrium case a more severe restriction on the timestep has to be regarded.

*supported by British-German ARC Grant, BMFT FKZ 03–KN7ERL

Introduction

The numerical simulation of the transport of solutes by dispersion and advection, which in addition can be adsorbed to the soil matrix by equilibrium/non–equilibrium kinetics, is one of the main problems in soil science and subsurface hydrology.

The aim of this paper is to develop an appropriate Lagrange–Galerkin discretization for a one–site transport model with either equilibrium or non–equilibrium adsorption sites. The considered model, in the case of non–equilibrium adsorption reactions, takes the following form. The unknowns u, v, the dissolved and the adsorbed concentration, have to fulfill:

$$
\begin{aligned}
\partial_t u + \partial_t v + \underline{q}\nabla u - \nabla\cdot(\underline{\underline{D}}\nabla u) &= 0 \quad in \quad (0,T]\times\Omega\subset(0,T]\times\mathbf{R}^N \\
\partial_t v &= k(\varphi(u) - v) \quad in \quad (0,T]\times\Omega \\
u(0,x) &= u_0(x) \quad , \quad x\in\Omega \\
v(0,x) &= v_0(x) \quad , \quad x\in\Omega,
\end{aligned}
\tag{1}
$$

and appropriate boundary conditions.

The water content and the bulk density are assumed to be constant and scaled to unity, and

$$
\begin{aligned}
\varphi \in C^r([0,\infty))\cap C^1_{loc}((0,\infty)) \quad & r\in(0,1) \\
\varphi(\cdot) &\geq 0 \\
\infty > \varphi'(s) &\geq 0 \quad \text{for } s>0 \\
\infty \geq \lim_{s\to+0}\varphi'(s) &\geq 0,
\end{aligned}
\tag{2}
$$

$$
k > 0, \tag{3}
$$

The diffusion/dispersion coefficient D and the flow field q are assumed to be smooth functions, such that:

$$q : (0,T] \times \Omega \to \mathbf{R}^N, \quad ||q|| \neq 0, \quad \nabla \cdot q = 0, \tag{4}$$

$$D : (0,T] \times \Omega \to \mathbf{R}^{N^2}, \quad \nu||\xi||^2 \leq \xi^T D(t,x)\xi \leq \mu||\xi||^2 \text{ for } \nu, \mu > 0. \tag{5}$$

The exact conditions under which (1) has weak or strong solutions are analysed in Knabner [3].

The equilibrium model can be obtained formally from (1) by the limiting process k to ∞, just leaving the dissolved concentration u as unknown.

We will consider Lagrange–Galerkin approaches for discretising the two models. For a linear instationary convection–diffusion equation, this type of method is characterised by writing the hyperbolic part as derivative along the characteristics, discretising this derivative by standard differences in time and dealing with the diffusive part implicitly in time. Finally the finite element discretization is applied to the weak formulation of the semi–discretised equation. The main advantage of this scheme are small truncation errors in time and thus large CFL–numbers.
A Lagrange–Galerkin approximation of the model considered should have this property as far as possible, too. Developing such a discretization has to handle three basic problems:

- the incorporation of the hyperbolic limit case $D = 0$, leading to discontinous solutions.
- the incorporation of degenerate nonlinearities of the Freundlich type ($\varphi'(0+) = \infty$) leading to solutions with a sharp front, which have much in common with the hyperbolic limit case (in particular the speed of propagation), cf. Knabner [3].
- the incorporation of kinetic adsorption reactions.

Throughout this paper we will restrict the derivation of the discretization to the one–dimensional case, i.e. $N = 1$, but it can be extended to two dimensions , as well. Furtheron we use linear finite elements with vertex quadrature to avoid instabilities, caused by quadrature formulas of higher order (see Morton et al. [4]).

Let $\Omega = [0, L]$, $L \in \mathbf{R}^+$ and for the sake of simplicity q a constant, say $q > 0$. ϕ denotes a differentiable approximation of φ. Just for the sake of simplicity uniform discretizations in time and in space are chosen.

$$0 = t^0 < \cdots < t^{n-1} < t^n < \cdots < t^{N_t} = T, \qquad \Delta t = t^i - t^{i-1}, 0 < i \le N_t, \tag{6}$$

$$0 = x_0 < \cdots < x_{j-1} < x_j < \cdots < x_{N_x} = L, \qquad h = x_i - x_{i-1}, 0 < i \le N_x. \tag{7}$$

From now on we will define characteristics not with respect to a given equation, but to a given flux. A Lagrange–Galerkin approach can result in small time truncation errors only, if the characteristics are computed sufficiently accurate. As consequence, a given flux b_j ($0 \le j \le N_x$), represented by its nodal values, will be approximated piecewise constant on each element and the characteristics are then calculated exactly with respect to this approximative flux. $\overline{x}_j$ denotes the point in space where the characteristic, starting at t^n at the node x_j, arrives at $t^n - \Delta t$ and is called 'backtrace'.
In formulas:

$$b_{j-1/2} := b(x_j), \tag{8}$$

$$\overline{x}_j = x_{j-m} + b_{j-m-1/2}\left[\sum_{k=1}^{m}\left(\frac{h}{b_{j+1/2-k}}\right) - \Delta t\right], \tag{9}$$

and $m \ge 0$ is the first integer for which

$$\sum_{k=1}^{m+1}\left(\frac{h}{b_{j+1/2-k}}\right) > \Delta t. \tag{10}$$

Furtheron the following notations will be used:

$$Q_j(x, U^{n-1}) := \begin{cases} \dfrac{\phi(U^{n-1}(x)) - \phi(U_j^{n-1})}{U^{n-1}(x) - U_j^{n-1}}, & \text{if} \quad U_j^{n-1} \ne U^{n-1}(x) \\ \phi'(U_j^{n-1}), & \text{otherwise.} \end{cases} \tag{11}$$

$$\delta(D\delta U)_j^n = \\ -\left[-(\frac{D_{j-1}+D_j}{2})U_{j-1}^n + (\frac{D_{j-1}+2D_j+D_{j+1}}{2})U_j^n - (\frac{D_j+D_{j+1}}{2})U_{j+1}^n\right]. \tag{12}$$

A first Lagrange–Galerkin discretization was already presented in Knabner, Kappmeier, Barrett [2]. Therefore we only summarize the derivation, restricted to the hyperbolic case $D = 0$, as the discretization of the diffusive term causes no problems and will be neglected therefore. There will be some small but important differences to the previous version.
The starting point is to introduce an ansatz–function $\alpha(t, x, u)$ for a given timestep and to define the characteristics with respect to the flux $b = \frac{q}{1+\alpha(t,x,u)}$. That is:

$$\begin{aligned} &(\partial_t\phi(u) - \alpha(t,x,u)\partial_t u) + \\ &(1+\alpha(t,x,u))\left[\partial_t u + \frac{q}{1+\alpha(t,x,u)}\partial_x u\right] = 0 \quad in\ (t^{n-1}, t^n) \\ &u(t,0) = g(t) \quad in \quad (t^{n-1}, t^n) \\ &u(t^{n-1}, x) = U^{n-1}(x). \end{aligned} \tag{13}$$

U^{n-1} denotes the computed solution of the last timestep. Here and in the following we use a Dirichlet boundary condition on the inflow boundary $x = 0$.
Known choices for α are $\alpha(t, x, u) = 0$ ('Pore–Velocity–Scheme') and $\alpha(t, x, u) = \phi'(u)$ ('Retarded–Velocity–Scheme' (RVS)). Choosing large CFL–numbers, the Pore–Velocity–Scheme tends to compute 'zig–zag–solutions', smears a lot in the case of steep profiles and the RVS fails, if shocks are present.
As a new approach we defined α on the discrete level in such a way, that a pure (discrete) derivative along the characteristic resulted. By this idea (see Knabner, Kappmeier, Barrett [2] for details) the following scheme is derived from (13). We denote this new scheme by FIS or by FISD in the case with diffusion:

$$U_j^n + \phi(U_j^n) = (1+\alpha_j)U^{n-1}(\overline{x}_j) - \alpha_j U_j^{n-1} + \phi(U_j^{n-1}), \tag{14}$$

where α_j is the infimum of all solutions $\beta \geq 0$ of

$$\beta = Q_j(\overline{x}_j, U^{n-1}), \tag{15}$$

for $\overline{x}_j$ defined by $\tilde{b}$

$$\tilde{b}_i = \frac{q}{1+\alpha_i} \qquad i \neq j, \tag{16}$$

$$\tilde{b}_j = \frac{q}{1+\beta}. \tag{17}$$

In this case

$$U_j^n = U^{n-1}(\overline{x}_j). \tag{18}$$

is equivalent to (14).

Note that in the present simple situation of one space dimension and a constant $q > 0$ the resolution of (14)–(17), which in general defines a nonlinear coupling between all values U_i^n and α_i, can proceed according to the numbering of the nodes. Then U_i^n, α_i, $i = 1, \cdots, j-1$ are already known and (14)–(17) define a nonlinear equation for α_j and U_j^n. In general for 'most' fluxes a 'downwind numbering' is possible, which preserves the property, that only known values α_i additionally to α_j are involved in the computation of the backtrace $\overline{x}_j$.
An open item in Knabner, Kappmeier, Barrett [2] was the existence of solutions β for the equations (15)–(17).

THEOREM 1 The computed solution U^{n-1} of the old time level is extended to **R**. *Furtheron $U^{n-1} \geq 0$, $\phi \in C^1([0,\infty))$, $q =$ const > 0 are assumed. $\alpha_i \in [0,\infty)$ are given for $0 \leq i < j$. The size of the timestep Δt from U^{n-1} to U^n is chosen.*
Then there exists a solution $\beta \geq 0$ for (15)–(17).

The solutions β for (15)–(17) can be computed by using the bisection method.

A speed up of convergence can be obtained by using a method of higher order (secant method, Brent's method). The main difference to the first version in Knabner, Kappmeier, Barrett [2] is the different piecewise constant approximation of the flux b, now defined by the downwind node (see (8)) and not by the midvalue of the flux at both nodes as in the previous paper. Now we have for $\Delta t \to 0$ that $b_{j-1/2} \to \frac{q}{1+\phi'(U_j^{n-1})}$ independent of the spatial discretization, i.e. the situation of the retarded–velocity–scheme.

The system (1) with the unknowns u, v can be transformed to an equivalent partial differential equation in the unknown u alone:

$$\begin{aligned} \partial_t u + \partial_t \varphi(u) + \underline{q}\nabla u - \nabla\cdot(\underline{\underline{D}}\nabla u) + & \\ k^{-1}\partial_t\left[\partial_t u + \underline{q}\nabla u - \nabla\cdot(\underline{\underline{D}}\nabla u)\right] = 0 & \quad in \quad (0,T]\times\Omega\subset(0,T]\times\mathbf{R}^N \\ u(0,x) = u_0(x) & \quad x\in\Omega \qquad (19) \\ k^{-1}\left[\partial_t u + \underline{q}\nabla u - \nabla\cdot(\underline{\underline{D}}\nabla u)\right](0,x) = v_0(x) - \varphi(u_0(x)) & \quad x\in\Omega. \end{aligned}$$

By (19) the non–equilibrium case can be interpreted as the equilibrium model amended by a pertubation term of the order $\frac{1}{k}$ in $\partial_t u$. Only in special cases this term can be rewritten as a purely spatial operator. In the one–dimensional case for constant coefficients we have for travelling wave solutions $\partial_t = -a\partial_x$ with a wave speed $a < q$ (cf. v.Duijn, Knabner [1]) and thus:

$$\begin{aligned} & k^{-1}\partial_t\left[\partial_t u + q\partial_x u - D\partial_{xx} u\right] = \\ & -k^{-1}\left[a(q-a)\partial_{xx} u - Da\partial_{xxx} u\right]. \end{aligned} \qquad (20)$$

In general an extension of the FIS/FISD to the non–equilibrium case will have an additional time truncation error due to the discretization of $\partial_t u$.

As it makes no additional difficulties to allow for diffusion, we will neglect diffusion during the analyis and derivation of the different schemes. But we formulate them for the case with diffusion.

Introducing a nonnegative ansatz function $\alpha(t,x,u,v)$, equation (1), restricted to (t^{n-1}, t^n) and with U^{n-1}, V^{n-1} as initial condition, becomes:

$$\begin{aligned} \partial_t v - \alpha(t,x,u,v)\partial_t u + & \\ (1+\alpha(t,x,u,v))\left[\partial_t u + \frac{q}{1+\alpha(t,x,u,v)}\partial_x u\right] = 0 & \quad in \quad (t^{n-1},t^n) \\ \partial_t v = k(\phi(u) - v) & \quad in \quad (t^{n-1},t^n) \qquad (21) \\ u(t,0) = g(t) & \quad in \quad (t^{n-1},t^n) \end{aligned}$$

$$u(t^{n-1}, x) = U^{n-1}(x)$$
$$v(t^{n-1}, x) = V^{n-1}(x).$$

In the equilibrium case the FIS chooses $\alpha(t, x, u)$, so that $\partial_t \phi(u) - \alpha(t, x, u)\partial_t u$ vanishes on the discrete level. The corresponding requirement for (21) is to define α by $\partial_t v - \alpha(t, x, u, v)\partial_t u = 0$ on the discrete level. However this idea leads to a flux b, which is possibly unbounded and changes sign.

The flux b should model the flux, or a part of it, of the dissolved concentration, α should depend on u. Furthermore b should incorporate the influence of the kinetics on the flux of u, by using the corresponding equilibrium reaction and a k–dependent correction term. Till the end of this Section u, v are assumed to be regular enough, to allow for the following transformations.

The ODE in (21) implies:

$$(\partial_t + k)(v) = k(\phi(u)), \tag{22}$$
$$(\partial_t + k)(\partial_t v) = k(\partial_t \phi(u)). \tag{23}$$

(23) can now be used in two ways:

1. By variation of the constant one gets:

$$\begin{aligned} \partial_t v(t) = \quad & \exp(-k(t - t^{n-1}))\partial_t v(t^{n-1}) + \\ & k \int_{t^{n-1}}^{t} \exp(-k(t-s))\partial_s \phi(u)\, ds \quad t \in [t^{n-1}, t^n), \end{aligned} \tag{24}$$

Replacing $\partial_t v$ in (21) by (24) and then $\partial_t v(t^{n-1})$ by means of (22) leads to:

$$\begin{aligned} & k \int_{t^{n-1}}^{t} \exp(-k(t-s))\partial_s \phi(u)\, ds - \alpha(t, x, u, v)\partial_t u \; + \\ & (1 + \alpha(t, x, u, v)) \left[\partial_t u + \frac{q}{1 + \alpha(t, x, u, v)} \partial_x u \right] \; = \\ & -k \exp(-k(t - t^{n-1}))(\phi(u(t^{n-1})) - v(t^{n-1})). \end{aligned} \tag{25}$$

To arrive at a Lagrange–formulation $\alpha(t, x, u, v)$ should be defined in such a way, that $k \int_{t^{n-1}}^{t} \exp(-k(t-s))\partial_s \phi(u)\, ds - \alpha(t, x, u, v)\partial_t u = 0$ (on the discrete level).

A possible scheme is determined by the chosen approximation of $k\int_{t^{n-1}}^{t}\exp(-k(t-s))\partial_s\phi(u)\,ds$.

2. Replacing $\partial_t v$ in (21) by means of (23) results in:

$$\begin{aligned}\partial_t\phi(u)-\alpha(t,x,u,v)\partial_t u-k^{-1}\partial_{tt}v+&\\ (1+\alpha(t,x,u,v))\left[\partial_t u+\frac{q}{1+\alpha(t,x,u,v)}\partial_x u\right]&=0. \qquad (26)\end{aligned}$$

Now $\alpha(t,x,u,v)$ will be defined according to the FIS, implying $\partial_t\phi(u)-\alpha(t,x,u,v)\partial_t u=0$ on the discrete level. This should restrict the time truncation error due to $\partial_t\phi(u)-\alpha(t,x,u,v)\partial_t u$.

The main difference to the equilibrium case is the pertubation term $k^{-1}\partial_{tt}v$. One way to handle this part is to replace it by $k^{-1}\partial_t\left[\partial_t u+\underline{q}\partial_x u\right]$ and to end up with equation (19). This second term can be written as a derivative in time of one along the characteristic of the underlying flowfield. Due to the derivative along the characteristic it may be expected, that this leads to a smaller time truncation error than the discretization of $k^{-1}\partial_{tt}v$ itself.

The scheme presented next is a consequence of (25), by approximating $k\int_{t^{n-1}}^{t}\exp(-k(t-s))\partial_s\phi(u)\,ds$ by $k\int_{t^{n-1}}^{t}\exp(-k(t-s))\,ds\ \partial_t\phi(u)=(1-\exp(-k(t-t^{n-1})))\partial_t\phi(u)$ and $\exp(-k(t-t^{n-1}))$ by $\frac{1}{1+k(t-t^{n-1})}$.

Discretising (25) : The resulting scheme for interior nodes is, with $b_j=\frac{q_j}{1+\alpha_j}$:

$$\begin{aligned}U_j^n-\frac{\Delta t}{h^2}\delta(D\delta U)_j^n+\frac{k\Delta t}{1+k\Delta t}\phi(U_j^n)=&\\ U^{n-1}(\overline{x}_j)+\frac{k\Delta t}{1+k\Delta t}\phi(U^{n-1}(\overline{x}_j))+\frac{k\Delta t}{1+k\Delta t}\left[V_j^{n-1}-\phi(U_j^{n-1})\right],& \qquad (27)\end{aligned}$$

where α_j is the infimum of all solutions $\beta\geq 0$ of

$$\beta=\frac{k\Delta t}{1+k\Delta t}Q_j(\overline{x}_j,U^{n-1}), \qquad (28)$$

for $\overline{x}_j$ defined by $\tilde{b}$

$$\tilde{b}_i = \frac{q}{1+\alpha_i} \qquad i \neq j, \tag{29}$$
$$\tilde{b}_j = \frac{q}{1+\beta}, \tag{30}$$

$$V_j^n = \frac{1}{1+k\Delta t}\left(k\Delta t\ \phi(U_j^n) + V_j^{n-1}\right) \quad \forall n \geq 1. \tag{31}$$

Discretising (19) : As consequence of (26) the resulting scheme for interior nodes is, with $b_j = \frac{q_j}{1+\alpha_j}$:

$$\begin{aligned} U_j^n - \frac{\Delta t}{h^2}\delta(D\delta U)_j^n + \frac{k\Delta t}{1+k\Delta t}\phi(U_j^n) = \\ \frac{k\Delta t}{1+k\Delta t}\left[U^{n-1}(\overline{x}_j) + \phi(U^{n-1}(\overline{x}_j))\right] + \frac{1}{1+k\Delta t}U^{n-1}(x_j - q\Delta t) \\ + \begin{cases} \frac{1}{1+k\Delta t}\left(U_j^{n-1} - \frac{\Delta t}{h^2}\delta(D\delta U)_j^{n-1} - U^{n-2}(x_j - q\Delta t)\right) & \text{if } n \geq 2 \\ \frac{k\Delta t}{1+k\Delta t}\left(v_0(x_j) - \phi(u_0(x_j))\right) & \text{if } n = 1, \end{cases} \end{aligned} \tag{32}$$

where α_j is the infimum of all solutions $\beta \geq 0$ of

$$\beta = Q_j(\overline{x}_j, U^{n-1}), \tag{33}$$

for $\overline{x}_j$ defined by $\tilde{b}$

$$\tilde{b}_i = \frac{q}{1+\alpha_i} \qquad i \neq j, \tag{34}$$
$$\tilde{b}_j = \frac{q}{1+\beta}, \tag{35}$$

V_j^n is defined as in (31).

It should be noticed, that both schemes ((32), (27)) reduce to the FIS/FISD for $k \to \infty$ and the linear finite element discretization of the instationary diffusion equation with adsorption, if $q = 0$.

Numerical Results

In the equilibrium case the FIS/FISD turned out to be superior to the Retarded–Velocity–Scheme or the Pore–Velocity–Scheme (see Knabner, Kappmeier, Barrett [2]). Therefore we take the Pore–Velocity–Scheme in the non–equilibrium case as benchmark, as well. For interior nodes the Pore–Velocity–Scheme (PVS) is:

$$\begin{aligned} U_j^n - \frac{\Delta t}{h^2}\delta(D\delta U)_j^n + \frac{k\Delta t}{1+k\Delta t}\phi(U_j^n) &= \\ U^{n-1}(x_j - q\Delta t) + \frac{k\Delta t}{1+k\Delta t}V^{n-1}(x_j) & \\ V_j^n = \frac{1}{1+k\Delta t}\left(k\Delta t\ \phi(U_j^n) + V_j^{n-1}\right) \quad \forall n \geq 1. & \end{aligned} \tag{36}$$

The three schemes ((32), (27), (36)) were tested in three situations:
For travelling waves with nonlinear isotherm $\varphi(u) = u^p$, $p \in (0,1)$ (situation 1), pulse with linear isotherm $\varphi(u) = K_d u$, $K_d > 0$ (situation 2), pulse with nonlinear isotherm $\varphi(u) = u^p$, $p \in (0,1)$ (situation 3).

In the equilibrium case (27) and (32) coincide with the FIS/FISD. For linear and nonlinear isotherms the FIS/FISD allows for larger CFL–numbers and smears much less than the Pore–Velocity–Scheme (36), if steep fronts are present. However strong refinement in space can be neccessary to conserve mass, in the case of degenerate nonlinear isotherms, sharp fronts and CFL–numbers larger than one. In summary the FIS/FISD seems to be the better choice.

In the non–equilibrium case the experiments show:
Scheme (27) requires smaller timesteps than the Pore–Velocity–Scheme (36), otherwise peaks and oscillations can be observed and therefore seems to be no good extension of the FIS/FISD.
In comparison to the equilibrium case a stronger restriction of the timestep size is necessary for Scheme (32), as well. Too large timesteps cause peaks and oscillations.
For $\frac{q}{1+\alpha}$ nearby q or small rate parameter k the Pore–Velocity–Scheme (36) allows for larger timesteps. In the other cases scheme (32) is superior.
The Pore–Velocity–Scheme (36) tends to smear more. In the case of too large timesteps

the Pore–Velocity–Scheme (36) computes zig–zag–solutions, resulting in a lot of smearing afterwards.

In consequence, scheme (32) seems to be an acceptable extension of the FIS/FISD to the non–equilibrium case.

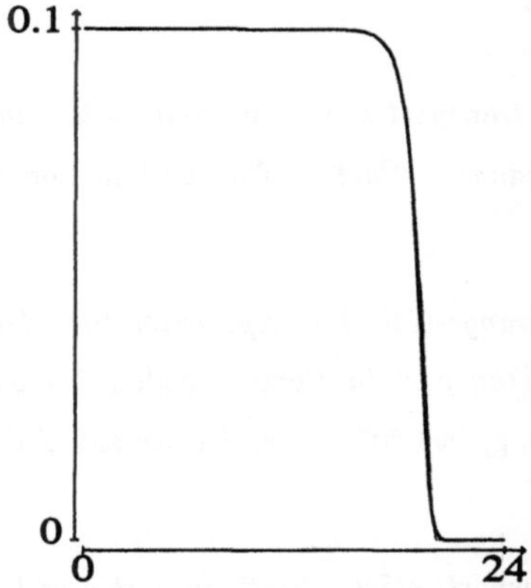

Figure 1: Scheme (32); $t = 20 \cdot \Delta t$

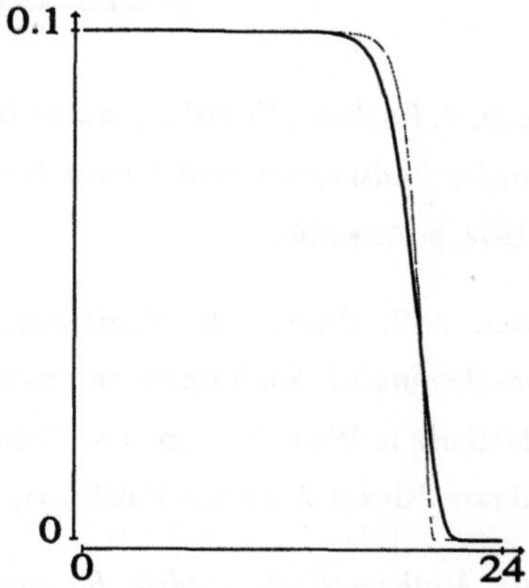

Figure 2: PVS (36); $t = 20 \cdot \Delta t$

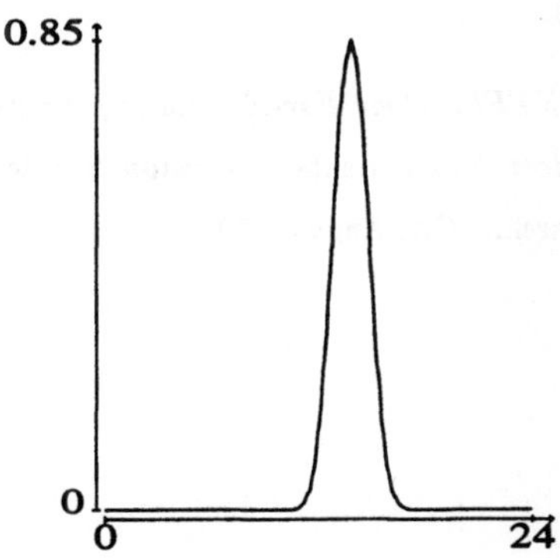

Figure 3: Scheme (32); $t = 35 \cdot \Delta t$

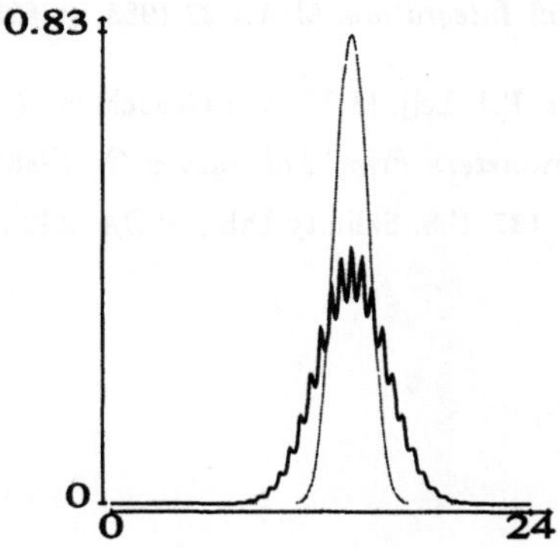

Figure 4: PVS (36); $t = 35 \cdot \Delta t$

The thick line displays the computed and the thin one the exact solution.

The figures 1, 2 present an example of situation 1 for $k = 1.0, p = 0.6, h = 0.05, \Delta t = 0.737777, D = 0, q = 1$.

The figures 3, 4 show an example of situation 2 for $k = 10.0, K_d = 1.0, h = 0.05, \Delta t = 8 \cdot 0.0737777, D = 10^{-8}, q = 1$.

The exact solutions of situation 2 were computed with the package of Toride et al. [5].

References

[1] C.J. v.Duijn, P. Knabner, *Travelling waves in the transport of reactive solutes through porous media: Adsorption and binary ion exchange – Part 2*, Transport in Porous Media 8 1992, pp.199-226.

[2] P. Knabner, J.W. Barrett, H. Kappmeier, *Lagrange-Galerkin Approximation For Advection-Dominated Nonlinear Contaminant Transport In Porous Media*, Computational Methods in Water Resources X., Volume 1, pp.299-307, Water Science and Technology Library, Kluwer Academic Publishers, 1994.

[3] P. Knabner, *Mathematische Modelle für den Transport gelöster Stoffe in sorbierenden porösen Medien*, Verlag Peter Lang, Frankfurt/M., 1991.

[4] K.W. Morton, A. Priestley, E. Süli : *Stability of the Lagrange–Galerkin Method with non–exact Integration*, M^2AN 22 1988, pp.625-653.

[5] N. Toride, F.J. Leij, M.Th. van Genuchten, *The CXTFIT Code For Estimating Transport Parameters From Laboratory Or Field Tracer Experiments – Version 2.0* Res. Rep. No. 137, U.S. Salinity Lab., USDA, ARS, Riverside, CA, August 1995.

A Parallel Adaptive Finite-Element Semi-Lagrangian Advection Scheme for the Shallow Water Equations

Jörn Behrens
Alfred-Wegener-Institut für Polar- und Meeresforschung
P.O. Box 12 01 61
D-27515 Bremerhaven

Summary

A numerical scheme for solving the shallow water equations in a plane periodical domain is presented. Adaptivity is utilized to resolve local phenomena adequately while saving computational cost. A semi-Lagrangian time discretization provides unconditional stability and allows large time-steps. Furthermore the semi-Lagrangian scheme is well suited for parallelization. An index array method is used to distribute data in a distributed virtual shared memory environment on a KSR-1 parallel computer. The scheme proves to be accurate and efficient. Parallelization and the use of adaptive methods make finite element methods competitive by the order of computing-time to finite difference methods in ocean or atmospheric modelling.

Introduction

A challenging topic in the field of modelling and computation in environmental sciences is the simulation of world ocean circulation. A working group at Alfred-Wegener-Institute focuses on the development and parallelization of ocean modelling software [7, 11]. The work presented here has been developed within this framework.

As a large range of scales plays an important role to the physics of ocean circulation [22], long time-series and high spatial resolution (e.g. for mesoscale phenomena) lead to large numbers of time-steps and unknowns simultaneously. Ocean modeling is therefore regarded as one of the so called grand challenges of computational physics [19].

Current ocean-models in general depend on standard numerical methods like finite-difference or spectral methods. These methods, however, suffer from several insufficiencies. Spectral methods are difficult to adjust to irregular domains. Finite-difference methods are not easy to adapt locally and are rather inflexible with respect to boundary conditions. The research project presented here aims at the following capabilities for ocean or atmospheric modeling software:

- High (local) spatial resolution.
- Parallel efficiency and scalability.
- Numerical efficiency and robustness.
- Flexibility with respect to boundary conditions and geometry.

These requirements have been addressed by different means. Adaptivity, in this case adapting mesh-sizes in a triangular grid, has been introduced in order to provide high local resolution maintaining the order of numerical operations. Finite element methods (FEM) have been utilized to achieve flexibility at the boundaries and provide a robust numerical scheme for the solution of the inherent elliptic boundary value problems. A semi-Lagrangian time discretization scheme has been chosen, because it is well suited for parallelization. Furthermore it is unconditionally stable providing large time-steps and numerical efficiency.

In this article we will focus on the presentation of the semi-Lagrangian discretization in combination with adaptive grid-refinement. Some aspects regarding parallelization will be mentioned and an implementation for the shallow water equations (SWE) will be briefly outlined. SWE have been chosen, because they are simple, yet they include the main difficulties occuring in more complicated situations. Furthermore, techniques applied to SWE can be generalized to primitive equation models.

The paper is organized in the following way: First an introduction to the semi-Lagrangian discretization will be given. Choosing the passive advection as an example for simplicity the combination of semi-Lagrangian discretization with grid adaptivity is demonstrated. Some results will be given. Second the application of these results to the shallow water equations will be shown. Results for a simple model-problem prove the numerical stability of the scheme. Finally concluding remarks will be given.

Semi-Lagrangian discretization of the passive advection equation

In order to give an introduction to the semi-Lagrangian method (SLM) of discretizing advection problems, the passive advection equation with a given wind is chosen. The one-dimensional equation reads

$$\frac{du}{dt} = \frac{\partial u}{\partial t} + \frac{dx}{dt}\frac{\partial u}{\partial x} = 0, \tag{1}$$

where the wind $\frac{dx}{dt} = a(x,t)$ and an initial condition $u(x, t = 0) = u^0(x)$ is given. The multi-dimensional form will be used in the sequel:

$$\frac{du}{dt} = \frac{\partial u}{\partial t} + \mathbf{a} \cdot \nabla u = 0. \tag{2}$$

Again $\mathbf{a}(\mathbf{x}, t)$ has to be given, where $\mathbf{x} = (x_1, x_2, \dots)$ and t are spatial and time coordinates respectively [17].

SLM discretization

The idea of the SLM is to discretize the Lagrangian form of equation (2) along trajectories [20]. At each time-step trajectories are chosen such that they end at grid-points of a prescribed grid. This scheme circumvents the disadvantage of purely Lagrangian schemes, i.e. a possible irregularity of the spatial distribution of the particle traces.

Equation (2) is discretized by the SLM in the following $\mathcal{O}(\Delta t^2)$-accurate way:

$$\frac{u(\mathbf{x}_m, t_n + \Delta t) - u(\mathbf{x}_m - \boldsymbol{\alpha}_m, t_n)}{2\Delta t} = 0. \tag{3}$$

Here $\mathbf{x}_m$ represents a grid-point ($m = 1, \ldots, N$, N number of grid-points), α_m represents the displacements to the upstream-points of the linear approximation to the trajectory corresponding to grid-point $\mathbf{x}_m$. Knowing the value of $u(\mathbf{x}_m - \alpha_m, t_n)$ at the upstream-point, one knows the new grid-value of u. A crucial part of the SLM, therefore, is the interpolation of the upstream-values which generally are located between grid-points.

Calculating the displacements is an easy task in the context of passive advection, where the wind is constantly given. The following iteration gives a good $\mathcal{O}(\Delta t^2)$ approximation to α_m after a few iteration steps (normally 3-4):

$$\alpha_m^{(k+1)} = \Delta t \; \mathbf{a}(\mathbf{x}_m - \frac{\alpha_m^{(k)}}{2}, t_n + \frac{\Delta t}{2}). \tag{4}$$

With equations (3) and (4) the algorithm of the SLM for passive advection is given by:

Algorithm 1 (semi-Lagrangian advection)

1. *Solve (4) for all* α_m.
2. *Interpolate* u *at upstream-points* $\mathbf{x}_m - \alpha_m$.
3. *Calculate new values of* u *according to equation (3).*

Different interpolation schemes have been investigated by several authors [8, 12, 14, 16, 24]. A good interpolation method, balancing the computational work with the gain in accuracy, is a cubic spline interpolation, which is of fourth order. As cubic interpolation is neither shape preserving nor monotonous nor mass preserving there are methods to force monotonicity and mass conservation [15, 8].

SLM and adaptivity

Triangular meshes are appropriate for loyal refinement. A grid generator based on red (regular) refinement with green closures is used for meshing [2]. Figure 1 shows the refinement procedure for loyal refinement avoiding hanging nodes. We call the mesh a triangulation and each cell an element (adopting terminology from FEM).

In order to introduce automatic adaptivity into the SLM, a criterion for refinement or coarsening is needed. As there are no sources or sinks in equation (2), accuracy of the numerical scheme depends only on the interpolation. The accuracy of interpolation can be improved by adaptivity in non-smooth areas of the function which is to be interpolated.

Thus the gradient of the function is used as a criterion for refinement, resp. coarsening. In each element the local gradient is calculated, a global maximum is determined, and refinement is performed where the local gradient is larger then a certain level. This is explicitly stated in:

Method 1 (Criterion for refinement) *Given a tolerance* θ_{fine}. *Refine an element* τ *of the triangulation* T, *if* $|\nabla u|_\tau \geq \theta_{fine} \cdot \max_{\tau \in T}(|\nabla u|_\tau)$, *where* $|\nabla u|_\tau$ *is the local gradient in one element.*

Analogously a coarsening criterion is defined:

Method 2 (Criterion for coarsening) *Given a tolerance* θ_{coarse}. *Coarse an element* τ, *if for at least three of four daughter elements the following inequality holds:* $|\nabla u|_{\tau_{daughter}} \leq \theta_{coarse} \cdot \max_{\tau_{daughter} \in T}(|\nabla u|_{\tau_{daughter}})$.

Besides the tolerances θ_{fine} and θ_{coarse} a water-mark $0 \le \rho \le 1$ is introduced which prevents the triangulation from being changed, if a fraction of triangles less than ρ is marked for adaption.

With these methods, using a bicubic spline interpolation for triangular irregular mesh-points [13], the following algorithm for the adaptive solution of the passive advection equation can be defined:

Algorithm 2 (Adaptive semi-Lagrangian Method)

1. *Perform initialization, preprocessing, input operations.*
2. *For all time-steps loop through the following steps:*
   (a) *Use old grid (from last time-step) as a initial grid for the new (current) time-step.*
   (b) *Perform semi-Lagrangian advection on the current grid (algorithm 1).*
   (c) *Adapt the grid according to methods 1 and 2 taking water-mark ρ into account.*
   (d) *If the grid has changed, go to step 2b of the algorithm* (inner iteration), *else:*
   (e) *Calculate diagnostics, plotting, etc. and increase time-step; go to step 2a of the algorithm* (outer iteration).
3. *Post-processing.*

Parallelizing the SLM

One of the main reasons for selecting the SLM for advection in this research project is its potential for parallelization. All calculations along trajectories (i.e. iterative calculation of displacements α_m, interpolation of upstream-values $u(\mathbf{x}_m - \alpha_m, t_n)$, and update of new values $u(\mathbf{x}_m, t_n + \Delta t)$) are independent and thus parallelizable.

However, adaptivity requires a dynamic data distribution scheme in order to provide good load balancing and efficient utilization of distributed memory. An index array method proves to be efficient on the KSR-1, a virtual shared memory computer. Index arrays are used to store the node indices belonging to a process. Each process then works exclusively on "his" nodes. Nodal data is fixed to determined memory locations belonging to corresponding processes [6].

An example of a loop changed for parallelization with index arrays is given below. Let the pseudo-code of a loop over all nodes of the triangulation be given by:

```
forall i ∈ {i node of the triangulation} do
    [perform calculation for node i]
enddo
```

Then this loop is changed to a nested loop where each process works exclusively on the nodes stored in the index array:

```
forall p = 1,...,#processes do parallel
    forall i ∈ index-array(p) do
        [perform calculation for node i]
    enddo
enddo parallel
```

In the calculation data may be read from remote storage locations. The corresponding data movement, however, will be executed by the virtual shared memory mechanism, the hardware implemented *AllCache Engine* of the KSR-1 [10]. Due to a natural node numbering most neighbouring nodal data will reside on the same memory *Subpage*, therefore the index array method minimizes data movement.

Results for a parallelized adaptive SLM implementation

Numerical stability and accuracy of the adaptive SLM are compared with results from several authors [8, 15]. A test case introduced by Zalesak [25] is used. Figure 2 shows the slotted cylinder test case at initial time and after six revolutions, forced by a circular wind-field. More detailed results are given in [5].

Numerical results only negligibly differ when comparing a calculation on a globally refined mesh with a locally adapted mesh with corresponding finest resolution. On the other hand a substantial amount of computing resources is saved by adaptivity. Figure 3 shows the locally adapted grid moving with the slotted cylinder in time.

Parallel efficiency for the SLM parts of the program is very high. Only due to a serial grid generation total efficiency for the locally adapted program decreases rapidly. A parallel grid generator is required. However, the development of parallel mesh generation is not subject of the work presented here. Parallel efficiency and speedup are shown in figure 4.

Adaptive semi-Lagrangian method for the shallow water equations

The shallow water equations in a plain domain, periodical in each direction, are given by:

$$\frac{\partial U}{\partial t} + U\frac{\partial U}{\partial x} + V\frac{\partial U}{\partial y} - fV + \frac{\partial \Phi}{\partial x} = 0, \tag{5}$$

$$\frac{\partial V}{\partial t} + U\frac{\partial V}{\partial x} + V\frac{\partial V}{\partial y} + fU + \frac{\partial \Phi}{\partial y} = 0, \tag{6}$$

$$\frac{\partial \Phi}{\partial t} + U\frac{\partial \Phi}{\partial x} + V\frac{\partial \Phi}{\partial y} + \Phi\left[\frac{\partial U}{\partial x} + \frac{\partial V}{\partial y}\right] = 0. \tag{7}$$

$\mathbf{V} = (U, V)$ is the velocity with the x- resp. y-component, Φ is the geopotential height and f denotes the Coriolis parameter [9]. Using the Lagrangian notation, the equations may be written in the simplified form:

$$\frac{dU}{dt} + \Phi_x - fV = 0, \tag{8}$$

$$\frac{dV}{dt} + \Phi_y + fU = 0, \tag{9}$$

$$\frac{d\Phi}{dt} + \Phi(U_x + V_y) = 0, \tag{10}$$

where $\frac{d}{dt} = \frac{\partial}{\partial t} + U\frac{\partial}{\partial x} + V\frac{\partial}{\partial y}$, subscripts $_x$ and $_y$ denote partial derivatives in x- resp. y-direction.

SLM discretization of the shallow water equations

Discretization of the SWE proceeds in two steps: First the discretization of equations (8), (9), and (10) by means of the SLM from the previous section is carried out; second an elliptic problem is formulated, using a semi-implicit formulation of equation (10).

A fully implicit discretization of equations (8), (9), and (10) is given by:

$$\begin{aligned} \frac{U^+ - U^\circ}{\Delta t} + \frac{\Phi_x^+ + \Phi_x^\circ}{2} - \frac{(fV)^+ + (fV)^\circ}{2} &= 0 \qquad (11)\\ \frac{V^+ - V^\circ}{\Delta t} + \frac{\Phi_y^+ + \Phi_y^\circ}{2} + \frac{(fU)^+ + (fU)^\circ}{2} &= 0 \qquad (12)\\ \frac{\Phi^+ - \Phi^\circ}{\Delta t} + \frac{\Phi^+(U_x^+ + V_y^+) + \Phi^\circ(U_x^\circ + V_y^\circ)}{2} &= 0. \qquad (13) \end{aligned}$$

Here the superscript $^+$ denotes the function-value at grid-points and at time $t + \Delta t$, $^\circ$ denotes function-values at upstream-points and time t.

In the process of deriving an elliptic problem, equation (13) is formulated semi-implicitly. For this purpose Φ is split into a constant mean $\bar{\Phi}$ and a deviation $\Phi' = \Phi'(x, y, t)$. Now equations (11), (12), and (13) can be reformulated:

$$\begin{aligned} \frac{U^+ - U^\circ}{\Delta t} + \frac{\Phi_x'^+ + \Phi_x'^\circ}{2} - \frac{(fV)^+ + (fV)^\circ}{2} &= 0 \qquad (14)\\ \frac{V^+ - V^\circ}{\Delta t} + \frac{\Phi_y'^+ + \Phi_y'^\circ}{2} + \frac{(fU)^+ + (fU)^\circ}{2} &= 0 \qquad (15)\\ \frac{\Phi'^+ - \Phi'^\circ}{\Delta t} + \Phi'^\circ(U_x^\circ + V_y^\circ) + \frac{\bar{\Phi}(U_x^+ + V_y^+) + \bar{\Phi}(U_x^\circ + V_y^\circ)}{2} &= 0. \qquad (16) \end{aligned}$$

Note that in equation (16) only the constant $\bar{\Phi}$ is treated implicitly.

The elliptic problem can be derived by first substituting and eliminating equations (14) and (15) to obtain

$$\begin{aligned} U^+ + \phi\frac{\Delta t}{2}\Phi_x'^+ + \varphi\frac{\Delta t}{2}\Phi_y'^+ &= [\text{known terms}], \qquad (17)\\ V^+ + \phi\frac{\Delta t}{2}\Phi_y'^+ - \varphi\frac{\Delta t}{2}\Phi_x'^+ &= [\text{known terms}]. \qquad (18) \end{aligned}$$

Using the divergence operator yields

$$U_x^+ + V_y^+ = \frac{\Delta t}{2}\left[\left(\phi\Phi_x'^+\right)_x + \left(\phi\Phi_y'^+\right)_y + \left(\varphi\Phi_y'^+\right)_x - \left(\varphi\Phi_x'^+\right)_y\right] + [\text{known terms}]. \qquad (19)$$

Finally substituting this into equation (16) gives an elliptic problem for Φ'^+:

$$\left(\phi\Phi_x'^+\right)_x + \left(\phi\Phi_y'^+\right)_y + \left(\varphi\Phi_y'^+\right)_x - \left(\varphi\Phi_x'^+\right)_y - \frac{4}{\bar{\Phi}(\Delta t)^2}\Phi'^+ = [\text{known terms}]. \qquad (20)$$

Building blocks for the implementation

With the discretization described above, an implementation for solving the SWE will consist of several components. First an algorithm for the adaptive SLM for the SWE is given below.

Algorithm 3 (Adaptive SLM for shallow water equations)

1. *Perform initialization, preprocessing, input operations.*
2. *For all time-steps loop through the following steps:*
   (a) *Use old grid (from last time-step) as a initial grid for the new (current) time-step.*
   (b) *Calculate the displacements* α_m.
   (c) *Interpolate and derive right hand side of equation (20).*
   (d) *Solve (20) by means of the FEM and obtain* Φ'^{+}.
   (e) *Estimate the local discretization error by an error estimator.*
   (f) *Adapt the grid in analogy to methods 1 and 2 using the error estimates.*
   (g) *If the grid has changed, go to step 2b of the algorithm* (inner iteration), *else:*
   (h) *Calculate velocities* (U, V) *from* Φ'^{+} *by equations (17) and (18).*
   (i) *Advect a tracer variable, calculate vorticity, etc. if required.*
   (j) *Calculate diagnostics, plotting, etc. and increase time-step; go to step 2a of the algorithm* (outer iteration).
3. *Post-processing.*

For calculating the displacements one needs to change the scheme introduced in (4), because of self-advection in this case. There are several extrapolation schemes to address this problem [21]. The right hand side of the elliptic problem is derived in a rather lengthy process. There are several interpolation steps involved as well as numerical gradient calculation. Bicubic spline interpolation is utilized as in the section above.

FEM with continuous, piecewise linear elements is chosen for solving the elliptic problem. The linear system of equations is solved by an iterative method for unsymmetric systems, namely BiCGSTAB [23]. It is preconditioned by simple Jacobian precondioning [4]. Error estimation is performed corresponding to results from Babuška/ Rheinboldt and Bank/ Weiser [1, 3].

In order to advect a passive tracer variable, the scheme for passive advection from the previous section is adopted. As displacements are known, the only operation required for tracer advection is an interpolation at each upstream point.

Parallelization

Parallelizing the SLM-parts of the implementation is as simple as in the previous section. All loops with nodal indexing are parallelized in the same way by index arrays.

The FEM is parallelized in the same way. Data access is performed in a nodal order wherever possible [6]. There are two exceptions: The element stiffness matrices are calculated in an element-wise order as well as the local error estimates. However, as no data dependency is involved in these two cases, automatic parallelization can be used.

All other loops are parallelized such that each processor works on the same data items within the whole time-step. Data redistribution takes place only when the grid has changed after adaption. Again the grid generator is serial. This results in some performance degradation.

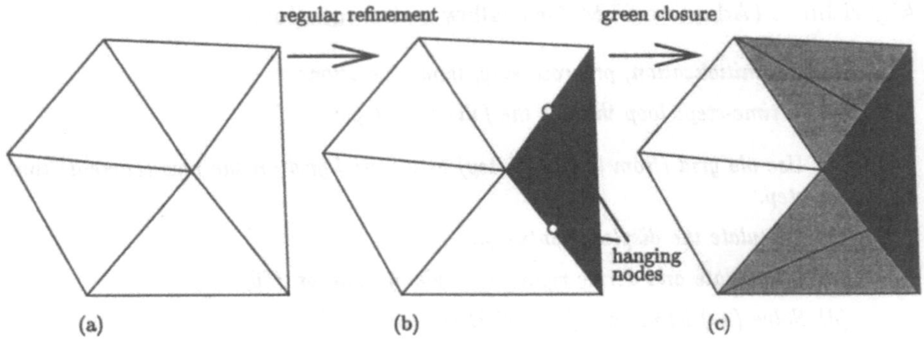

Figure 1: An initial coarse triangulation (a) is locally refined by regular refinement (b). This leads to hanging nodes, which are not admissible in a triangulation. Triangles containing hanging nodes are then bisected by a *green closure* (c) in order to re-establish admissible triangulation.

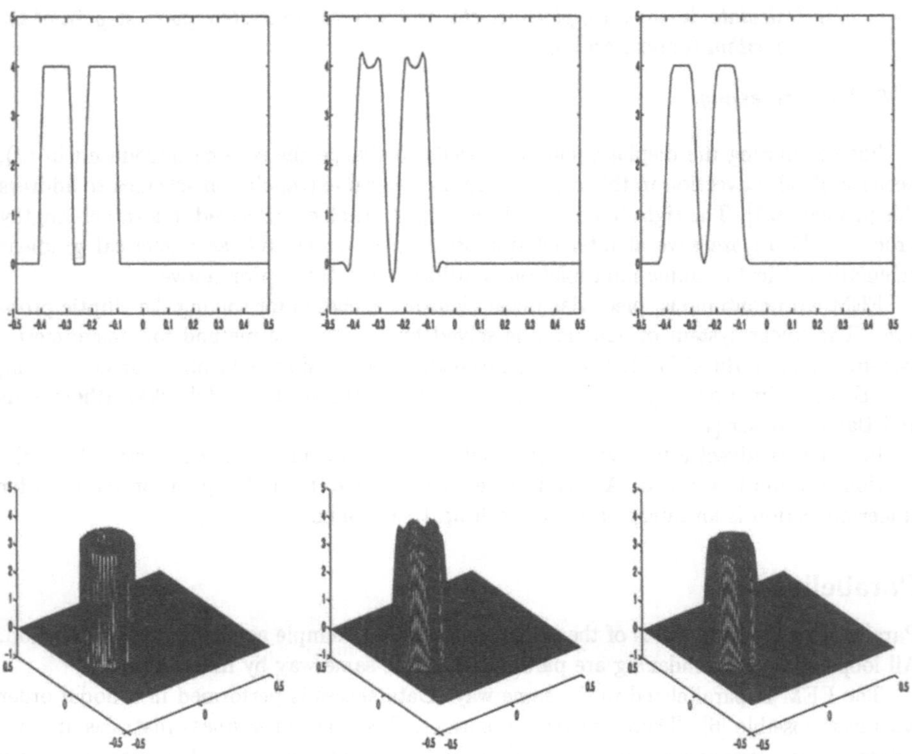

Figure 2: The slotted cylinder test case at initial state, after six rotations wit plain bicubic spline interpolation, and with quasi-conservative interpolation (from left).

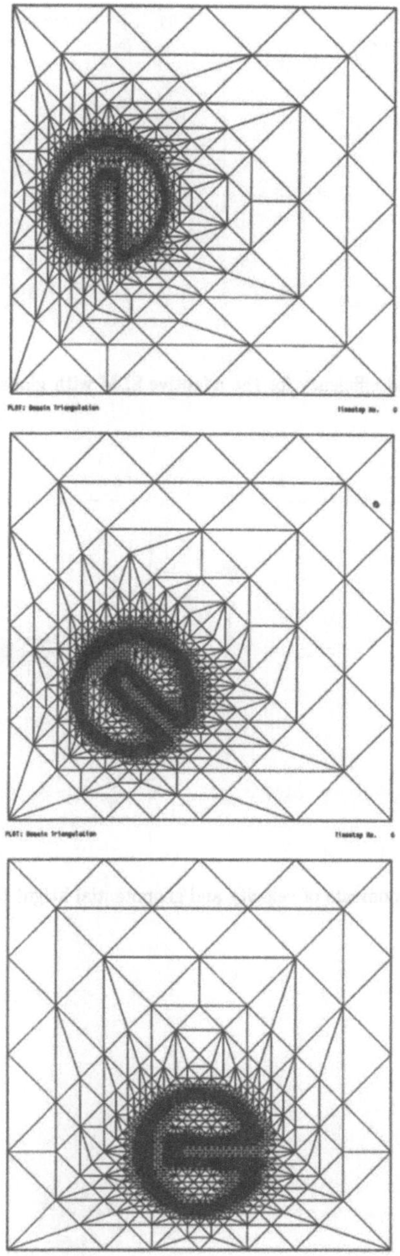

Figure 3: **Locally adapted grid moving with the slotted cylinder. Screen-shots at initial time, after 1/8 and 1/4 revolutions.**

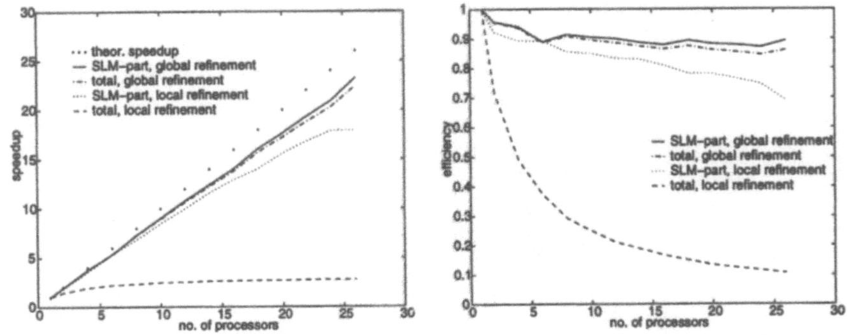

Figure 4: Parallel speedup and efficiency for the adaptive SLM with global and local refinement, serial grid generation

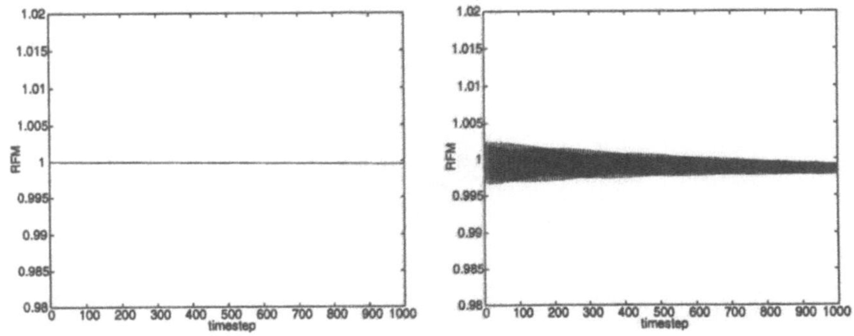

Figure 5: First moments of velocity and geopotential height for the test-case.

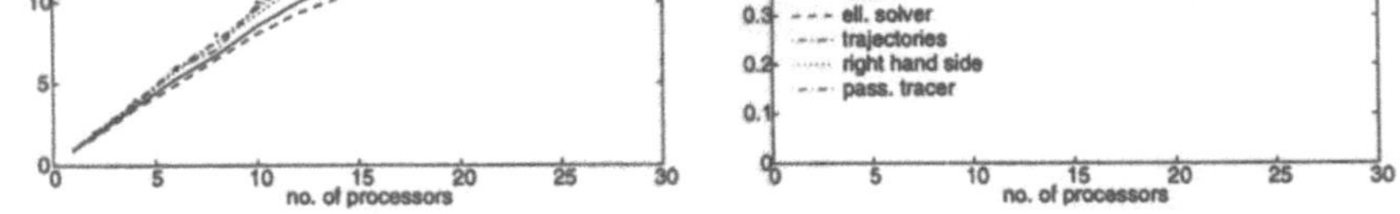

Figure 6: Parallel speedup and efficiency for the adaptive SLM applied to the shallow-water equations.

Results

Results are shown for a model problem adopted from [18]. Stationary Rossby waves are used to determine numerical dissipation of the scheme. Initial conditions are given by

$$\Psi_0 = K_1 \sin x \sin y, \tag{21}$$
$$\Phi_0 = \pi^2 K_2(\cos 2x + \cos 2y) + \bar{\Phi}. \tag{22}$$

U and V are derived from (21) by

$$U = -\frac{\partial \Psi_0}{\partial y}, \quad V = \frac{\partial \Psi_0}{\partial x}. \tag{23}$$

First and second moments of the velocity $\mathbf{V} = (U, V)$ and the geopotential height Φ are taken as diagnostic values. Ideally these values should be conserved. Because of numerical dissipation no conservation can be achieved. However, the loss of energy (first moment of velocity) and mass (first moment of geopotential height) is very small. Figure 5 shows losses of less than 0.1 percent after 1000 time-steps.

Parallel efficiency again is high, especially for all SLM related functions. Efficiency rates of above 90% for up to 26 Processors on a KSR-1 can be achieved. Due to the serial grid generation and the iterative solver which does not scale optimally, the efficiency rate for the total program ranges between 90% and 65% for 4 to 26 processors, shown in figure 6.

Concluding remarks

An adaptive semi-Lagrangian method for the solution of the shallow water equations has been developed and parallelized. The implementation consists of several components, including grid generation, error estimation, solution of an elliptic boundary value problem by finite element methods, semi-Lagrangian time discretization, and diagnostics. Parallelization is efficient and scalable.

Some features still need to be developed. A parallel grid generator is crucial to provide scalability up to a large number of processors. The flexibility of FEM with respect to the domain geometry has not been exploited yet. A problem still not addressed is determination of criteria for refinement in the presence of tracers. A completely mathematical error estimator based on the discretization error of the elliptic solver might not be sufficient, as the physical properties of the fluid are not included. Finally the extension to three-dimensional applications including adaptive meshes is still a field of research with many unsolved problems.

However, it could be shown that SLM and adaptivity go well together. Parallelization of the program makes FEM, which are numerically more demanding, competitive to finite difference methods even for grand challenge problems like atmospheric or ocean modelling.

References

[1] I. Babuška and W. C. Rheinboldt. Error estimates for adaptive finite element computations. *SIAM J. Numer. Anal.*, 15(4):736–754, 1978.

[2] R. E. Bank. The efficient implementation of local refinement algorithms. In I. et. al. Babuska, editor, *Adaptive computational methods*, pages 74–81, Philadelphia, 1983. SIAM.

[3] R. E. Bank and A. Weiser. Some a posteriori error estimators for elliptic partial differential equations. *Math. Comp.*, 44, No. 170:283–301, 1985.

[4] R. Barrett, M. Berry, T. F. Chan, J. Demmel, J. Donato, J. Dongarra, V. Eijkhout, R. Pozo, C. Romine, and H. van der Vorst. *Templates for the Solution of Linear Systems*. SIAM, Philadelphia, 1993.

[5] J. Behrens. An adaptive finite element semi-Lagrangian advection scheme and its parallelization. Report 62, Alfred-Wegener-Institut, Bremerhaven, 1995. AWI Berichte aus dem Fachbereich Physik.

[6] J. Behrens. Parallelization strategies for matrix assembly in finite element methods. In A. Ferreira and J. D. P. Rolim, editors, *Parallel Algorithms for Irregular Problems: State of the Art*, pages 3–24, Dordrecht, 1995. Kluwer Academic Publishers.

[7] J. Behrens, B. Fritzsch, W. Hiller, and H. P. Kersken. Vergleich eines 2-D Mehrgitter Helmholtzlösers mit einem FFT-basierten direkten Löser in QG-Modellen. Report 37, Alfred-Wegener-Institut, Bremerhaven, 1993. AWI Berichte aus dem Fachbereich Physik.

[8] R. Bermejo and A. Staniforth. The conversion of semi-Lagrangian advection schemes to quasi-monotone schemes. *Mon. Wea. Rev.*, 120:2622–2632, 1992.

[9] A. E. Gill. *Atmosphere-Ocean Dynamics*. Academic Press, London, 1982.

[10] Kendall Square Research Corp., Waltham, MA. *KSR/Series Principles of Operation*, 1994. Revision 7.0.

[11] H.-P. Kersken, B. Fritzsch, O. Schenk, W. Hiller, J. Behrens, and E. Krauße. Parallelization of large scale ocean models by data decomposition. In W. Gentzsch and U. Harms, editors, *High-Performance Computing and Networking*, number 796 in Lecture Notes in Computer Science, pages 323–328, Berlin, 1994. Springer-Verlag.

[12] J. D. McCalpin. A quantitative analysis of the dissipation inherent in semi-Lagrangian advection. *Mon. Wea. Rev.*, 116:2330–2336, 1988.

[13] L. B. Montefusco and G. Casciola. Algorithm 677: C^1 surface interpolation. *ACM Trans. on Math. Softw.*, 15(4):365–374, 1989.

[14] A. Oliveira and A. M. Babtista. A comparison of integration and interpolation Eulerian-Lagrangian methods. *Int. Journ. Num. Meth. Fluids*, 21:183–204, 1995.

[15] A. Priestley. A quasi-conservative version of the semi-Lagrangian advection scheme. *Mon. Wea. Rev.*, 121:621–629, 1993.

[16] P. J. Rasch and D. L. Williamson. On shape-preserving interpolation and semi-Lagrangian transport. *SIAM J. Sci. Stat. Comput.*, 11(4):656–687, 1990.

[17] P. Roache. *Computational Fluid Dynamics*. Hermosa Publishers, Albuquerque NM, 1972.

[18] Robert Sadourny. The dynamics of finite-difference models of the shallow-water equations. *J. Atmos. Sci.*, 32:680–689, 1975.

[19] H. D. Simon, editor. *Parallel computational fluid dynamics – implementations and results*. Scientific and Engineering Computation. The MIT Press, Cambridge Massachusetts, 1992.

[20] A. Staniforth and J. Côté. Semi-Lagrangian integration schemes for atmospheric models - a review. *Mon. Wea. Rev.*, 119:2206–2223, 1991.

[21] A. Staniforth and C. Temperton. Semi-implicit semi-Lagrangian integration schemes for a barotropic finite-element regional model. *Mon. Wea. Rev.*, 114:2078–2090, 1986.

[22] K. E. Trenberth, editor. *Climate System Modeling*. Cambridge University Press, Cambridge, 1992.

[23] H. A. van der Vorst. Bi-CGSTAB: A fast and smoothly converging variant of Bi-CG for the solution of nonsymmetric linear systems. *SIAM J. Sci. Stat. Comput.*, 13(2):631–644, 1992.

[24] D. L. Williamson and P. J. Rasch. Two-dimensional semi-Lagrangian transport with shape-preserving interpolation. *Mon. Wea. Rev.*, 117:102–129, 1989.

[25] S. T. Zalesak. Fully multidimensional flux-corrected transport algorithms for fluids. *Jou. Comput. Phys.*, 31:335–362, 1979.

Practical Aspects of the Simulation of Viscous Flow using Lattice Boltzmann Automata

J. Bernsdorf and M. Schäfer
Department of Fluid Mechanics (LSTM)
University of Erlangen – Nürnberg
Cauerstr. 4, D–91058 Erlangen, Germany

SUMMARY

A relatively new approach in computational fluid dynamics is the simulation of fluid microdynamics with cellular automata. Binary fluid particles with unit mass (FHP) or fluid densities (Lattice Boltzmann) are propagated with unit velocity on a hexagonal Bravais lattice in discrete time steps. Simple collision rules result after suitable time and space averaging, in a Navier - Stokes like behavior of the 'computer fluid'. Several recent results indicate that cellular automata can be a good working tool for the numerical simulation of fluid flow. Compared to the well established "traditional" finite elements (FE) or finite volume (FV) methods, especially for highly complex geometries, the cellular automata seem to be a very promising approach, since geometrical complexity has nearly no influence on computing time during the simulation.

Some results comparing FV and LBA approaches are presented for viscous fluid flow in geometries with increasing complexity. As a first step towards practical application, a simulation through a structure obtained from a digitized electron microscope picture of a sedimentary layer is considered.

Introduction

Detailed information about transport parameters in complex geometries, as for instance porous media, is of importance in a wide range of scientific and engineering problems. While some macroscopic results can be obtained from synthetic model systems [19], a microscopic pressure and velocity distribution in complex flow regions is always the result of numerical simulations (or of some experimental work).

Besides the well known finite-volume and finite-element techniques, a relatively new approach, based on the numerical solution of a simplified Boltzmann–type equation, is more and more used [6, 1]. In recent years, these cellular automata like numerical tools are developped to a stage, which makes them interesting for practical application in numerical fluid dynamics [18, 2, 4, 8, 20]. For highly complex geometries, they appear to have some advantages compared to the "classical" methods, due to the possibility of a relatively simple implementation of obstacle structures and that the CPU–times are nearly independent from the complexity of these structures.

In the following, we first give some introduction to the lattice gas and lattice Boltzmann automata from a practical point of view. Afterwards, advantages and disadvantages of the lattice Boltzmann and finite volume approaches are discussed and quantitative comparative results related to accuracy, convergence properties, and computing times are given for test cases in simple and complex geometries. Finally, we illustrate a relatively easy way to get flow simulation results from "real" porous structures.

Introduction to FHP and Lattice Boltzmann Automata

The basic idea of lattice gas and lattice Boltzmann automata is the numerical simulation of a simplified molecular dynamics of the fluid. This is done by evaluating a time and space discrete Boltzmann equation, the so called lattice Boltzmann equation [6]. After appropriate time and space averaging, macroscopic values like pressure and velocity can be obtained. It can be shown, that these macroscopic values, if some assumptions are fulfilled, have a behavior similar to that governed by the Navier–Stokes equations [10, 12, 21].

FHP – Automata

The FHP automata was first proposed by Frisch, Hasslacher and Pomeau in 1986 [6]. The basic principle of this approach is that binary "particles" with unit mass are propagated on a hexagonal Bravais lattice (see Fig.1, an orthogonal lattice would cause aniotropic flow) in discrete time steps with unit velocity.

The update of the lattice takes place in three steps:

- Propagation
- Find collision states and perform collisions
- Acceleration

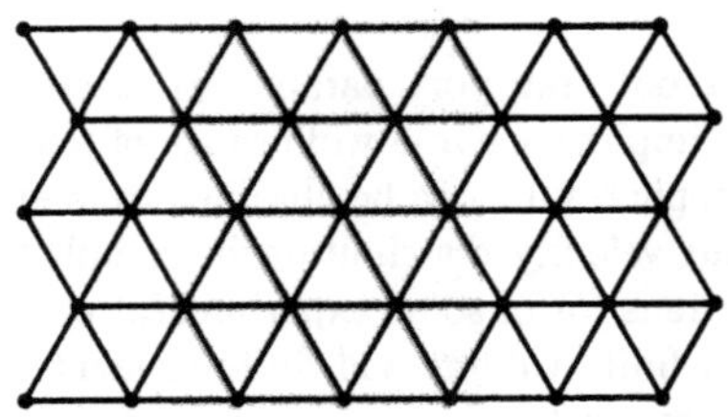

Figure 1: Hexagonal Bravais lattice.

Propagation is only possible between the six nearest neighbours along the lattice connection lines. A particle position and (discrete) velocity vector is indicated by its lattice coordinates and its position at the node. Every node splits up into six *cells* (see Fig.2), and the six possible velocity vectors $\vec{c}_j$ are:

$$\vec{c}_j = [\cos(2j\pi/6), \sin(2j\pi/6)], j = 1, ..., 6. \tag{1}$$

An exclusion principle is imposed that does not allow the location of more than one particle per cell at the same time. The propagation step consists of moving all particles

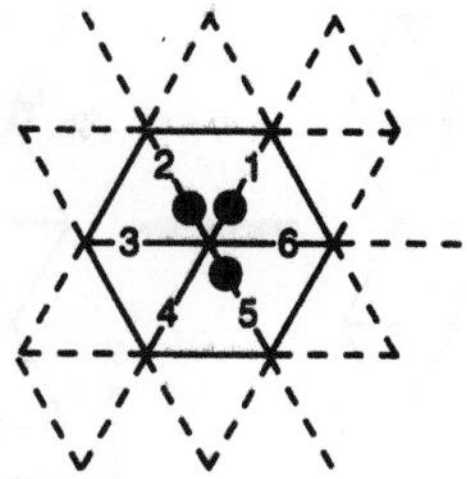

Figure 2: Lattice node with particles at cell 1, 2 and 5.

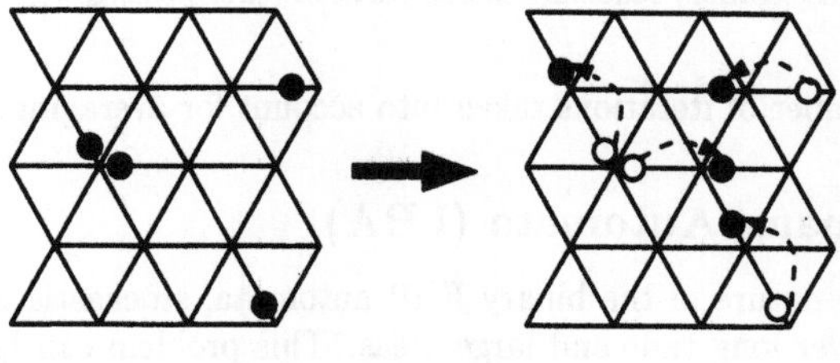

Figure 3: Propagation: particles moving to new positions.

to the next node in direction of its velocity vector (see Fig.3). Depending on the model (FHP I, II or III), more or less complicated collision rules may change the velocity vectors of those particles, entering a *collision state* after the propagation step. A minimum set of collision rules to reproduce realistic flow are the head–on and three particle collisions which are used in FHP I (see Fig.4). Particle mass and momentum is conserved during collisions. To get some directed flow (and a pressure gradient), the velocity vectors of a

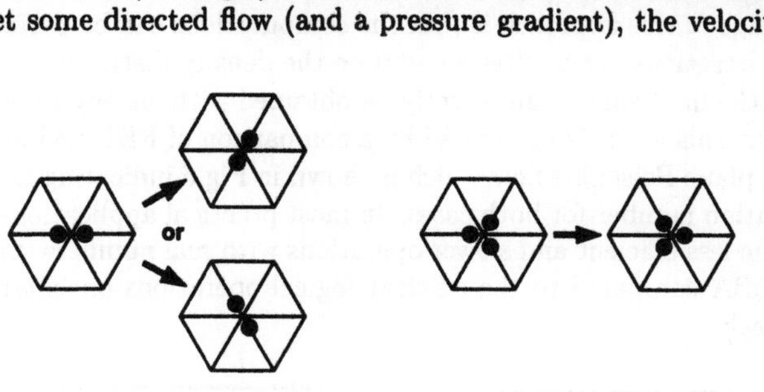

Figure 4: Head–on and three particle collisions.

certain amount of randomly chosen particles have to be changed in every iteration. To get a flow in (positive) x–direction, for example, this is done by moving particles from cell 3 to cell 6 at the corresponding nodes (see Fig.5). This local disturbance is propagated through the lattice, and after a number of iterations, an equilibrium between acceleration and viscous forces imposed by collisions and boundary effects (see below) evolves. In order to obtain macroscopic values, like pressure and velocity, from the particle distributions, appropriate time and space averaging procedures have to be carried out.

Denoting by $n_i(\vec{r}_*, t_*) \in (0,1)$ the binary particle density at cell $i \in (1,2,...,6)$ of lattice node $\vec{r}_*$ at time t_*, for example, the mean velocity can be obtained by:

$$< u >= \frac{1}{M} \sum_{t_*=1}^{M} \frac{\sum_i \vec{c}_i \sum_{\vec{r}_*} n_i(\vec{r}_*, t_*)}{\sum_i \sum_{\vec{r}_*} n_i(\vec{r}_*, t_*)}, \tag{2}$$

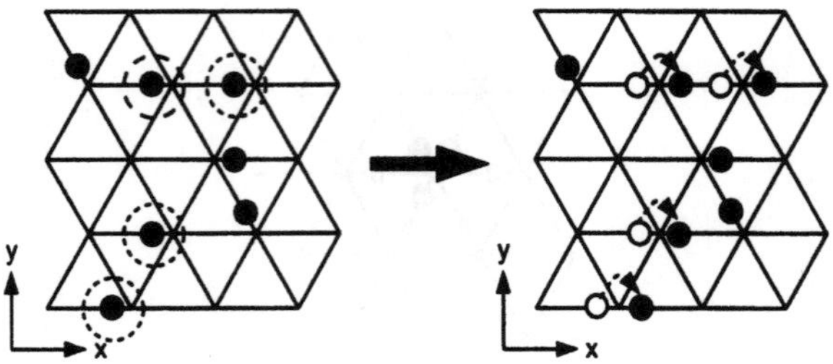

Figure 5: Acceleration: randomly chosen particles (circle) change velocity vector.

where M is the number of iterations taken into account for averaging.

Lattice Boltzmann Automata (LBA)

Due to the discrete nature of the binary FHP automata, stochastic noise makes it necessary to average over long time and large areas. This problem can be avoided by using particle densities as ensemble average of the Boolean population. The particle collision rules are replaced by matrix multiplications which produce the new particle distributions after the collisions:

$$N_i(\vec{r}_* + \vec{c}_i, t_* + 1) = N_i(\vec{r}_*, t) + \Delta_i(N), \tag{3}$$

where the exact definition of the collision operator $\Delta_i(N)$ depends on the details of the employed model. (For the two most common see [17, 21]). The densities N_i are the ensemble averages of the binary particle densities n_i.

The advantage of this approach is that the computations can be performed on smaller grids with less iterations, since after some time the density distribution does not change any more, and the final values can directly be obtained without any time and space averaging processes. This effect is illustrated by a comparison of FHP and lattice Boltzmann automata for a plane Poiseuille flow which is shown in Fig.6 indicating the mean flow rate versus the iteration number for both cases. In most practical applications, this advantage compensates the less efficient and slower operations with real numbers (particle densities) necessary for LBA compared to the efficient logical operations on binary particles with an FHP approach.

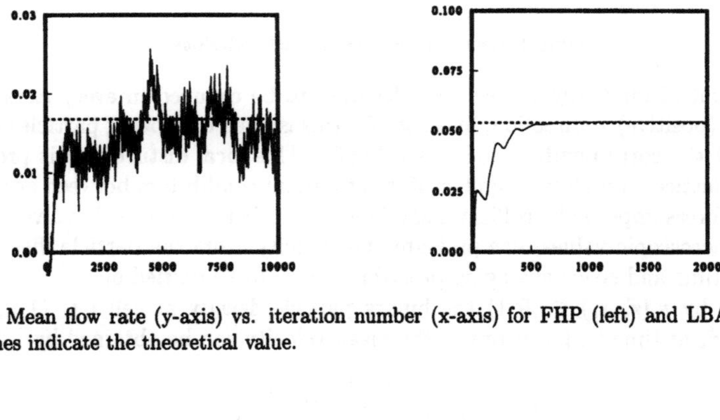

Figure 6: Mean flow rate (y-axis) vs. iteration number (x-axis) for FHP (left) and LBA (right). The dashed lines indicate the theoretical value.

Obstacles and Channel Walls

In both cases, FHP as well as lattice Boltzmann automata, arbitrary obstacle structures can be implemented by occupying lattice nodes. Particles or particle densities which are moved to occupied nodes are simply bounced back (see Fig.7). This method makes

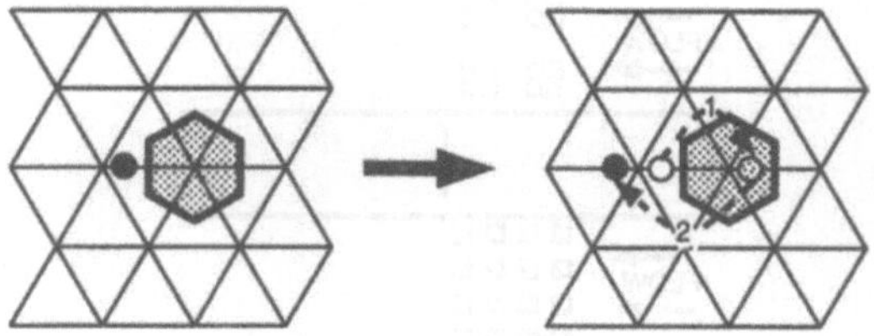

Figure 7: Particle moved on occupied node (1) is bounced back (2).

it easy to implement complex structures or to change the obstacle structure during the computation, which is necessary for problems with time–varying flow geometry. A possible disadvantage is, that real structures can only be approximated by hexagons. Also, one has to take care that the smallest structure elements are large enough to avoid the so called finite–size–effect, which may lead to unphysical results [13, 14].

The same bounce–back principle is applied to the channel walls (for this and other possible boundary conditions see [16]). All particles tending to leave the lattice are reflected towards their *incoming* direction (see Fig.8), resulting in a zero mean velocity at the boundary (no slip boundary condition).

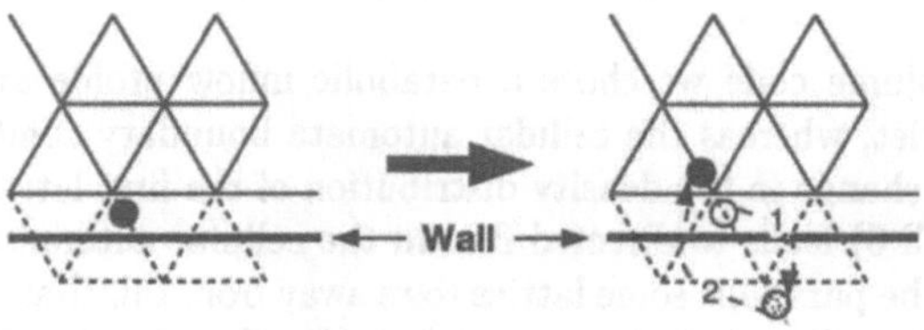

Figure 8: Channel walls: leaving particles (1) are bounced back (2).

Comparison FV and LBA Methods

To get some information about the efficiency of the lattice Boltzmann method discussed above in complex geometries, we compared it with the commercial finite volume (FV) code FASTEST. As lattice Boltzmann method we used the enhanced collision type as it was proposed by Higuera and Succi [11].

The underlying numerical method for the FV code is based on a fully conservative 2nd order finite volume discretization on blockstructured grids with collocated arrangements of variables, an iterative pressure-correction method of SIMPLE type for coupling of variables and linearization, an ILU method for the solution of the sparse linear systems, and a nonlinear multigrid method for convergence acceleration. A detailed description of the employed methodology can be found in [5]. The code has proven to work with high numerical efficiency for the considered type of problems, and thus, a comparison with it can be viewed as a severe criterion for the valuation of the capabilities of the

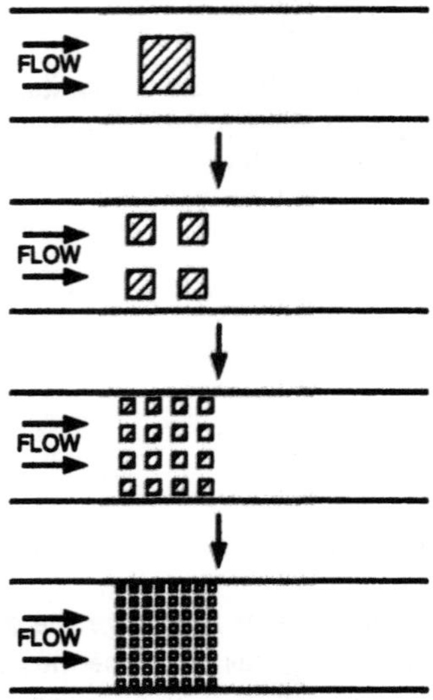

Figure 9: Increasing complex obstacle structure.

lattice Boltzmann approach. As test cases we consider a $lx = 4\,ly$ sized channel with different numbers of square obstacles, which are regularly placed in the second quarter of the channel (see Fig.9), representing a geometry of increasing complexity. The sizes of the obstacles are chosen in a way that the space occupied by them is the same for all cases.

For the finite volume code we chose a parabolic inflow profile and a zero gradient condition at the outlet, whereas the cellular automata boundary conditions are periodic in flow direction. A change in the density distribution of the first lattice row (decrease in cell 3, increase in cell 6) leads to directed flow in the cellular automata. The flow profile already appeared to be parabolic some lattice rows away from this first "acceleration" row. The region downstream the obstacle area, at least for the considered Reynolds number $Re = 0.1$, is long enough to avoid any influence of the outlet to the flow profile.

As reference quantities the pressure difference between the cross–sections at $x = 1/8\,lx$ and $x = 7/8\,lx$ of the channel (see Fig.10) and the maximum velocity were computed. In Fig.11 these values are indicated for the 2 x 2 obstacle case for both methods, when

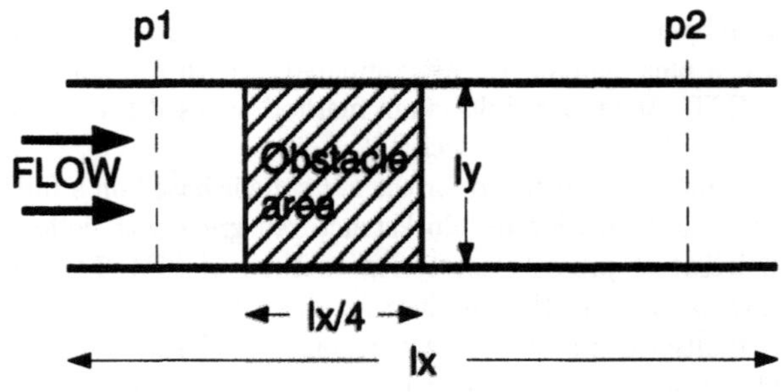

Figure 10: Channel with obstacle area. Pressure was measured at p1 and p2.

using lattice / grid sizes of increasing fineness ranging from 64 x 16 cells (grid 1) to 1024 x 256 cells (grid 5). The results show very good agreement in the high resulution limit

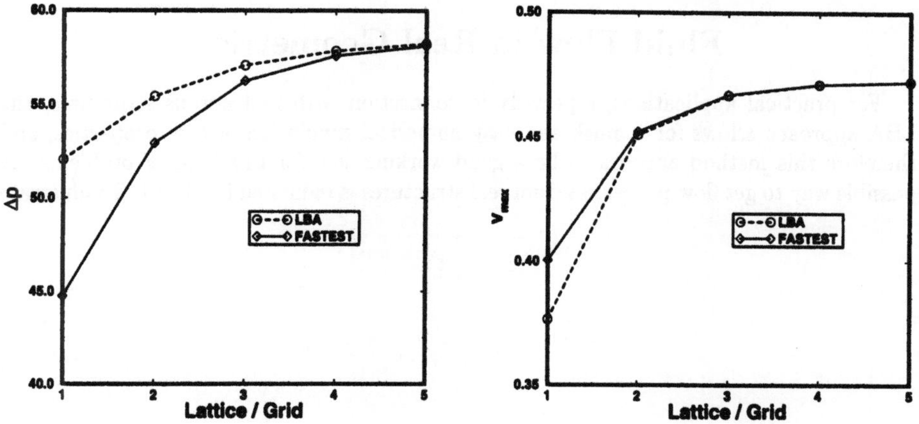

Figure 11: Pressure drop (left) and maximum velocity (right) in LBA and FASTEST.

for both pressure drop and maximum velocity. Actually, the pressure drop on the coarser grids is closer to the fine grid value for the LBA, whereas the finite volume code gives closer values for the velocity. For the other obstacle configurations very similar results concerning the reference quantities were found. The CPU–times for the computations

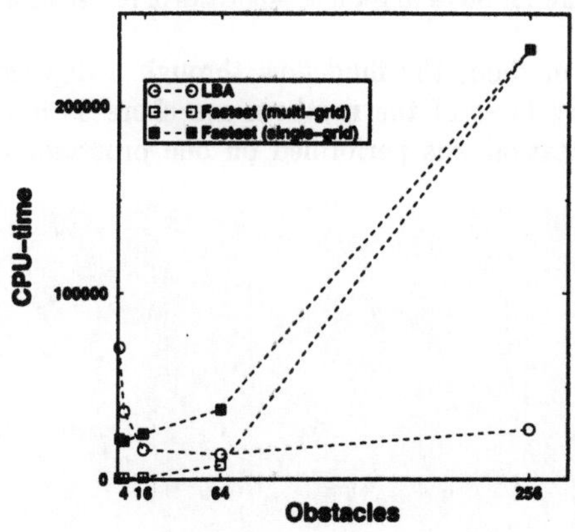

Figure 12: CPU - times vs. number of obstacles for LBA and FASTEST (single- and multigrid).

with 1 to 16 x 16 obstacles and a lattice / grid with 512 x 128 cells are shown in Fig.12 for both methods, where for the FV code timings are given for multigrid and single–grid computations (all computations were done on a SUN10 workstation). For the multigrid case with every increase in the number of obstacles one grid level less could be used for computation, because the coarsest grid has to be still fine enough to fully resolve the obstacle structure. So, in the limit (grid 5) the multi- and singlegrid methods coincide. Comparing CPU–times, an increase for the FV code with increasing number of obstacles is evident. For the lattice Boltzmann automata, the necessary CPU-time decreases with increasing number of obstacles and finally becomes almost independent on the number of obstacles.

Fluid Flow in Real Geometries

For practical applications, especially in connection with real porous structures, the LBA approach allows for a quick and easy numerical simulation of flow properties, and therefore this method appears to be a good working tool for this kind of problems. A possible way to get flow properties from real structures is indicated in Fig.13. To illustrate

Sample preparation

⇓

(Microscope) sample picture

⇓

Digitize picture

⇓

Digitized picture ⇒ obstacle file

⇓

Numerical calculation with LBA

Figure 13: Steps to get flow properties of real structures.

the capabilities of the method, the fluid flow through a digitized electron microscope picture of a sedimentary layer of the northern sea shore as it is shown in Fig.14 was simulated. The computation was performed on one processor of a Cray-YMP. 10000

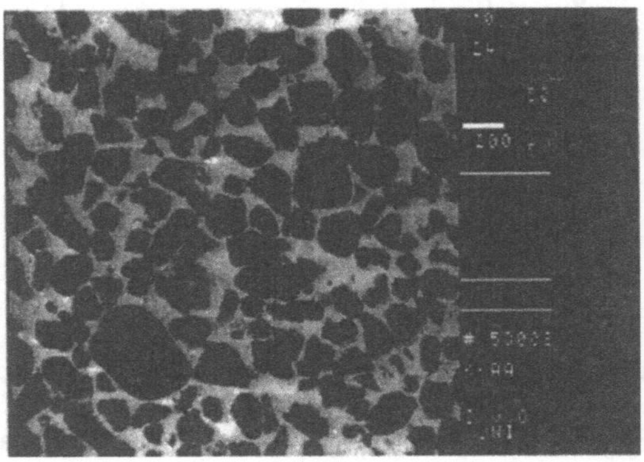

Figure 14: Sedimentary layer, from [15].

iterations were needed for a 500 x 500 lattice (380 x 500 for the obstacle area, 2 x 60 before and behind the obstacle) to reach the steady state. This corresponds to 1200 s of total CPU-time where $1.47 * 10^6$ site updates per second can be performed with a performance of 190 MFLOPS (from 330 theoretical peak performance). This indicates the possibility of a straightforward vectorization of the lattice Boltzmann cellular automata computations, which constitutes another advantage of this method.

Fig.15 shows the computed velocity vectors, and an enlarged section can be seen in Fig.16. The pictures illustrate that, regardless the complexity of the pores, the flow features are well captured.

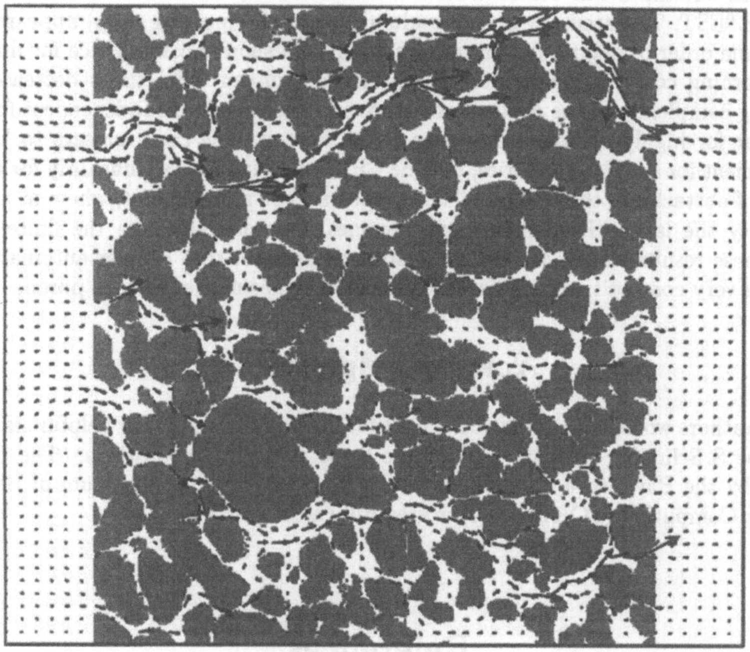

Figure 15: Numerically simulated flow. Every tenth velocity vector (in both dimensions) is plotted.

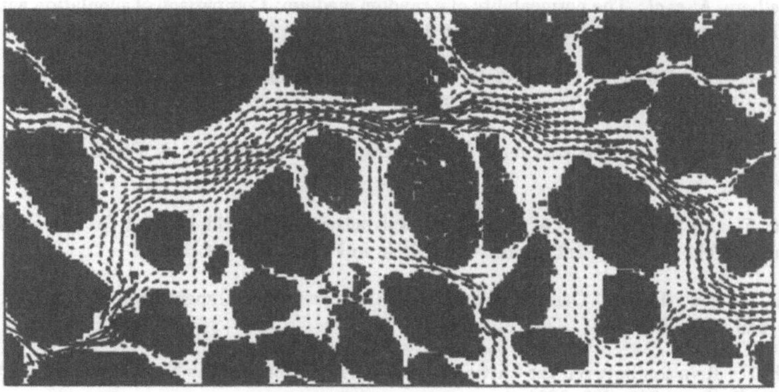

Figure 16: Section from the simulated area. Every third velocity vector (in both dimensions) is plotted.

Conclusions

To show some aspects interesting for practical application of CFD in complex geometries, we presented a fast method based on cellular automata to get flow profiles of "real" porous structures (at least in two dimensions). The quantitative behaviour of the lattice Boltzmann automata was demonstrated in comparison to the commercial finite volume code FASTEST. The accuracy of the results (velocity and pressure drop) appeared to be in the same order of magnitude for both methods and showed convergence for fine grids.

Concerning complex geometries, the lattice Boltzmann CPU–times decreased with increasing complexity of the obstacle structure and became almost independent from this parameter for highly complex structures. The finite–volume code appears to be much more efficient for simple geometries, but the CPU–times increase significantly for a higher number of obstacles in the channel. There appears to be a breakpoint, where at a certain complexity the lattice Boltzmann code becomes more efficient than the finite–volume code. These results strengthen the often stressed opinion, that cellular automata based codes are very competitive tools for CFD in complex geometries such as porous media.

Acknowledgements

The authors would like to thank F. Durst for his support and many helpful discussions as well as S. Schreiber for carrying out the finite volume computations. The financial support by the Bayerische Forschungsstiftung in the Bavarian Consortium of High Performance Computing (FORTWIHR) is greatfully acknowledged.

References

[1] Benzi, R. *et al.*: The Lattice Boltzmann Equation: Theory and Applications, in: Phys. Rep. 222, No.3 pp. 145 - 197, 1992.

[2] Cancelliere, A. *et al.*: The permeability of a random medium: Comparison of simulation with theory, in: Phys. Fluids A 2, pp. 2085 - 2088, Dec. 1990.

[3] Chen, H. *et al.*: Recovery of the Navier - Stokes equations using a lattice - gas Boltzmann method, in: Phys. Rev. A, Vol. 45 Nr.: 8, April 1992.

[4] Chen, S. *et al.*: Lattice Gas Automata for Flow through Porous Media, in: Physica D 47, pp. 72 - 84, 1991.

[5] Durst, F. and Schäfer, M.: A Parallel Blockstructured Multigrid Method for the Prediction of Incompressible Flows, in: Int. J. for Num. Meth. in Fluids, in press, 1996.

[6] Frisch, U. *et al.*: Lattice - Gas Hydrodynamics in two and three Dimensions, in: Complex Systems 1, pp. 649 - 707, 1986.

[7] Frisch, U.: Relation between the Lattice Boltzmann Equation and the Navier Stokes Equations, in.: Physica D 47, pp.231 - 232, 1991.

[8] Hasslacher, B.: Discrete Fluids, in.: Hao Bai - Lin (ed), Directions in Chaos, Vol. 2, Singapore, 1988.

[9] Hayot, F. *et al.*: Implementation and Performance of a Binary Lattice Gas Algorithm on Parallel Processor Systems, in: Journal of Computational Physics 80, pp. 277 - 287, 1989.

[10] Higuera, F.J *et al.* : Boltzmann Approach to Lattice Gas Simulations, in: Europhys. Lett., 9 (7), pp. 663 - 668, 1989.

[11] Higuera, F.J. *et al.*: Lattice Gas Dynamics with Enhanced Collisions, in: Europhys. Lett., 9 (4), pp. 345 - 349, 1989.

[12] Kingdon, E.D. *et al.*: A Lattice Boltzmann model for the simulation of fluid flow, in: J. Phys. Math., Gen. 25, pp. 3559 - 3566, 1992.

[13] Kohring, G.A.: Effect of Finite Grain Size on the Simulation of Fluid Flow in Porous Media, in: J. Phys. II 1, pp. 87 - 90, 1991.

[14] Kohring, G.A.: Limitations of a finite Mean Free Path for Simulating Flows in Porous Media, Universität Köln, April 1991.

[15] Kropp, J.: Analyse der inhomogenen Elementverteilung in Wattsedimenten, Universität Oldenburg, Diplomarbeit, 1992.

[16] Lavallée, P. *et al.*: Boundaries in Lattice Gas Flows, in: Physica D 47, pp. 233 - 240, 1991.

[17] Qian, Y.H.: Lattice BKG Models for Navier - Stokes Equation, in: Europhys. Lett., 17 (6), pp. 479 - 484, 1992.

[18] Rothmann, D.H.: Cellular-automaton fluids: A model for flow in porous media, in: Geophysics, Vol. 53. No. 4, pp. 509 - 518, April 1988.

[19] Sahimi, M: Flow and Transport in Porous Media and Fractured Rock, Weinheim 1995.

[20] Succi, S. *et al.*: Three - Dimensional Flows in Complex Geometries with the Lattice Boltzmann Method, in: Europhys. Lett., 10 (5), pp. 433 - 438, 1989.

[21] Succi, S. *et al.*: The Lattice Boltzmann Equation: A New Tool for Computational Fluid - Dynamics, in: Physica D 47, pp. 219 - 230, 1991.

FLOW CHANNELING IN UNSATURATED POROUS MEDIA OF STRONG HETEROGENEITY

Jens Birkhölzer and Chin-Fu Tsang
Lawrence Berkeley Laboratory, Earth Sciences Division, MS 50E
1 Cycloton Rd., Berkeley CA 94720

SUMMARY

Numerical simulations have been performed to study unsaturated/saturated flow in a two-dimensional vertical column of strongly heterogeneous soil. Different saturation scenarios were applied varying from fully saturated conditions to low saturations, corresponding to different infiltration rates into the soil. After solving the flow field, a particle tracking procedure was performed to model the transport of contaminants throughout the model region. It was found that particles (or solutes) travel along preferred flow paths, which are called channels. This is the case in saturated as well as in unsaturated media. The grade of channeling and the location of channels is dependant on the overall saturation. Strong channeling effects were obtained in the saturated case and in cases of low saturation. Within intermediate saturation ranges, however, the system exhibits a more homogeneous flow pattern; channeling effects are less significant.

INTRODUCTION

The variation of permeability in strongly heterogeneous porous soil causes strong variations in the fluid velocities. The fluid seeks to flow in the least resistive pathways, which are called channels. Particles (or solutes) travelling along those channels may be transported many times faster than the mean time for the bulk of the fluid. In practical problems concerning the transport of radioactive or toxic wastes, this earlier arrival of contaminants may be of vital importance.

In the last decade flow channeling phenomena have been intensively studied with regard to *saturated* media. TSANG & TSANG [1], MORENO et al. [2,3] or MORENO & TSANG [4] investigated the effects of channeling in single fractures with variable apertures. They pointed out that the hydraulic properties of the channels are different from those of the global porous medium. It was found that the channel properties are invariant regardless of the actual location of the channels and the overall direction of flow. Thus, the channel properties can be used to characterize the flow and transport behaviour in the whole system. TSANG & TSANG [5] or MORENO & TSANG [6] extended their results to solute transport through two- and threedimensional porous media with strongly variable hydraulic properties, finding that flow channeling also occurs in such systems.

In this paper we address the question of flow channeling in *unsaturated* soil. It is obvious that channels in saturated media develop along areas of large permeability. In unsaturated soil, however, large permeabilities are correlated to larger pore volumes, which would drain faster than the smaller pores. This can lead to strongly reduced flow rates along zones of high permeabilities; therefore the grade of channeling and the location of channels might be very different from the saturated case. To study such behaviour, we perform numerical simulations of fluid flow and particle transport in a two-dimensional vertical column of heterogeneous soil. Different saturation scenarios are applied varying from fully saturated conditions to low saturations.

FLOW AND TRANSPORT SIMULATION

In our study we simulate flow and transport processes in unsaturated porous media of strong heterogeneity. A quadratic model area is chosen with a side length of 10.0 m. Figure 1 shows the model area representing a vertical cross section of a porous medium. However, it may as well represent a vertical single fracture with strongly varying aperture distribution.

The heterogeneity is introduced by dividing the area in 40 x 40 quadratic cells or grid elements. A value of permeability is assigned to each element from a permeability distribution, which is assumed to be lognormally distributed and correlated in space. The permeability distribution has a geometric mean of 10^{-12} m^2 and a standard deviation of 1.737 in log10 (4.0 in natural log). The correlation in space is isotropic with a correlation length of 1.0 m. An exponential covariance function is used. Three different realizations of heterogeneous fields are generated. We assume that the porosity in each element is linearly correlated to the log permeability. A porosity of 0.35 is associated to the mean log10 permeability of -12; values of 0.30 and 0.40 are associated to log10 permeabilities of -16 and -8, respectively.

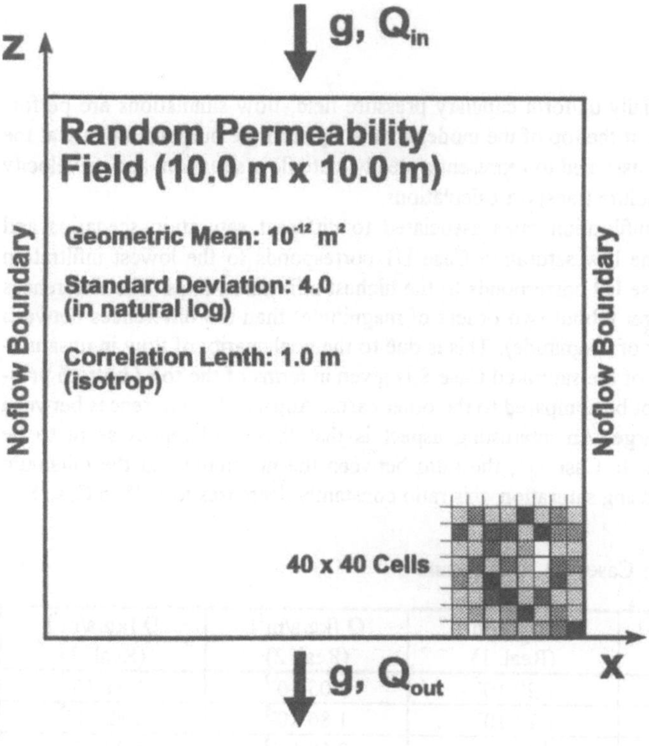

Fig.1 Model Area

Fluid flows with gravity from top to the bottom, the right and left boundaries are closed. At the top (inflow) boundary we prescribe certain infiltration rates into the model area (see Fig. 2). We do that in a way that saturations at the inflow cross section turn out to be constant with values of 0.3 (Case U1), 0.5 (Case U2) and 0.7 (Case U3), respectively. In all cases the bottom boundary condition is kept constant; it is held at a specified saturation of 0.7. For the sake of comparison, a fully saturated flow field is calculated as well (Case S). In that case, we simply apply a pressure gradient $\Delta P = P_1 - P_2$ between the top and the bottom boundary; no gravity forces are considered.

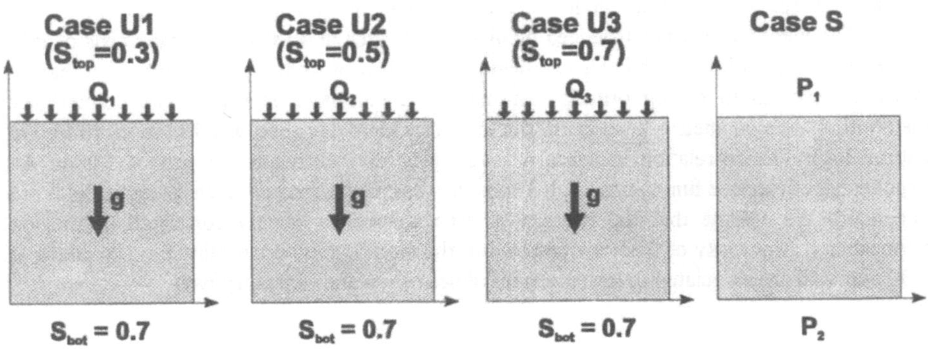

Fig. 2 Boundary conditions

Starting from an initially uniform capillary pressure field, flow simulations are performed until the infiltration rate at the top of the model area is equal to the output flow rate at the bottom. The last time step is assumed to represent a steady-state flow situation, and its velocity field serves as input for the solute transport calculations.

Table 1 shows the infiltration rates associated to different saturation scenarios and realizations. As expected, the low saturation Case U1 corresponds to the lowest infiltration rates; the high saturation Case U3 corresponds to the highest infiltration rates. The differences between U1 and U2 are larger (about two orders of magnitude) than the differences between U2 and U3 (about one order of magnitude). This is due to the nonlinearity of flow in unsaturated soil. The infiltration rate of the saturated Case S is given in terms of the top / bottom pressure difference ΔP and cannot be compared to the other cases. Apparently, differences between the realizations are quite large. An interesting aspect is that these differences seem to be depending on the saturation. In Case U1, the ratio between the maximum and the minimum flow rate is 2.40. With increasing saturation, this ratio constantly decreases to 1.85 in Case S.

Table 1 Infiltration rates for Cases U1, U2, U3 and S

Case	S_{top} [-]	S_{bot} [-]	Q [kg/s/m^2] (Real. 1)	Q [kg/s/m^2] (Real. 2)	Q [kg/s/m^2] (Real. 3)
U1	0.3	0.7	1.27 10^{-7}	3.05 10^{-7}	1.81 10^{-7}
U2	0.5	0.7	1.23 10^{-5}	1.86 10^{-5}	0.82 10^{-5}
U3	0.7	0.7	7.34 10^{-5}	9.59 10^{-5}	5.02 10^{-5}
S	1.0	1.0	ΔP x 5.21 10^{-8} Pa^{-1}	ΔP x 7.27 10^{-8} Pa^{-1}	ΔP x 3.92 10^{-8} Pa^{-1}

Flow Simulation

The flow simulations are performed with the Integrated Finite Difference Code TOUGH2, developed by Karsten Pruess at the Lawrence Berkeley Laboratory (PRUESS, [7] and [8]). TOUGH2 is a widely used simulator for nonisothermal flow of multicomponent, multiphase fluids in porous and fractured media. In our study only a small part of TOUGH2's capabilities is used, since only one phase (liquid) and one component (water) under isothermal conditions are taken into account. It is assumed that the air phase is immobile, so only the movement of water has to be calculated. The pressure in the air phase is constant (reference pressure P_r). Under these assumptions, the flow of water in the unsaturated zone can be described by the Richard's equation as follows

$$\frac{\partial}{\partial t}\left(\phi\ S\ \rho\right)+\nabla\left\{k\frac{k_r}{\eta}\rho\left[\nabla P-\rho g \delta_z\right]\right\}=0\,. \tag{1}$$

Here ϕ is porosity, S is saturation, ρ is density, k is absolute permeability, k_r is relative permeability, η is viscosity, $P = P_r + P_{cap}$ is pressure in the water (sum of reference pressure P_r and capillary pressure P_{cap}) and g is gravitational acceleration.

Capillary pressures and relative permeabilities depend strongly on the saturation. The lower is the water content in the soil pores, the higher is the capillary pressure, but the smaller is the relative permeability. Many workers have introduced functional relationships for capillary pressures and relative permeabilities, and a number of those are available within TOUGH2. In our study, we use VAN GENUCHTEN's capillary pressure and relative permeability expressions, which are given as follows (VAN GENUCHTEN, [9])

$$P_{cap}=\frac{1}{\alpha}\left\{\left(S_e\right)^{-\frac{1}{m}}-1\right\}^{\frac{1}{n}} \tag{2}$$

$$k_r=\left(S_e\right)^{\frac{1}{2}}\left\{1-\left(1-\left(S_e\right)^{\frac{1}{m}}\right)^{m}\right\}^{2} \tag{3}$$

with the effective liqid saturation

$$S_e=\frac{S-S_r}{1-S_r}\,. \tag{4}$$

Here S_r is the residual saturation; $1/\alpha$, n and m are parameters characterizing the pore geometry. The first parameter $1/\alpha$ denotes the minimum pressure needed to drive air into a water saturated soil (air entry pressure); large capillary forces are represented by large values for $1/\alpha$. The second parameter n characterizes the distribution of pore sizes (1 for very heterogeneous pore sizes, infinity for constant pore sizes). The last parameter m is defined as (1-1/n) (MUALEM, [10]).

We shall assume a residual saturation of 0.2 and a pore size distribution coefficient of 2.0. The air entry pressure related to the mean absolute permeability of 10^{-12} m^2 is chosen to -10000 Pa. However, as already mentioned, areas of high absolute permeability are expected to drain faster due to large pores sizes and less capillary strength. Likewise, areas of low absolute permeability are expected to have larger water saturations, due to smaller pore sizes and stronger capillary suction. LEVERETT ([11]) proposed a model to calculate the capillary pressure of a soil with varying absolute permeability as follows

$$P_{cap} = \overline{P}_{cap}\sqrt{\frac{\overline{k}}{k}} \qquad (5)$$

where k is an arbitrary permeability value, $\overline{k}$ is the mean permeability, and $\overline{P}_c$ is the capillary pressure associated to this mean value. A number of studies have shown in the past that this relationship holds for a variety of different soils (e.g. DAVIES, [12]; WANG, [13]).

Given a certain liquid saturation of the soil, the mean capillary pressure is obtained from Eq. (2) and, according to the absolute permeability of the element, corrected by applying LEVERETT's scaling model using Eq. (5). Thus, after scaling, each cell of our model region has a different capillary pressure function. In our example, a capillary pressure of -25000 Pa is associated to a water saturation of 0.5 for a cell with mean permeability. However, for a low permeability cell ($k = 10^{-15}$ m^2) the same pressure value is associated to almost full saturation; for high permeabilities ($k = 10^{-9}$ m^2) the saturation value drops almost down to the residual saturation. A special TOUGH2 version was developed to simulate this heterogeneity in capillary suction (FINSTERLE et al., [14]).

Particle Tracking Procedure

After solving the flow field and calculating particle velocities in each cell, a particle tracking procedure is performed to model the transport of contaminants through the porous medium. A total number of 300 particles is released at the inflow cross section of the model region. Some simulation runs have been performed with a higher number of particles to check the sensitivity of the model results. It has been shown, however, that a total number of 300 particles is sufficient to guarantee stable results. We use a particle tracking technique implemented in the interactive plotting software TECPLOT (TECPLOT, [15]), featuring a predictor-corrector integration scheme to trace particles through the vector field.

The number of particles introduced in each cell is proportional to the specific discharge into that cell. Within each cell, particles are randomly distributed along the horizontal line across the cell. Each particle is then followed through the model area until it reaches the bottom. During the particle tracking process, a number of parameters are recorded along the trace, such as absolute and relative permeabilities, saturations, pressures, flow rates or velocities. These data are used to calculate channel statistics and compare them to the statistics of the global porous medium. Finally, particles are collected at the outlet surface, and the travel time for each particle is calculated. Following MORENO & TSANG ([6]), all these calculations are based on only the fastest 90% of particles. This is reasonable because (1) particles transported through the fast channels are more significant, because (2) particles may decay and, in field tests, would not be recorded anymore after long travel times, and because (3) particles which get stuck in low velocity regions might bias the statistical results. The latter effect may sometimes be physically correct; however, it can as well be created by inaccuracies in the tracking procedure.

RESULTS AND DISCUSSION

Flow Field

The results of the flow simulations shall first be presented for one particular realization of the random permeability field. We choose Realization 1 in Table 1 which seems to represent an average permeability field with regard to the infiltration rates. Again, Cases U1, U2 and U3 are the three unsaturated cases with bottom saturation of 0.7 and top saturations of 0.3, 0.5 and 0.7; Case S is the fully saturated case. The distribution of absolute permeabilities is the same for all four cases; it is shown in Figure 3.

First, we look at the magnitude of particle velocities within the flow region. This allows for a visual interpretation with regard to the grade of channeling and the location of channels. Figure 4 shows the relative velocity distribution for the saturated medium Case S. As expected from previous studies (e.g. TSANG & TSANG, [5]; MORENO & TSANG, [6]) the majority of flow takes place in one or two selected pathways while a major part of the model area exhibits very low velocities. It seems reasonable to refer to these preferred flow paths as one-dimensional channels, and to describe the flow field in the model area based on the properties of those channels rather than on the properties of the entire porous medium. A comparison of Figure 3 and Figure 4 indicates that the channels develop along interconnected zones of high absolute permeabilities; the fluid seeks to flow in the least resistive pathways.

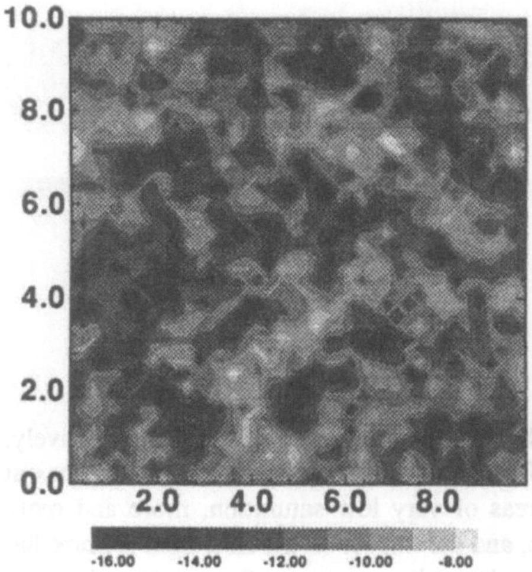

Fig. 3 Random Permeability Field (in log10)

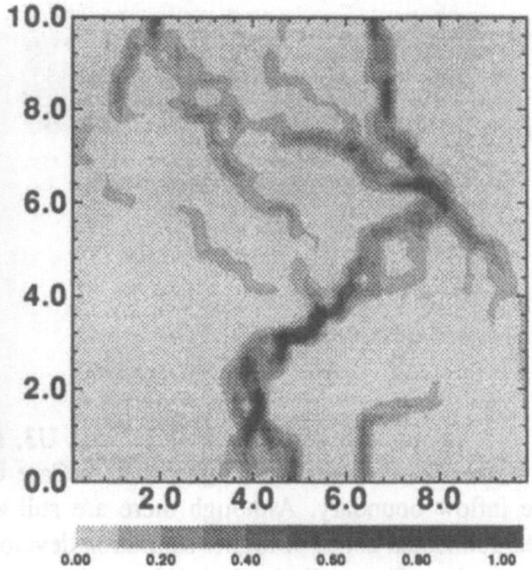

Fig. 4 Rel. Particle Velocities for Case S

Figure 5 shows the results of the first unsaturated Case U1. The calculated saturations in Figure 5a correlate to the absolute permeability field in Figure 3. Low absolute permeability zones are associated to saturation values close to 1.0 while high absolute permeability zones exhibit saturations just above the residual saturation. We forced such a behaviour in our simulation by providing different capillary pressure functions for each cell of the model area.

It is apparent from Figure 5b that the saturation distribution in the model area has a strong impact on the particle velocity field. Again, the majority of flow follows preferred paths which may be called channels, and the grade of channeling seems similar to Case S. However, the location of channels is totally different. While in Case S preferred flow paths correlate to high absolute permeabilities, channels now develop along interconnected zones of higher saturations which correspond to zones of low absolute permeabilities. Apparently, the relative permeabilitiy values according to equation (3) drop so rapidly with decreasing saturations that eventually the product of absolute and relative permeabilities (effective permeabilities) becomes very small and velocities in low saturations zones become almost negligible, despite of high absolute permeabilities.

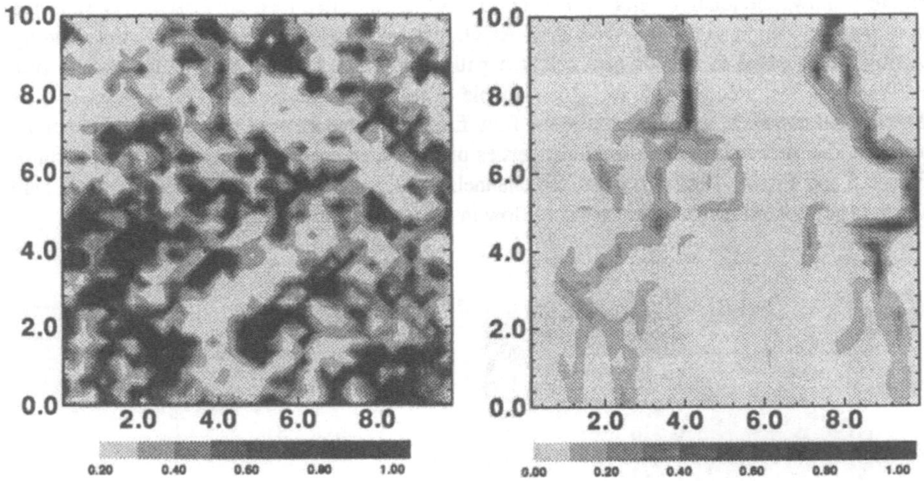

Fig. 5 Case U1 (S_{top}=0.3, S_{bot}=0.7)
a. Saturation
b. Relative Particle Velocities

Let us now discuss Cases U2 and U3, shown in Figure 6 and Figure 7, respectively. The overall saturations are higher than in Case U1, according to the higher infiltration rates at the inflow boundary. Although there are still areas of very low saturation, more and more interconnected zones of higher saturation develop, and the impact on the flow field is immediately evident. Especially in Case U3 we observe an almost homogeneous flow pattern. It is not possible to identify distinct channels and the variation of particle velocities seems smaller.

Apparently, intermediate to high saturations can have a natural smoothing effect on flow in strongly heterogeneous soil. For the low saturation case, however, the impact of drastically dropping relative permeabilities is so strong that the effective permeability field seems as heterogeneous as the absolute permeability field. To support this intuitive assumption we calculate the standard deviation of the effective permeabilities in the model area, presented in Figure 8 (in log10). Note that for the saturated Case S (S_{top}=1.0) the effective permeability is equals to the absolute permeability. Realization 1 exhibits the dashed line with square symbols. Realization 2 and 3 are added to get a more complete picture. The standard deviations of Case

U1 (S_{top}=0.3) and Case S are in the same range; both represent a very heterogeneous permeability field which allows for flow channneling. For Case U2 (S_{top}=0.5) and U3 (S_{top}=0.7), however, the standard deviations drop down significantly. Especially Case U3 represents an intermediate range of heterogeneity whère channeling effects are not evident.

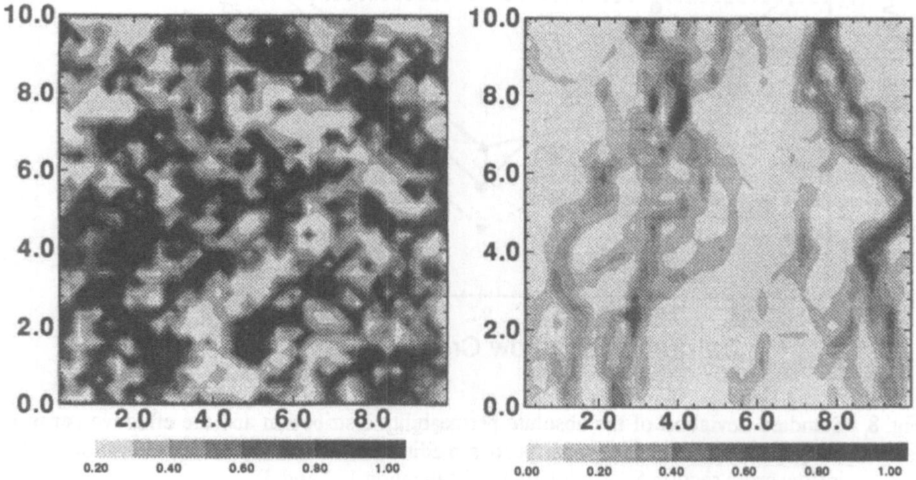

Fig. 6 Case U2 (S_{top}=0.5, S_{bot}=0.7)
a. Saturation
b. Relative Particle Velocities

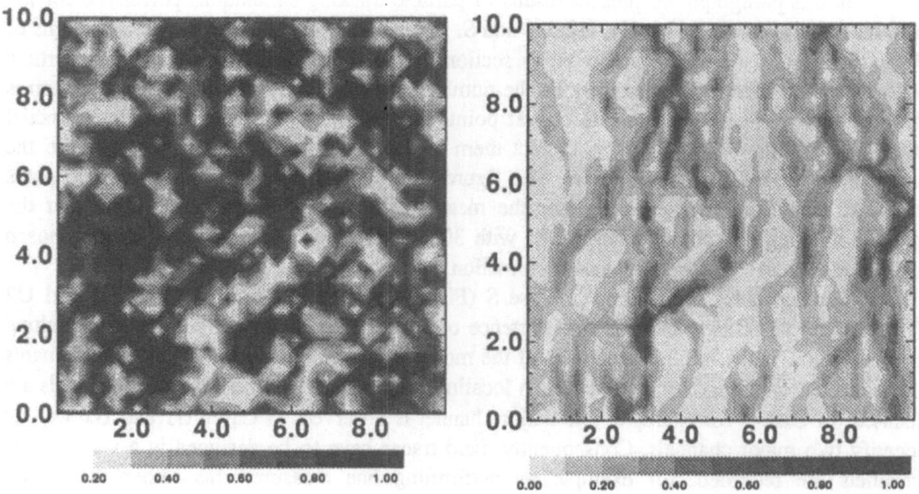

Fig. 7 Case U3 (S_{top}=0.7, S_{bot}=0.7)
a. Saturation
b. Relative Particle Velocities

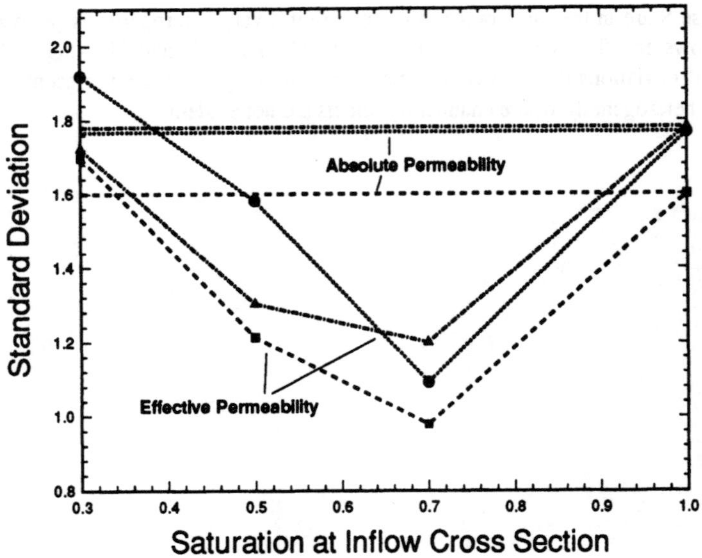

Fig. 8 Standard deviation of the absolute permeability distribution and the effective permeability distribution of the global porous medium, as a function of the saturation at the inflow cross section S_{top} (in log10, for Realization 1, 2 and 3)

Particle Tracking Simulation

In this paragraph we present results of particle tracking simulations performed for the different saturation scenarios U1, U2, U3 and S. Figures 9,10,11,12 show the particle paths of 50 particles released at the inflow cross section of the model area. Flow field and particle velocities correspond to Realization 1. The number of particles introduced at each location is proportional to the infiltration rate at that point. We follow the particles through the model area, record data along the trace, collect them at the outlet cross section and measure the travel time needed for each particle. The figures represent a snapshot of the particle paths taken at time T(Mean) which denotes the mean travel time of all particles. Note that the particle tracking procedure is performed with 300 or more particles; 50 particles are chosen only in the figures for a better visual presentation.

The saturated simulation run Case S (Fig. 9) and the unsaturated Cases U1 and U2 (Fig. 10 and 11) clearly indicate the existence of channels. Particles (or solutes) travel along preferred flow paths, while large zones of the model area are not affected. Thus, contaminants cross the bottom boundary only at certain locations, depending on where the flow channels are located. In Case S for example, one major channel is observed, in Cases U1 and U2 we can identify two major channels. Consequently, field tracer have to be designed in a way that all channels are recorded, for example by performing line measurements rather than point measurements. Compared to these results Case U3 (Fig. 12) exhibits a very different picture: Here the particle paths are widely spread, and distinct channels can hardly be identified.

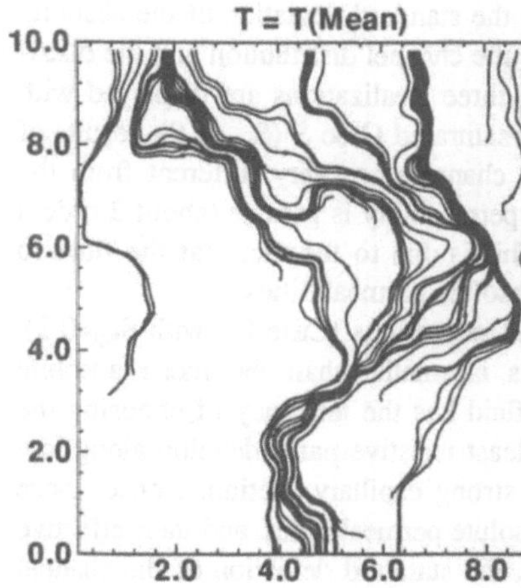

Fig. 9 Particle Tracks for Case S

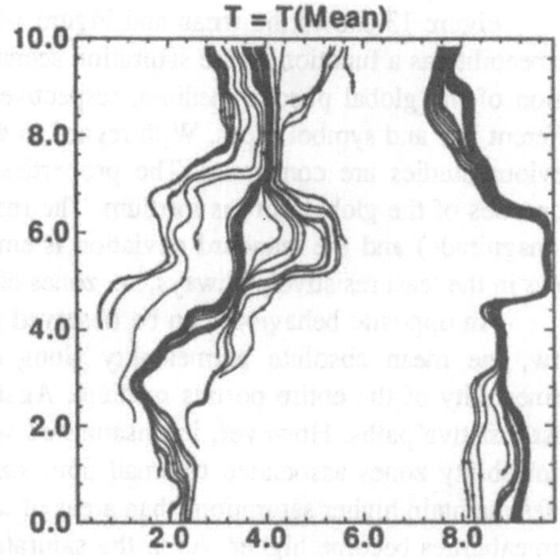

Fig. 10 Particle Tracks for Case U1

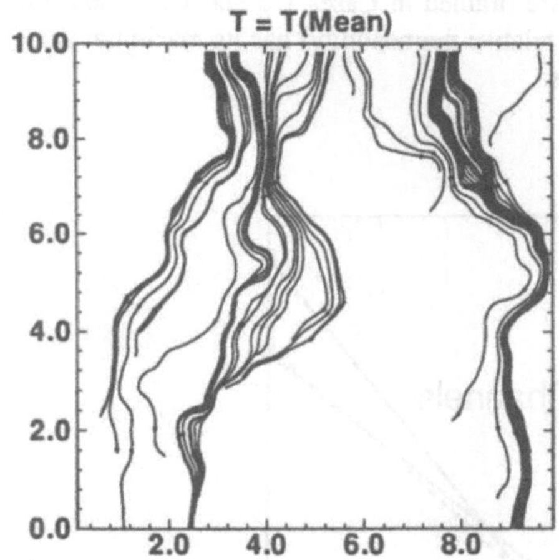

Fig. 11 Particle Tracks for Case U2

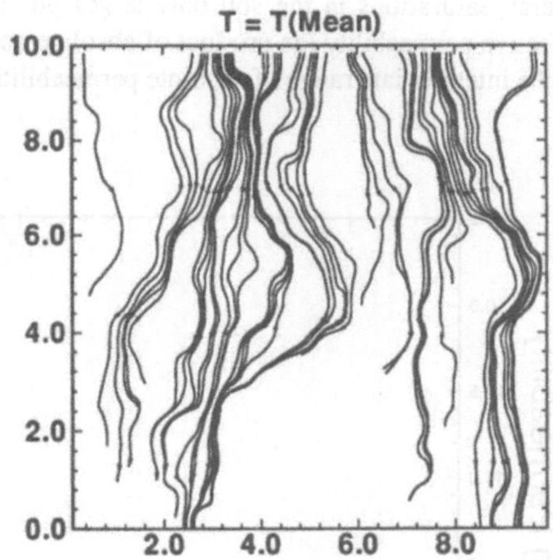

Fig. 12 Particle Tracks for Case U3

In the next paragraph we focuse on the hydraulic properties of the channels compared to the hydraulic properties of the entire porous medium. Namely, we are interested in the absolute permeability distribution along the channels. An interesting question is wether the mean and standard deviation of this distribution is a function of the saturation scenario. This issue is studied by collecting the absolute permeabilities of the 90% fastest particles travelling through the flow field. From these data, a new distribution of absolute permeabilities is obtained (the channel distribution), which is then compared to the absolute permeability distribution of the global porous medium.

Figure 13 shows the mean and Figure 14 shows the standard deviation of the absolute permeability as a function of the saturation scenario, for the channel distribution and the distribution of the global porous medium, respectively. The three Realizations are displayed with different line and symbol types. With regard to the fully saturated Case S (S_{top}=1.0), results of previous studies are confirmed. The properties of the channels are very different from the properties of the global porous medium. The mean log permeability is greater (about 2 orders of magnitude) and the standard deviation is smaller. This is due to the fact that the fluid to flows in the least resistive pathways, i.e. zones of high absolute permeabilities.

An opposite behaviour can be observed for small saturations (Case U1 with S_{top}=0.3). Now, the mean absolute permeability along channels is smaller than the mean absolute permeability of the entire porous medium. Again, the fluid has the tendency of choosing the least resistive paths. However, in unsaturated soil the least resistive paths develop along low permeability zones associated to small pore sizes and strong capillary suction. Hence, these zones maintain higher saturations than areas of large absolute permeabilities, and their effective permeabilities become higher. As in the saturated case, the standard deviation of the channel permeability distribution is smaller than the standard deviation of the global porous medium.

For intermediate ranges of saturation (Cases U2 with S_{top}=0.5 and U3 with S_{top}=0.7), the mean absolute permeabilities of the channels shift to higher values, while the standard deviations are still as small as in Case U1. Due to the higher infiltration rates and the increasing overall saturations in the soil only larges pores are drained in Cases U2 and U3. Then, the effective permeability (as product of absolute and relative permeability) has its maximum value in the intermediate range of absolute permeabilities.

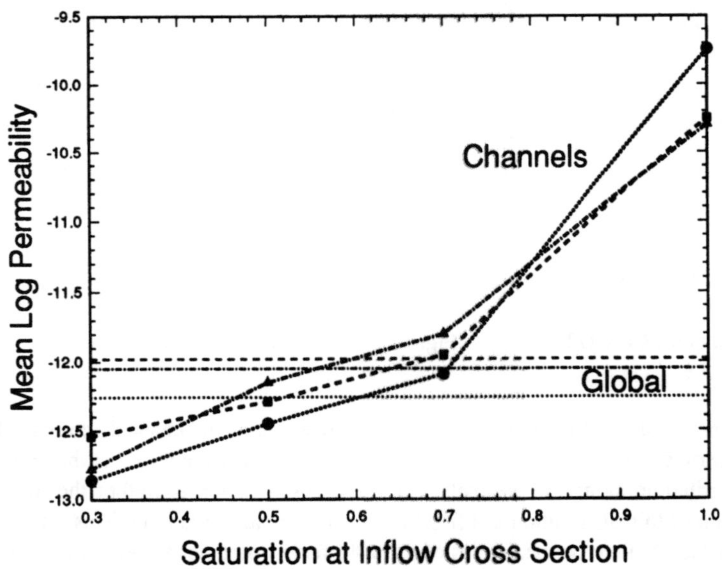

Fig. 13 Difference between the mean absolute permeability of the channels and the global porous medium, as a function of the saturation at the inflow cross section S_{top} (in log10, for Realization 1, 2 and 3)

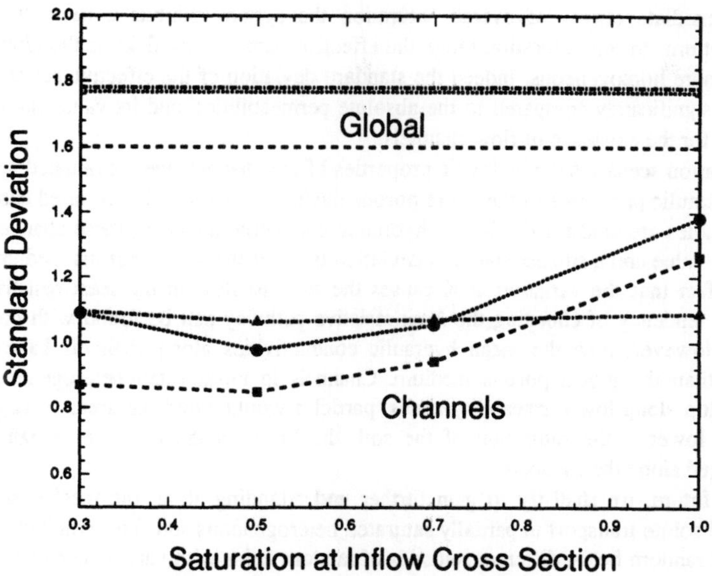

Fig. 14 Difference between the standard deviation of the absolute permeability distribution of the channels and the global porous medium, as a function of the saturation at the inflow cross section S_{top} (in log10, for Realization 1, 2 and 3)

An interesting aspect is that all realizations exhibit the same tendency of increasing mean log permeability with increasing overall saturation, despite of the fact that the location of channels is totally different from Realization to Realization. Apparently, the channel properties are independent of the specific location and geometry, provided that the same saturation scenario is considered. Thus, it seems possible to characterize the solute transport through the entire porous medium by only providing the statistical properties of the channel distribution.

CONCLUSIONS

Numerical simulations have been performed to study flow and solute transport in a two-dimensional vertical column of heterogeneous soil. Different saturation scenarios were applied varying from fully saturated conditions to low saturations. It was found that in strongly heterogeneous soil particles (or solutes) travel along preferred flow paths, which may be called channels. This is the case in saturated as well as in unsaturated media; however, the grade of channeling and the location of channels is different for different scenarios. The channel properties are independent of the specific location and geometry, provided that the same saturation scenario is considered.

A distinct channeling effect was obtained in the saturated case and in the case of low saturation. The effective permeability field of the low saturation case (product of absolute and relative permeabilites) is as heterogeneous as the absolute permeability field which is relevant for the saturated case. The standard deviation of the effective permeabilities is in the same range as the standard deviation of the absolute permeabilities.

Within intermediate ranges of overall saturation the system exhibits a more two-dimensional flow pattern. In this saturation range the effective permeability field of the global porous medium is more homogeneous. Indeed the standard deviation of the effective permeabilities drops down significantly compared to the absolute permeabilities, and its value can be used as an indicator for the existence of flow channels.

For all saturation scenarios the hydraulic properties of the channels were evaluated and compared to the hydraulic properties of the entire porous medium. For the fully saturated case, results of former studies are confirmed. The log-hydraulic conductivities along these channels have a greater mean value and a smaller standard deviation than for the global porous medium. This is due to the fact that the variation in K causes the fluid to flow in the least resistive pathway. The same tendency of choosing the least resistive pathway can be found with partially saturations. However, now the mean hydraulic conductivities along channels have a lower mean value than the global porous medium. Channels in unsaturated heterogeneous media tend to develop along low permeability zones, particles would avoid the areas of large permeabilities. The lower is the saturation of the soil, the lower is the absolute hydraulic permeability measured along the channels.

In the near future, we shall try to gain further understanding about the mechanisms controlling flow and solute transport in partially saturated heterogeneous soil. This will include more realizations of random fields, different statistics of random fields with varying correlation length or standard deviation and eventually a more thorough study on the impact of different capillary pressure and relative permeability functions.

REFERENCES

[1] TSANG, Y.W., C.F. TSANG (1987), Channel model of flow through fractured media, *Water Resour. Res.*, 25(9), 2076-2080.

[2] MORENO, L., Y.W. TSANG, C.F. TSANG, F. HALE, I. NERETNIEKS (1988), Flow and transport in a single fracture: A stochastic model and ist relation with field observations, *Water Resour. Res.*, 24(12), 2033-2048.

[3] MORENO, L., C.F. TSANG, Y.W. TSANG, I. NERETNIEKS (1990), Some anomalous features of flow and solute transport arising from fracture aperture variability, *Water Resour. Res.*, 26(10), 2377-2391.

[4] MORENO, L.; TSANG, C.F. (1991): Multiple peak response to tracer injection tests in single fractures: A numerical study, Water Resources Reseach, Vol. 27, No. 8, pp. 2143-2150.

[5] TSANG, Y.W.; TSANG, C.F. (1989): Flow channeling in a single fracture as a wwo-dimensional strongly heterogeneous permeable medium, Water Resources Research, Vol. 25, No. 9, pp. 2076-2080.

[6] MORENO, L., C.F. TSANG (1994), Flow channeling in strongly heterogeneous porous media: A numerical study, *Water Resour. Res.*, 30 (5), 1421-1430.

[7] PRUESS, K. (1987), TOUGH user's Guide, Lawrence Berkeley Laboratory report LBL-20700, Berkeley, CA.

[8] PRUESS, K. (1991), TOUGH2 - A general-purpose numerical simulator for multiphase fluid and heat flow, Lawrence Berkeley Laboratory report LBL-29400, Berkeley, CA.

[9] VAN GENUCHTEN, M.T. (1980), A closed-form equation for predicting the hydraulic conductivity of unsaturated soil, *Soil Sci. Soc. Am. J.*, 892-898.

[10] MUALEM, Y. (1978), A new model for predicting the hydraulic conductivity of unsaturated porous media, *Water Resour. Res.*, 12 (3), 513-522.

[11] LEVERETT, M.C. (1941), Capillary behaviour in porous solids, *AIME Trans.*, Vol. 142, 1941.

[12] DAVIES, P.B. (1991), Evaluation of the role of threshold pressure in controlling flow of waste-generated gas into bedded salt at the waste isolation plant, Sandia National Laboratories report SAND90-3246, Albuquerque, NM.

[13] WANG, J.S.Y. (1992), Variations of hydrological parameters of tuff and soil, Proc. 3rd Int. Conf. on High Level Radioactive Waste Management at Las Vegas, NV, 727-731.

[14] FINSTERLE, S., G.J. MORIDIS, K. PRUESS (1994), A TOUGH2 equation-of-state module for the simulation of two-phase flow of air, water and a miscible gelling liquid, Lawrence Berkeley Laboratory report LBL-36086, Berkeley, CA.

[15] TECPLOT (1995), TECPLOT - Version 6 - User's Manual, Amtec Engineering, Bellevue, WA.

BLOCK SSOR PRECONDITIONERS FOR 3-D GROUNDWATER FLOW PROBLEMS

D. Braess and C. König
Faculty of Mathematics and Faculty of Civil Engineering
Ruhr-Universität Bochum, D-44780 Bochum, Germany

SUMMARY

In the simulation of 3-dimensional groundwater flow problems often meshes are generated for which the horizontal mesh size is much larger than the distances in the vertical direction. The matrices which arise from the discretization of differential equations are known to have large condition numbers. Here, the geometry leads to an additional increase of the condition number. While standard preconditioners as SSOR and ICC can cope with the differential operators, they are less suited for the geometrical difficulties. We show that here block-SSOR preconditioners are an appropriate tool.

INTRODUCTION

In groundwater flow simulations great systems of linear equations are to be solved. Stationary problems lead to the discretization of elliptic differential equations of second order. If Darcy's law

$$v = -K \operatorname{grad} h$$

is inserted in the continuity equation

$$\operatorname{div} v = q$$

then an elliptic equation of second order arises

$$-\operatorname{div}(K \operatorname{grad} h) = q.$$

When time-dependent problems are solved by implicit integration schemes, then equations with a similar structure are created.

In the modeling of 2-dimensional problems with several thousands of unknowns, preconditioned conjugate gradient methods (PCG-methods) have turned out to be efficient. In particular, the standard preconditioners which can be used as black-box codes are successful in most cases. There are no big differences between preconditioning by incomplete Cholesky decomposition (ICC) and symmetric successive overrelaxation (SSOR), when computing times are measured. There is a small preference for the ICC method [1, 4, 5].

The computing times are more sensitive to the question whether meshes with well-shaped triangles (and quadrilaterials) or with slim triangles have been created. Meshes are usually generated so that sources and wells become nodes in the mesh and (small) rivers become edges. Since there are these restrictions from a practical point of view, one is prepared to have a slightly larger number of iteration steps than in similar problems of other applied sciences. Indeed, this is the situation when 2-dimensional models are treated.

The situation turned out to be less favorable when 3-dimensional problems were first attacked. The iteration with ICC-preconditioner broke down in many cases and the iterations with SSOR-preconditioner suffered from a dramatic increase of the number of steps. This was our experience, and the same was reported by other groups.

It is our aim to understand this phenomenon and to find a remedy.

SOME CHARACTERISTICAL PROPERTIES OF 3-DIMENSIONAL MODELS

When 3-dimensional models for groundwater flow problems are performed, one has typically domains whose projection onto a horizontal plane covers an area of several km^2. On the other hand, the aquifer, i.e. the zone in which groundwater is transported, has a thickness of about 10 to 100 m. To achieve a good model of the geological situation, the domain is divided into 3 to 20 layers.

As a consequence, the 3-dimensional meshes have quite different meshsizes in horizontal and vertical direction. Typical are mean values of meshsizes of

100 m – 500 m in the horizontal direction,

1 m – 10 m in the vertical direction.

Since the coefficients of the equations which arise from a differential equation of second order are proportional to h^{-2}, we get an anisotropy of

$$10^3 \text{ to } 10^4.$$

Here we have already taken into account that the permeability in the horizontal direction is greater than the vertical one.

The standard preconditioners SSOR and ICC are well suited to cope with the properties of the matrix which are due to the discretization of an elliptic equation. Also ICC can deal with problems with jumps in the dispersion coefficient since it is less point-oriented than SSOR. On the other hand it cannot compensate the increase of the condition number of the matrix induced by the anisotropy of the grid.

We will see that block preconditioners are suited [1, 2, 6].

The 3-dimensional meshes which are encountered in groundwater, usually have a special structure. They can be understood as copies of identical horizontal layers. The number of layers is between 3 and 20. Thus in the vertical direction there is a very regular structure while the meshes within every 2-dimensional layer may be a very irregular.

A MODEL PROBLEM

In order to describe the phenomena more clearly and to simplify the representation, we turn to a model problem with the same structure. We consider the Poisson equation on a rectangle

$$\begin{aligned} -\Delta u &= f \text{ on } [0, n_x\Delta x] \times [0, n_z\Delta z] \\ u &= 0 \text{ on the lateral boundary,} \\ \frac{\partial u}{\partial n} &= 0 \text{ on the top and on the bottom.} \end{aligned}$$

The typical stencil for the discretization, i.e. the typical coefficients in the matrix associated to an interior point is given by

$$\begin{pmatrix} & -1 & \\ -\varepsilon & 2+\varepsilon & -\varepsilon \\ & -1 & \end{pmatrix}.$$

Here $\varepsilon = (\Delta z/\Delta x)^2$. In order to have a comparable model, we consider $\varepsilon \approx 10^{-3}$, $n_x = 40$ and $n_z = 10$. The coupling in the vertical direction is much higher than in the horizontal one.

Now it is clear that a point relaxation scheme cannot be successful. The point Gauss-Seidel scheme can be understood as follows. Each update of a component of the vector is performed such that the induced quadratic form is minimized. A difference to the vertical neighbours, however, increases the quadratic term by a large amount and is penalized by a large factor. The updates are damped by this large factor. A slow-down of the iteration is the consequence.

The ICC method has a different shortcoming. The matrix has a block structure and the diagonal blocks

$$\begin{pmatrix} 1+\varepsilon & -1 & & & \\ -1 & 2+\varepsilon & -1 & & \\ & \ddots & \ddots & \ddots & \\ & & -1 & 2+\varepsilon & -1 \\ & & & -1 & 1+\varepsilon \end{pmatrix}$$

are almost singular. The truncation errors which are caused by the incompleteness of the decomposition meet with a nearly singular matrix.

Table 1. Number of iteration steps for model problem and three preconditioners. ($n_x = 40$, $n_z = 10$, accuracy= 10^{-5})

anisotropy ε	1	10^{-1}	10^{-2}	10^{-3}	10^{-4}
preconditioner					
ICC	26	35	42	74	94
SSOR	26	41	105	161	> 200
block-SSOR	19	18	17	17	16

Candidates for overcoming the difficulties are block-SSOR and block-ICC methods. Specifically the first method turns out to be successful. When the updates of all the points on a vertical line are done simultaneously, the strong coupling is no longer a disadvantage.

The results in Table 1 support the considerations above.

In principle, the same holds for block ICC provided that the inversion of the diagonal block matrices is performed exactly. In this case, all the blocks of the matrix which contain only one nonzero entry are completely filled. To avoid this drawback, block-ICC is usually combined with some approximations of the inverses. Inverses of tridiagonal matrices are approximated by tridiagonal ones. These approximations are here not appropriate for the same reason as mentioned for point ICC.

Table 2. Three examples of matrices in groundwater simulations. Number of iteration steps k and computing time t for an accuracy of 10^{-5}. (nc = no convergence)

region	Düsseldorf		Heidelberg		Ruhr-Lippe	
variables	9246		16464		45666	
non-zero entries	85499		161645		546000	
horizontal layers	6		12		6	
area in km^2	4.5		0.71		152.8	
	k	t	k	t	k	t
preconditioner						
ICC	49	7.8	*nc*		*nc*	
point SSOR	183	24.1	620	150	750	280
block SSOR	30	4.6	278	83	150	60
AMG with block SSOR	7	5.4	128	105	30	81

NUMERICAL RESULTS

We have examined several preconditioners for 7 real-life groundwater flow problems. Table 2 contains the results for three of them. Here the number of iteration steps is listed together with the computer time (in seconds) on a workstation with 7 MFlops. Table 3 contains only the iteration count. It is obvious that the block-SSOR preconditioner leads to the fastest iteration and is the most reliable. The reduction of the number of iteration steps is always significant. In some cases it is more than a factor of 5. Specifically, the difference bewtween the preconditioners are very great in those cases in which a region with a great area is modelled.

We emphasize that the codes are (black-box) general purpose codes. No individual programming was performed in connection with these investigations.

In Table 2 also results for preconditioners with algebraic multigrid methods (AMG) is shown [3]. The costs for one cycle are now as high as for about 4 steps with the other preconditioner. Therefore, we do not save computer time although the number of iteration steps was reduced once more.

Table 3. Four examples of matrices in groundwater simulations. Number of iteration steps for an accuracy of 10^{-5}.

region	*Brandenb.*	*Anhalt*	*CokePlant*	*Dorsten*
variables	24000	24260	75816	34074
non-zero entries	290000	292000	987000	412000
horizontal layers	6	17	13	6
area in km^2	23.5	98.9	0.49	113.0
preconditioner				
ICC	1411	859	128	173
point SSOR	2369	1149	184	282
block SSOR	155	112	86	81

REFERENCES

[1] Axelsson, O.: "Iterative Solution Methods", Cambridge University Press (1994).

[2] Axelsson, O.: "Incomplete block matrix factorization preconditioning methods." The ultimate answer? J. Comp. Appl. Math. 12, 13, 3–18 (1985).

[3] Braess, D.: "Towards algebraic multigrid for elliptic problems of second order", Computing 55, 379–393 (1995).

[4] Meyerink, A., van der Vorst, H.A.: "An iterative solution method for linear systems of which the coefficient matrix is a symmetric M-matrix", Math. Comp. 31, 148–162 (1977).

[5] Schmid, G., Braess, D.: "Comparison of fast equation solvers for groundwater flow problems". In "Groundwater Flow and Quality Modelling" (Hrsg.: E. Custodio u.a.) pp. 173–188. Reidel (1988).

[6] Wittum, G.: "On the robustnes of ILU-smoothing", SIAM J. Sci. Stat. Comput. 10, 699–717 (1989).

A DISCRETISATION FOR TRANSPORT PROBLEMS WITH DOMINANT CONVECTION USING CHARACTERISTICS AND FINITE ELEMENTS

Carola Blömer
GKW Ingenieurgesellschaft m.b.H.
Am Birkenbusch 9a,
D-44803 Bochum, Germany

SUMMARY

In this paper a discretisation for two-dimensional steady state transport processes in porous media is presented. To guarantee stability and good accuracy of the discrete solution even in the convection dominated case two mean concepts are used: The convective part of the equation is discretised by using the method of characteristics. Mixed finite element / finite volume techniques are applied to the resulting modified equation. The discretisation can be applied also to the hyperbolic limit problem of transport without dispersion. Some examples are presented. They demonstrate the stability of the discretisation and the sensitivity with respect to the choice of the steplength parameter for the discrete characteristics. For convection dominated problems the linear systems achieved by this discretisation are singular perturbations of subdiagonal systems, if a 'flow oriented' numbering of the finite element nodes is used. Gauss-Seidel and SOR-algorithm with overrelaxation can be used as efficient solvers.

INTRODUCTION

The calculation and prognosis of the spreading of substances dissolved in groundwater can be done efficiently with numerical models. The following problem arises in practical applications: Standard finite element methods require frequently very fine discretisations to avoid numerical dispersion and guarantee stability of the numerical solution of the transport equation. Due to this restriction finite element grids, which are fine enough to give good accuracy of the discrete solution for the groundwater flow problem, often can not be used for the numerical solution of the transport problem. The disadvantage of standard finite element methods lies in the fact, that on coarse grids the transport equation is a singular perturbation of the hyperbolic limit problem of convective transport. Standard discretisation methods can not be applied to this limit problem. Consequently they have stability problems in the singular perturbed case of convection dominated transport. This paper presents a new discretisation [1], starting with the construction of a discretisation for the hyperbolic limit problem. The resulting method is combined with finite element / finite volume techniques to construct a discretisation for the convection dominated transport problem.

POLLUTANT TRANSPORT IN GROUNDWATER: A SINGULAR PERTURBED PROBLEM

When sorption, chemical and biological reactions and diffusion is neglected, the distribution of a pollutant dissolved in groundwater is determined by convection and hydromechanical dispersion (see e.g. [2]). The partial differential equation for steady state transport in two dimensions is:

$$\begin{aligned} -\nabla(\mathrm{D}\nabla c)+\mathrm{v}\cdot\nabla c+qc=f &\quad \text{in } \Omega\subset\mathbb{R}^2,\\ c=\bar{c} &\quad \text{on } \Gamma_0,\\ (\mathrm{D}\nabla c-\mathrm{v}c)\cdot\mathrm{n}=\mathrm{g}_1 &\quad \text{on } \Gamma_1,\\ (\mathrm{D}\nabla c)\cdot\mathrm{n}=\mathrm{g}_2 &\quad \text{on } \Gamma_2, \end{aligned} \qquad \text{(P)}$$

with $\Gamma=\Gamma_0\cup\Gamma_1\cup\Gamma_2=\delta\Omega$. n is the outward normal to Γ and

c concentration of pollutant,
$f=q\,c^*$ solute sources and sinks,
c^* concentration of sources and sinks,
q sources and sinks of groundwater,
$\mathrm{v}=(\mathrm{v}_x,\mathrm{v}_y)$ velocity vector, satisfying Darcy's law: $\nabla\cdot\mathrm{v}=q$.

The hydromechanical dispersion tensor is defined by

$$\mathrm{D}=\frac{1}{\sqrt{\mathrm{v}_x^2+\mathrm{v}_y^2}}\begin{pmatrix}\alpha_L\mathrm{v}_x^2+\alpha_T\mathrm{v}_y^2 & (\alpha_L-\alpha_T)\mathrm{v}_x\mathrm{v}_y\\ (\alpha_L-\alpha_T)\mathrm{v}_x\mathrm{v}_y & \alpha_T\mathrm{v}_x^2+\alpha_L\mathrm{v}_y^2\end{pmatrix}, \text{ with}$$

α_L longitudinal dispersion coefficient,
α_T transversal dispersion coefficient, with $\alpha_T \ll \alpha_L$.

In practical problems no restrictions can be made with respect to the smoothness of the boundary conditions $\bar{c}$ or the source term qc^*. In the convection dominated case, this means $\|\mathrm{D}\| \ll \|\mathrm{v}\|$, discontinuous data distribution lead to internal layers in the solution of (P). These layers follow the direction of the characteristics of the velocity field v. The solution of the convection dominated problem is determined by the hyperbolic convective part of the partial differential equation. In this case the transport problem (P) can be understood as a singular perturbation of the hyperbolic problem:

$$\begin{aligned} \mathrm{v}\cdot\nabla c+qc=f &\quad \text{in } \overline{\Omega}\backslash\Gamma_{in},\\ c=\bar{c} &\quad \text{on } \Gamma_{in}, \end{aligned} \qquad (\mathrm{P}_0)$$

where Γ_{in} denotes the inflow part of the boundary.

The stability of standard discretisation techniques for transport problem (P) is determined by the Peclet number *Pe*. The stability condition is given by:

$$Pe = \frac{h\|\mathrm{v}\|}{2\|\mathrm{D}\|} = \frac{h}{2\alpha_T} < 1,$$

where h is the local mesh size of a given triangulation of Ω.

DISCRETISATION OF THE LIMIT PROBLEM USING CHARACTERISTICS

As mentioned above the solution of the convection problem (P_0) is determined by the characteristics of the velocity field v. In the following characteristics are used to construct a discretisation for the convection problem (see [3], [5]). The characteristic $X(x,.)$ through a point $x \in \Omega$ is defined as the solution of

$$\frac{d}{dt} X(x,t) = \mathrm{v}(X(x,t)), \quad X(x,0) = x.$$

By using this definition the term $\mathrm{v}\cdot\nabla c$ can be rewritten as a derivative of c in flow direction v:

$$(\mathrm{v}\cdot\nabla c)(x) = \left.\frac{d}{dt}\right|_{t=0} c(X(x,t)). \tag{1}$$

With this relation problem (P_0) can be formulated as follows:

$$\begin{array}{ll} c(x) = c(X^{\Delta t(x)}(x)) + I(f-qc,x) & x \in \overline{\Omega}\backslash\Gamma_{in}, \\ c(x) = c_{in}(x) & x \in \Gamma_{in}, \end{array} \qquad (\tilde{\mathrm{P}}_0)$$

with $X^s(x) = X(x,-s)$ and $I(f-qc,x) = \int_0^{\Delta t(x)} (f-qc)(X^s(x))ds$.

For a given triangulation of Ω let $(x_i)_i$ denote the meshnodes and $(\varphi_i)_i$ denote the basis functions for piecewise linear finite elements: $\varphi_i(x_j) = \delta_{i,j}$. The discrete approximation of $(\tilde{\mathrm{P}}_0)$ is defined as follows:

Find $c_h = \Sigma c_i\ \varphi_i$, with

$$\begin{array}{ll} c_i = c_h(X_h^{\Delta t_i}(x_i)) + I_h(f-qc,x_i) & x_i \in \overline{\Omega}\backslash\Gamma_{in}, \\ c_i = c_{in}(x_i) & x_i \in \Gamma_{in}. \end{array} \qquad (\tilde{\mathrm{P}}_0^h)$$

The steplength parameter Δt_i for the calculation of the discrete characteristics are determined by the rule that the length of the discrete characteristic $X_n^s(x_i)$, $s \in [0, \Delta t_i]$, should be proportional to the local meshsize h:

$$\int_0^{\Delta t_i} \|\mathrm{v}(X^s(x_i))\| ds = \tau\, O(h). \tag{2}$$

Figure 1 shows an example for $\tau = 6$. The choice of the stepsize parameter τ is discussed in the following chapters.

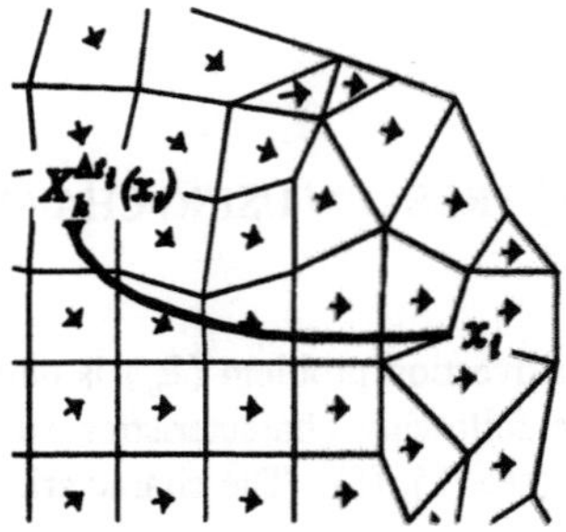

Figure 1: Discrete characteristic for $\tau = 6$.

The integrals over the source term $f - qc$ are calculated by using finite volume interpolations of the source functions: Let $(\overline{\varphi}_i)_i$ denote the finite volume basis functions being piecewise constant on cells Ω_i attached to the meshnodes x_i:

$$\overline{\varphi}_i(x) = \begin{cases} 1 & x \in \Omega_i, \\ 0 & x \notin \Omega_i. \end{cases}$$

The interpolation operator is defined as

$$u \to \overline{u}_h, \quad \overline{u}_h(x) = \Sigma u(x_i)\overline{\varphi}_i$$

and the integral over the source term is defined as

$$I_h(f - qc, x_i) = \int_0^{\Delta t_i} (\bar{f}_h - \bar{q}_h \bar{c}_h)(X_h^s(x_i))ds.$$

The solution of the discrete problem $(\tilde{P}_0^h)$ requires the evaluation of the piecewise linear function c_h at the endpoints of each characteristic:

$$X_h^{\Delta t_i}(x_i) = \sum_{\mu=1}^{n(i)} \lambda_{k_\mu(i)} x_{k_\mu(i)}, \text{ with } \sum_{\mu=1}^{n(i)} \lambda_{k_\mu(i)} = 1.$$

The node-set $S(i) = \{k_\mu(i), \mu = 1, \ldots, n(i)\}$ can be used to construct a natural numbering of the nodes, beginning at the inflow boundary and numbering in flow direction. If this enumeration is used, the matrix of the linear system for $(\tilde{P}_0^h)$ is nearly subdiagonal. The Gauss-Seidel algorithm can be applied as an efficient iterative solver.

DISCRETISATION OF THE CONVECTION DOMINATED TRANSPORT PROBLEM

By using the relation (1) in the same way as for the hyperbolic limit problem (P_0) a new formulation of the transport problem (P) can be obtained:

$$c(x)=c(X^{\Delta t(x)}(x))+I(f-qc,x)+I(\nabla(\mathrm{D}\nabla c),x) \quad x\in\Omega,$$
$$+ \text{ boundary conditions.}$$

In the first discretisation step the integral over the characteristic of the dispersion part is approximated by:

$$I(\nabla(\mathrm{D}\nabla c),x) \approx \Delta t(x)\,\nabla(\mathrm{D}\nabla c)(x).$$

After some rearrangements this approximation leads to the following modified transport problem:

$$-\nabla(\mathrm{D}\nabla c)(x)+\frac{c(x)-c(X^{\Delta t(x)}(x))}{\Delta t(x)}=I(f-qc,x) \quad x\in\Omega, \qquad (\tilde{\mathrm{P}})$$
$$+ \text{ boundary conditions.}$$

The finite difference $\left(c-c\circ X^{\Delta t(.)}\right)/\Delta t(.)$ can be interpreted as an upwind discretisation of the convection term $\mathrm{v}\cdot\nabla c$. The integral over the characteristic $I(f-qc,.)$ can be understood as an upwind weighting of the source term $f-qc$.

To avoid numerical dispersion caused by linear interpolation of the hyperbolic parts of the equation a mixed finite element / finite volume technique is used to discretise this problem. The hyperbolic part of the equation is discretised using finite volume techniques where as finite element basis- and testfunctions are applied to the symmetric dispersion part of the equation. The discrete problem is defined as follows:

Find $c_h=\Sigma c_i\varphi_i$ with

$$\int_\Omega \nabla\varphi_i(\mathrm{D}\nabla c_h)dx+\int_\Omega \frac{\bar{c}_h-\overline{(c_h\circ X_h^{\Delta t})}}{\overline{\Delta t}}\bar{\varphi}_i dx=\int_\Omega \frac{\overline{I_h(f-qc_h,.)}}{\overline{\Delta t}}\bar{\varphi}_i dx \quad \forall\varphi_i. \qquad (\tilde{\mathrm{P}}^{\mathrm{h}})$$

REMARKS:

1. For $\mathrm{D}=0$ the discrete problem $(\tilde{\mathrm{P}}^{\mathrm{h}})$ is equivalent to $(\tilde{\mathrm{P}}_0^{\mathrm{h}})$.
2. If a natural numbering of the nodes is used, the linear system for $(\tilde{\mathrm{P}}^{\mathrm{h}})$ in the convection dominated case is a singular perturbation of a nearly subdiagonal matrix. The SOR algorithm with overrelaxation can be applied as an efficient iterative solver.
3. The accuracy of the discrete solution depends on the choice of the steplength parameter τ, which determines the parameter $\Delta t(.)$ (see (2)):
   - In the convection dominated case ($Pe>1$) τ has to be large to avoid numerical dispersion caused by the linear interpolation of the concentration at the endpoints of each discrete characteristic.
   - In the elliptic case ($Pe<1$) small steplength parameters have to be chosen.

The discretisation method is implemented in the program package SICK 100, version 20.1, developed by the Arbeitsgruppe Grundwassermodelle of the Ruhr Universität Bochum [4]. The code for the CHAracteristic Finite Element discretisation is called CHAFE.

Results for some examples are presented in the following:

EXAMPLE 1: PARALLEL FLOW

Consider a parallel flow problem on the region $\Omega = [0,200] \times [0,200]$.$\Omega$ is discretised into 400 triangles and a grid-direction nearly perpendicular to the flow direction v (see figure 2). The boundary conditions are fixed to be $\bar{c} = 100$ on the lower right edge of the inflow boundary $\Gamma_0 = \{(x,0),\ x \in [180,200]\} \cup \{(200,y),\ y \in [0,20]\}$. The discrete solutions obtained with CHAFE are compared with the discrete solutions obtained with bilinear finite elements on a rectangular grid with 200 × 200 elements [6].

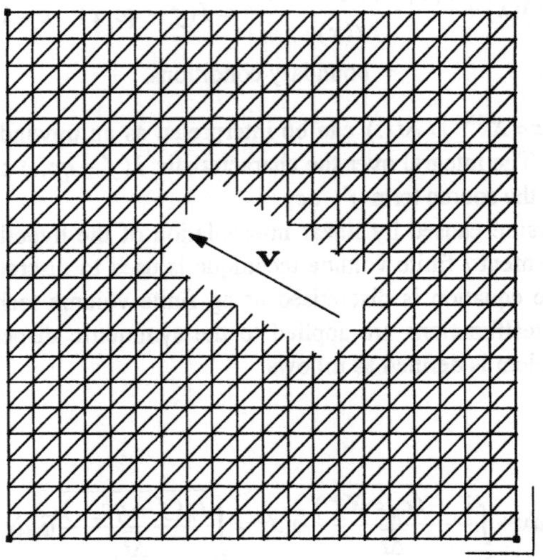

Figure 2: Triangulation of $\Omega = [0,200] \times [0,200]$ for parallel flow problem.

For the elliptic case with dispersion coefficients $\alpha_L = 100$, $\alpha_T = 10$ and a Peclet number of 0.5 the discrete solutions for steplength parameters $\tau = 1$ and $\tau = 2$ are shown in figure 3. Compared to the solution of standard finite elements on the fine grid the discretisation with the new method shows substantial numerical dispersion. This can be explained by the unfavourable grid direction. Comparing the discrete solutions for $\tau = 1$ and $\tau = 2$ we observe that larger steplength parameters τ lead to less numerical dispersion perpendicular to the flow direction but increase the numerical dispersion in flow direction.

For the convection dominated case with dispersion coefficients $\alpha_L = 1$, $\alpha_T = 0.1$ and a Peclet number of 50 stable discrete solution are obtained with CHAFE (see figure 4). Good accuracy is achieved, even for the unfavourable triangulation, if the steplength parameter τ is taken large enough. For smaller parameters of τ additional numerical dispersion is caused by the linear interpolation at the endpoints of the discrete characteristic.

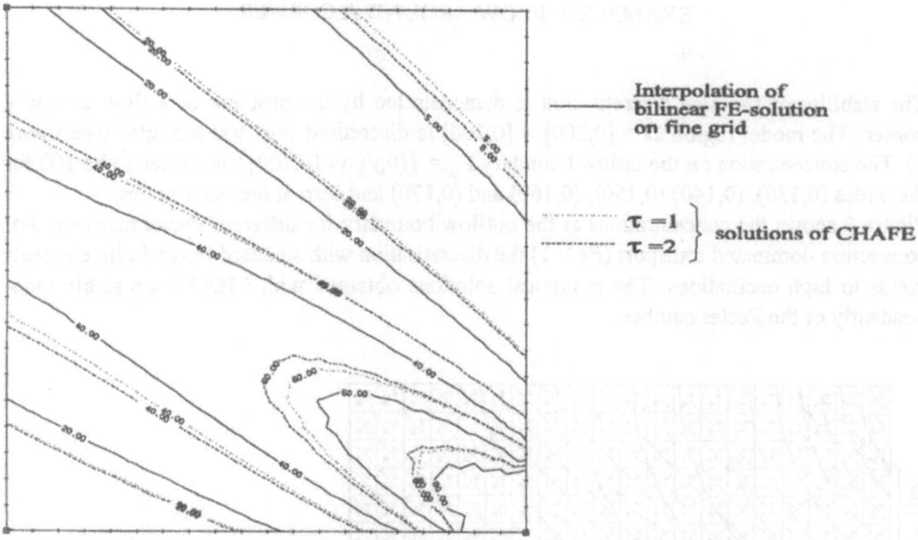

Figure 3: Discrete solutions for parallel flow problem with $\alpha_L/\alpha_T = 100/10$ and $Pe = 0.5$.

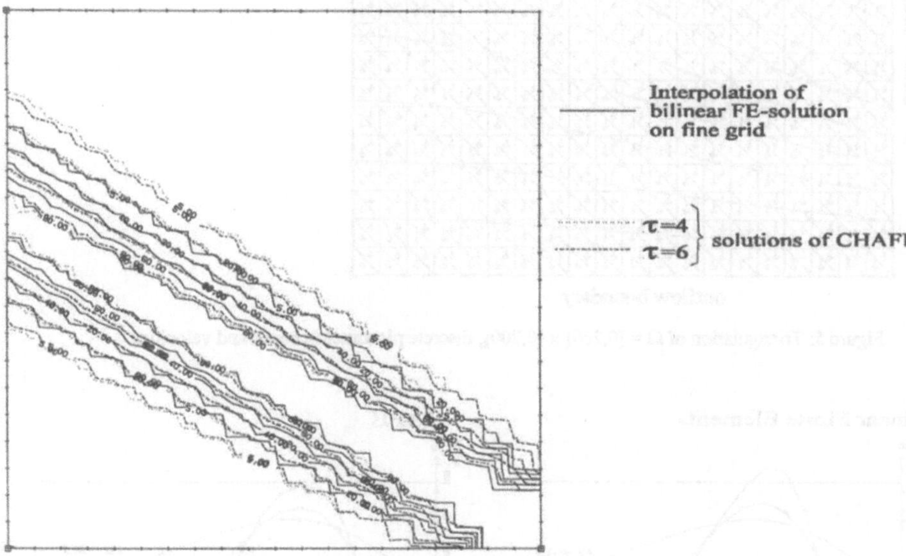

Figure 4: Discrete solutions for parallel flow problem with $\alpha_L/\alpha_T = 1./0.1$ and $Pe = 50$.

The stability of the new discretisation is demonstrated by the problem of a flow around a corner. The model region $\Omega = [0,200] \times [0,200]$ is discretised into 400 triangles (see figure 5). The concentration on the inflow boundary $\Gamma_{in} = \{(0,y),\ y \in [0,200]\}$ is chosen to be 100 for the nodes (0,130), (0,140) (0,150), (0,160) and (0,170) and zero at the other nodes.
Figure 6 shows the concentrations at the outflow boundary for different Peclet numbers. For convection dominated transport ($Pe > 1$) the discretisation with standard linear finite elements yields to high oscillations. The numerical solutions obtained with CHAFE are stable independently of the Peclet number.

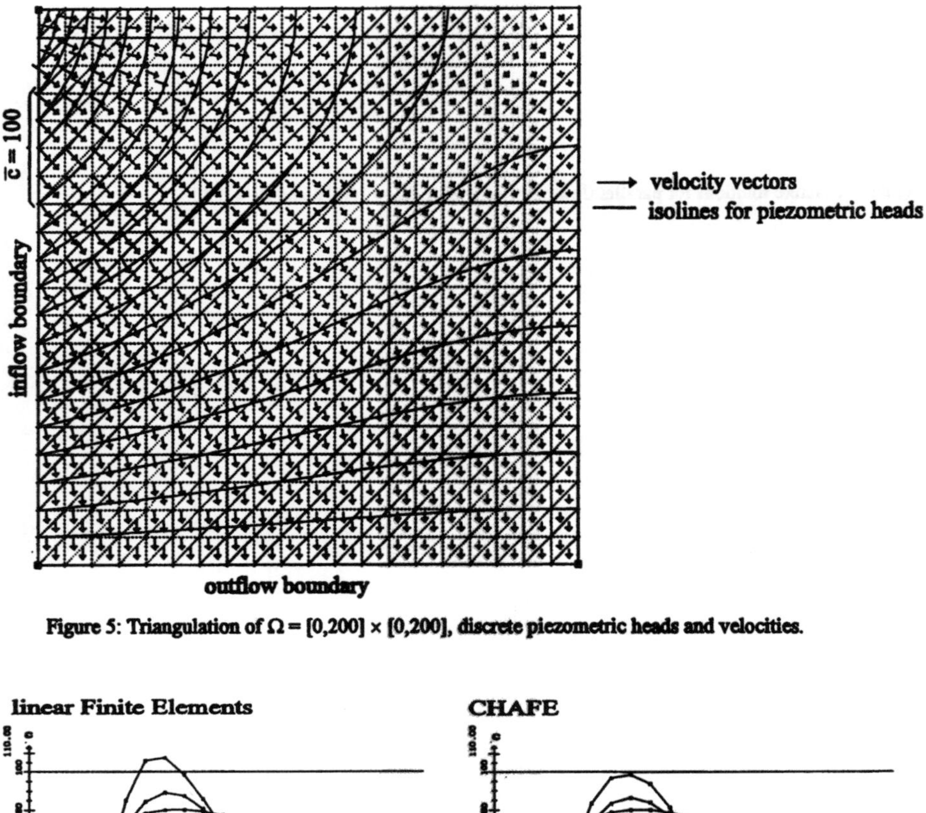

Figure 5: Triangulation of $\Omega = [0,200] \times [0,200]$, discrete piezometric heads and velocities.

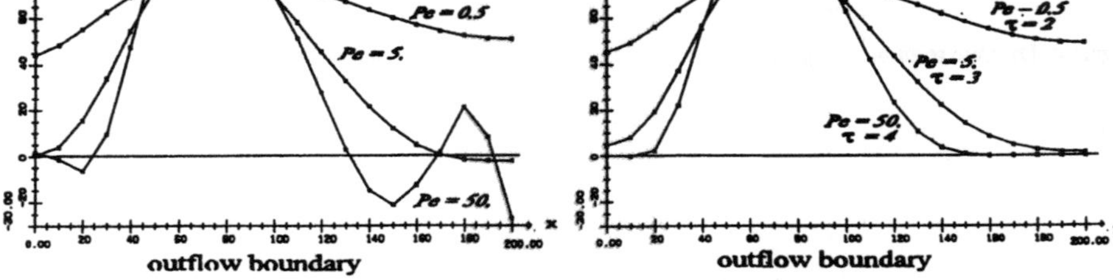

Figure 6: Concentrations at the outflow boundary for different Peclet numbers.

EXAMPLE 3: A PRACTICAL PROBLEM

This example shows a practical transport problem. For the given region with nearly parallel flow direction some pollutant sources are known from measurements (see figure 7). The concentration of these sources vary from 0.047% to 100%. The task is to calculate the outspreading of pollutant with an accuracy of 0.01%.

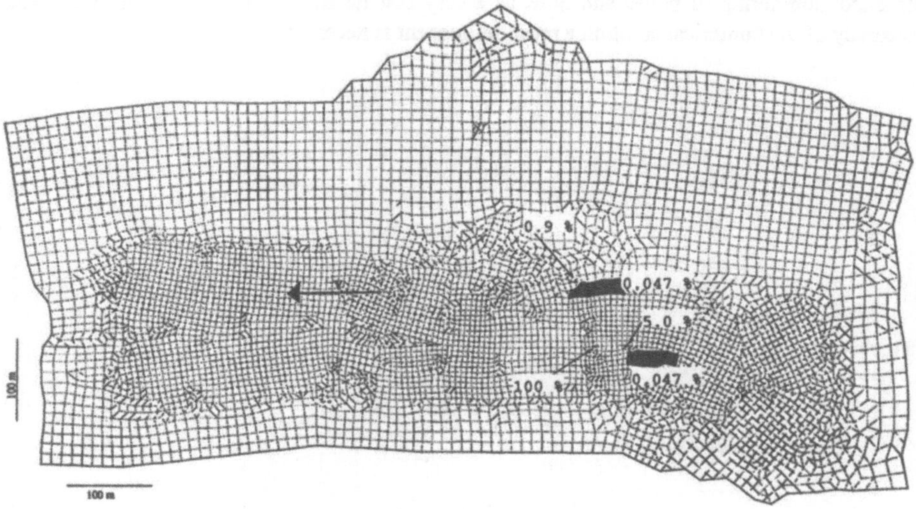

Figure 7: FE-mesh and solute sources in %.

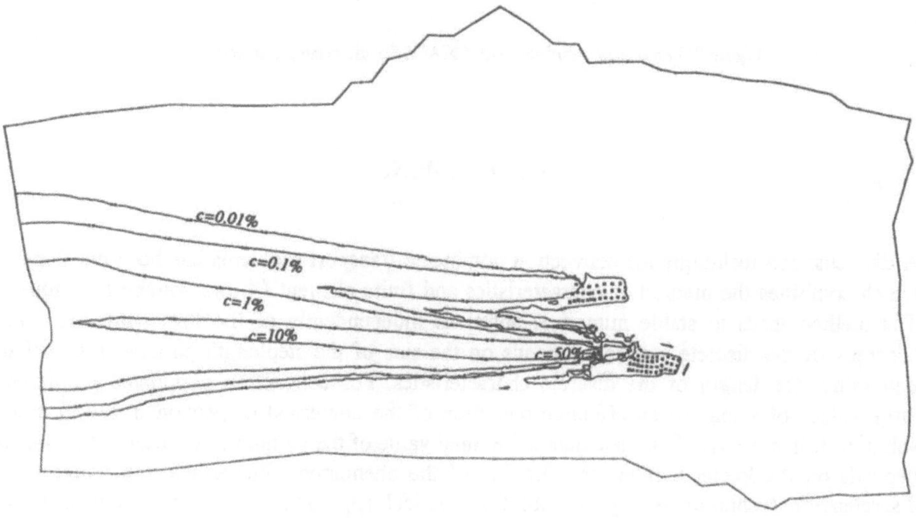

Figure 8: Numerical solution with bilinear finite elements.

For this problem the stability condition for standard finite element discretisation is not satisfied. The average Peclet number is $Pe \approx 5$. The discrete solution obtained with bilinear finite elements can not achieve the required accuracy (see figure 8). The discrete solution shows oscillations up to peak values of −13% in the concentration.

With CHAFE stable numerical solutions are obtained. With steplength parameters large enough (e.g. $\tau = 8$) the numerical dispersion can be reduced to a small value (see figure 9). The accuracy of the numerical solutions is good enough to give a prognosis of the concentration plum. This result can be obtained with little CPU time using the feature of 'flow oriented' numbering of nodes and SOR as a very fast iterative solver. To achieve a higher accuracy of the numerical solution a mesh refinement is necessary.

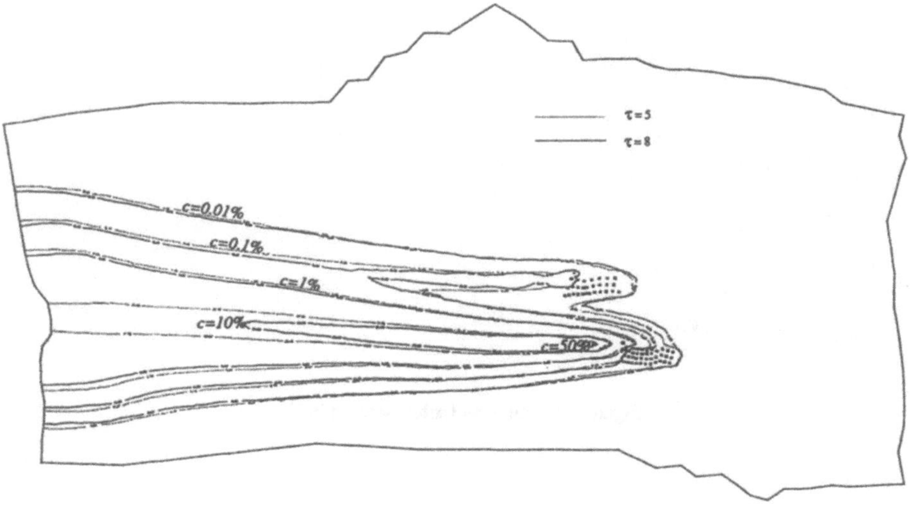

Figure 9: Numerical solutions with CHAFE for different parameters of τ.

CONCLUSIONS

A discretisation technique for convection dominated transport problems has been developed, which combines the method of characteristics and finite element / finite volume techniques. The method leads to stable numerical solutions independently of the Peclet number. The accuracy of the discrete solution depends on the size of the steplength parameter τ, which determines the length of the discrete characteristics. For convection dominated problems, large values of τ lead to an effective reduction of the numerical dispersion in the discrete solution. It is not possible to calculate an optimal value of the parameter τ, because this value depends on the Peclet number and the size of the phenomena. Nevertheless the presented discretisation technique is a good tool for practical applications to predict concentration plums on realistic grids. The discretisation results in an optimal numbering of the mesh nodes. With this enumeration numerical results can be calculated in reasonable CPU times by using standard SOR algorithms as iterative solvers.

REFERENCES

[1] Blömer, C.: "Eine neue Diskretisierung für zweidimensionale Transportprobleme mit dominanter Konvektion", Thesis, Ruhr Universität Bochum (1992).

[2] Kinzelbach, W., "Numerische Methoden zur Modellierung des Transports von Schadstoffen im Grundwasser", R. Oldenburg Verlag (1987).

[3] Pironneau O.: "On the transport-diffusion algorithm and its application to the Navier-Stokes equations", Num. Math. 38, 309-332 (1982).

[4] "SICK 100 Benutzerhandbuch", Arbeitsgruppe Grundwassermodelle, Ruhr-Universität Bochum (1992).

[5] "Süli, E.: "Convergence and nonlinear stability of the Lagrange-Galerkin method for the Navier Stokes equations", Num. Math. 53, 459-483 (1988).

[6] Voss, C. I.: "SUTRA, Saturated-Unsaturated Transport, A Finite-Element simulation model for saturated-unsaturated, fluid density-dependent groundwater flow with energy transport or chemically-reactive single-species solute transport", U.S. Geological Survey Water Resource Investigation Report 84-4369 (1984).

Comparison of Approaches for the Coupling of Chemistry to Transport in Groundwater Systems

Olaf Cirpka & Rainer Helmig

Institut für Wasserbau, Universität Stuttgart, Pfaffenwaldring 61, D-70550 Stuttgart, Germany
olaf@iws.uni-stuttgart.de , rainer@iws.uni-stuttgart.de

Summary

The basic requirements for numerical schemes tackling multi-component reactive transport are stated. Constraints for the discretization of transport using different approaches for coupling of chemistry to transport are summarized. Operator-split, an iterative two-step approach and direct coupling of the reaction-advection-diffusion equation are compared in an application of reductive dechlorination in a one-dimensional system.

1 Introduction

In the last decade, numerical simulation of multi-component reactive transport has become an active field of research in subsurface hydrology. These models couple mass transfer processes, abiotic transformations and/or microbial activity to advective-dispersive transport, which is dependent on the flow field. Basic processes are illustrated in figure 1. Practical application of the models is limited by high uncertainty in the quantitative descriptions of subsurface processes. For hydraulics the uncertainties concern mainly parameter fields such as hydraulic conductivity, for reactive processes uncertainties arise even at the conceptual level.

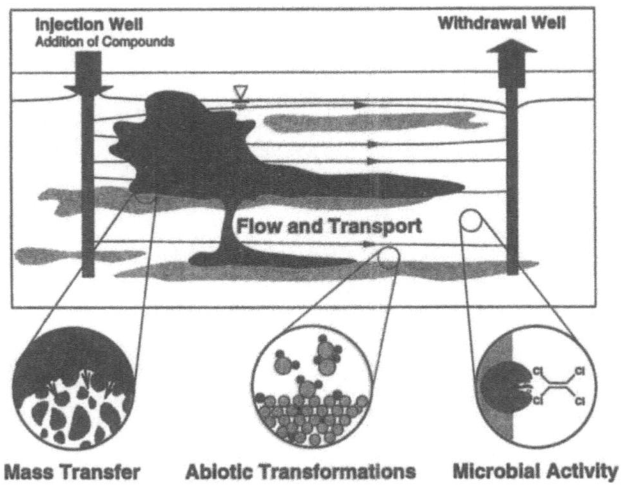

Fig. 1: Processes of interest.

As a consequence, a number of studies on multi-component reactive transport focus on the identification, parametrization and quantification of reactive processes, using standard techniques for the solution of the transport equation and for coupling. However, numerical model results may be dependent on the chosen numerical schemes, therefore numerical errors could be missinterpreted as effect of physical or chemical processes.

The objective of this study is to identify requirements for the different parts of a general numerical scheme for multi-component reactive transport. Different approaches for coupling and for numerical treatment of transport will also be compared and discussed. Specific reaction processes will not be addressed in detail.

2. Types of Equations

2.1. Transport Equation

For a given flow field the transport of solutes in groundwater can be calculated by solving (1) for each mobile compound *i*:

$$\phi\frac{\partial c_i}{\partial t} + \vec{\nabla}\cdot\left(\vec{v}c_i - \mathbf{D}\vec{\nabla}c_i\right) - \phi r_i = 0 \qquad (1)$$

where ϕ is the porosity, $\vec{v}$ is the DARCY velocity, **D** is the dispersion tensor following the SCHEIDEGGER parametrization, and r_i is a source-sink term due to reactive processes.

The reactive source-sink term is the only nonlinear term in (1). For a conservative compound (1) becomes a linear partial differential equation of parabolic type. Nevertheless the numerical solution may be non-trivial, since

- the hyperbolic term dominates over the parabolic term,
- soil properties are highly variable or even discontinuous in space, reflecting geological structures, and
- discontinuities occur in the solution, as a result of discontinuous soil properties, reactive processes and boundary conditions.

Refinement of the spatial discretization would yield smooth solutions because of the dispersion term. However, for practical application discretization at the length-scale of local dispersivities would be too computationally expensive. Therefore numerical schemes must be capable of handling discontinuous solutions.

2.2 Reactive Processes

2.2.1 Fast Reactions

The description of fast reactions is based on the assumption that thermodynamic equilibrium is reached immediately after introduction of a disturbance (e.g. due to transport). In the context of reactive transport this is known as the local equilibrium assumption. All equations describing equilibrium must be coupled to the mass balance equation (2) for chemical elements:

$$\sum_{\alpha=1}^{n_{phases}}\sum_{i=1}^{n_{species}} \theta_\alpha Y_i^{el} c_i^\alpha \Big|_{before\ equilibrium} = \sum_{\alpha=1}^{n_{phases}}\sum_{i=1}^{n_{species}} \theta_\alpha Y_i^{el} c_i^\alpha \Big|_{at\ equilibrium} \qquad (2)$$

where θ_α is the volumetric fraction of phase α, Y_i^{el} is the mass fraction of the chemical element *el* in species *i* and c_i^α is the volumetric concentration of species *i* in phase α.

Chemical equilibrium yields systems of nonlinear algebraic equations (AES). For homogeneous reactions of type (3) the law of mass action (4) holds:

$$aA + bB \leftrightharpoons cC + dD \tag{3}$$

$$\frac{[C]^c[D]^d}{[A]^a[B]^b} = K_{aA+bB \leftrightharpoons cC+dD} \,. \tag{4}$$

Precipitation and dissolution of minerals (5) is commonly described by the solubility product (6):

$$aA + bB \leftrightharpoons A_aB_b\downarrow \tag{5}$$

$$[A]^a[B]^b = K_{A_aB_b} \,. \tag{6}$$

Note that (6) is valid only as long as the mineral phase A_aB_b is present.

The equilibrium coefficients in (4) and (6) can be computed from the free formation enthalpy of the reacting species. It should be mentioned that the activities $[A]$, $[B]$, ... in (4) and (6) are not equal to concentrations in the general case. Activity coefficients must be introduced especially for systems with high ionic strength.

Mass transfer between a sorbing and the aqueous phase is commonly described by sorption isotherms such as the LANGMUIR and FREUNDLICH isotherms or in the simpliest case by linear partitioning. Mass transfer between the gaseous and liquid phases as well as between the different liquid phases are mostly described by HENRY's law for idealized solutes, and by RAOULT's law for idealized solvents [26].

2.2.2 Slow Reactions

Slow reactions are controlled by reaction kinetics. This yields systems of nonlinear ordinary differential equations (ODES). In many applications kinetics are described by first order laws, where the reaction velocity is proportional to the deviation from equilibrium (7). This is valid namely for first order mass transfer kinetics.

$$\frac{\partial c}{\partial t} = -k(c - c_{eq}) \,. \tag{7}$$

Biotic transformations are typical examples for catalyzed reactions, where the reaction velocity is dependent on the amount of catalyst and on the number of free reaction sites on the catalyst. The latter leads to nonlinear terms of the MICHAELIS-MENTEN type, resulting in first- to zeroth-order kinetics dependent on the concentration. For most biotic transformations no reverse reactions must be taken into account.

An example for biotic transformations may be aerobic degradation of a substrate (8), as applied in [15]:

$$\begin{aligned} v_{reac} &= v_{\max}\frac{c_O}{c_O+K_O}\frac{c_S}{c_S+K_S}X & \frac{\partial c_O}{\partial t} &= -Y_O v_{reac} \\ \frac{\partial c_S}{\partial t} &= -Y_S v_{reac} & \frac{\partial X}{\partial t} &= Y_X v_{reac} - k_{dec}X \end{aligned} \tag{8}$$

where v_{max} is the maximum reaction velocity per unit biomass, K_O and K_S are the MICHAE-

LIS-MENTEN coefficients of oxygen and the substrate respectively, X is the biomass, Y_O and Y_S are stoichiometric coefficients, Y_X is the yield coefficent and k_{dec} is a first order decay coefficient for the biomass.

Note that the choice of coefficients, namely the yield coefficient, determines the stiffness of the ODES defined in (8).

2.2.3 Coupled Fast and Slow Reactions

In many application fast reactions are coupled with slow reactions. Typically the following assumptions are made about the kinetics:

- Acid-base reactions are at equilibrium.
- Microbial activity is controlled by kinetics.
- Abiotic redox reactions are "in between".

In most studies abiotic redox reactions are treated as fast reactions although kinetics should be considered. The lack of kinetic parameters leads to the application of equilibrium models even for precipitation and dissolution processes.

The combination of fast and slow reactions yields a system of nonlinear differential-algebraic equations (DAES), which is the limiting case for very stiff ODES.

3. Solution of the Reactive Sub-Problem

For the general case, where reactive processes are described by a nonlinear DAES, the following requirements must be met by the numerical scheme:

- Nonlinearity requires linearization, which will be a NEWTON type method in the most cases.
- Stiffness requires fully implicit integration in time.
- Accuracy requires a multistep approach for integration to achieve a high order of consistency in a fully implicit scheme.
- Robustness, stability and efficiency requires adaptivity in the choice of the timestep and the order of approximation.

GEAR's stiff method is a possible approach, which meets the requirements listed above. In order to solve DAES, the original scheme has to be modified.

A frequently used and freely available solver for DAES is DASSL [22]. A detailed description of DASSL is given in [2]. Recently DASSL has been modified to include sparse matrices and iterative solvers. DASPK [3] is a modification of DASSL applying KRYLOV subspace techniques.

DASSL is based on a variable timestep, variable order, fixed leading coefficient multistep method for integration in time. It applies polynomial approximation up to fifth order. Linearization is done by a modified NEWTON method. Adaptivity in time stepping and choice of order leads to a very stable and accurate scheme.

The user of DASSL must supply a definition of the DAES and its JACOBIAN. The initial conditions, including time derivatives, must be consistent with the DAES. This leads to additional computational effort if the algebraic equations in the DAES are nonlinear. If transport and reactive processes are decoupled in the overall method for reactive transport, application of a DAESsolver like DASSL might be inefficient, since the nonlinear algebraic equations must be solved twice, once in the initialization of the reactive sub-problem and once in the DAES solver itself.

An alternative approach decouples the AES and the ODES; this has been applied by various authors [4,15,16]. For this approach either the AES is solved first and the resulting concentrations are taken as initial conditions for the solution of the ODES, or vice versa. A great advantage of this approach is the oppurtunity to use specialized solvers for equilibrium chemistry such as PHREEQE [20]. However, sequential coupling decreases the time integration scheme to first order consistency.

3.1 Requirements of Chemistry to the Transport Scheme

Choosing a transport scheme, one must consider additional constraints which result from the simulation of reactive processes:

- Negative concentrations would cause serious stability problems in the simulation of reactive processes, therefore oscillations must be avoided. This constraint will be met if the transport scheme is monotonicity preserving.
- Only mixed compounds may react, therefore artificial diffusion caused by the discretization of transport must be minimized. The transport scheme should be of at least second order consistency in space and time.

These two requirements, monotonicity preservation and high order of consistency, are contradictory in the framework of linear EULERIAN methods for advection dominated transport, especially if discontinuities occur. Nevertheless linear EULERIAN methods are frequently applied to multi-component reactive transport. Note that schemes of the LAX-WENDROFF type applied in [29] are unstable for PECLET numbers exceeding 2, wheras linear upwinding schemes applied in [7, 15] suffer from excessive artificial diffusion.

The latter might be tolerable if mixing of the compounds is dominated by other processes such as chromatographic effects or sorption kinetics. However, if compounds of comparable mobility are interacting, artificial diffusion may lead to significant overestimation of mixing and hence of reaction rates. Therefore more sophisticated methods for transport should be applied in the context of multi-component reactive transport.

4. Coupling Techniques

4.1 Operator Split Approach

The basic idea of the operator split approach introduced in a general context by YANENKO [1971] is to split the coupled problem into a transport problem and a reactive problem, both of which are solved independently and sequentially (9):

$$c(t_{n+1}) = c(t_n) + \Delta c_{trans} + \Delta c_{chem} \quad (9)$$

This yields the following scheme:

1. Take $c(t_n)$ as initial condition and solve:

$$\tilde{c}_{n+1} = \int_{t_n}^{t_{n+1}} transport\ without\ chemistry\ dt \, .$$

2. Take $\tilde{c}_{n+1}$ as initial condition and solve:

$$\tilde{c}_{n+1} = \int_{t_n}^{t_{n+1}} chemistry\ without\ transport\ dt\,.$$

3. Goto next timestep.

In the transport step one linear PDE must be solved per mobile compound, in the chemistry step one nonlinear DAES per node or cell.

It should be mentioned that variations of the method exist, namely the three-step approach in [27], where a transport step over the half timestep is followed by a chemistry step over the entire timestep and a second transport step over the half timestep.

The operator split approach has some obvious advantages. It is easy to implement, computational costs are low, and the transport scheme may be optimized independent of the scheme for reactive processes.

The main problem of this approach lies in the introduction of an unknown coupling error. The modelled processes are interacting and occurying in parallel, but treated independently in a sequence. Unstable solutions may be avoided with the use of sufficiently small timesteps. To the knowledge of the authors, the corresponding error estimators do exist only for simplified situations such as linear decay or single MICHAELIS-MENTEN kinetics [14,19].

The coupling condition in the operator split approach concerns only the exchange of results between the two half-steps. Spatial discretization may differ between transport and chemistry scheme, as long as the results are redistributed.

4.2 Iterative Two-Step Method

The introduction of an iteration into the operator split approach may limit the coupling error of the scheme. In [15] a reactive source-sink term was applied in the transport step, which is iteratively updated by the chemistry step. As in the operator split approach, the intermediate concentration $\tilde{c}_{n+1}$ was used as initial condition for the chemistry step.

This method was modified in [7], such that the concentration at the previous timestep is used as initial condition for both half-steps. This requires the definition of an explicit transport rate for the chemistry step. The basic principle may be explained by (10):

$$c_i(t_{n+1}) = \int_{t_n}^{t_{n+1}} \tau_i + r_i \, dt \qquad \textbf{(10)}$$

where τ_i is the transport rate for compound i at the node of interest and r_i is the reactive source-sink term. In the transport step τ_i is calculated implicitly and r_i explicitly, in the chemistry step τ_i is an explicit zeroth order term and r_i is calculated implicitly. This is leading to the following scheme:

1. For initialization of the new timestep take the old reactive source-sink term as initial guess:

$$r_{i,n+1}^{(0)} = r_{i,n}\,.$$

2. Take $c_i(t_n)$ as initial condition for the transport step:

$$c_{i,n+1}^{trans,(\nu)} = \int_{t_n}^{t_{n+1}} trans - r_{i,n+1}^{(\nu-1)}\, dt \,.$$

3. Calculate an explicit transport rate for iteration step ν:

$$\tau_{i,n+1}^{(\nu)} = \frac{c_{i,n+1}^{trans,(\nu)} - c_i(t_n)}{t_{n+1} - t_n} + r_{i,n+1}^{(\nu-1)} \,.$$

4. Take $c(t_n)$ as initial condition for the chemistry step:

$$c_{i,n+1}^{chem,(\nu)} = \int_{t_n}^{t_{n+1}} chem + \tau_{i,n+1}^{(\nu)} \,.$$

5. Calculate an explicit reactive source-sink term for iteration step ν:

$$r_{i,n+1}^{(\nu)} = \frac{c_{i,n+1}^{chem,(\nu)} - c_i(t_n)}{t_{n+1} - t_n} - \tau_{i,n+1}^{(\nu)} \,.$$

6. Check convergence:

$$\left\| c_{i,n+1}^{chem,(\nu)} - c_{i,n+1}^{trans,(\nu)} \right\| < \varepsilon_i \,?$$

7. Go back to transport step with $\nu = \nu + 1$ or to next timestep respectively.

As in the operator split approach, for each transport step one linear PDE per mobile compound must be solved, for each chemistry step one nonlinear DAES per node or cell.

The main advantage over the operator-split method is the control over the coupling error. This must be paid by multiple evaluation of both half-steps, leading to higher computational effort. The two methods can still be optimized almost independent of each other. Implementation is quite simple, therefore most existing EULERIAN transport model can easily be extended to include reactive multi-component transport.

A key problem of the approach is that convergence, checked in step 6 of the scheme, is not guaranteed for an arbitrary choice of timestep. Besides the CFL condition, which is stated for many conservative transport schemes, the size of the timesteps is also limited by the convergence behaviour of coupling. For optimal efficiency this should be done in an adaptive way. A possible criterion for reduction of the timestep, could be a maximum number of iterations, or a minimum rate of convergence.

Note that the iterative two-step method includes two linearizations, one for coupling and one included in the chemistry scheme. If a nonlinear scheme is applied in the transport step a third linearization occurs. Under certain conditions this may be less efficient than a direct coupling scheme with consistent linearization of the entire set of equations.

Compared to the operator split approach, coupling conditions are extended. The transport scheme must allow spatially distributed source-sink terms independent of the flow field, the chemistry scheme must consider the additional zeroth order transport rate. Spatial coincidence is required over the entire timestep, not only for exchange of results.

4.3 Direct Coupling

In the direct coupling approach, the PDE for transport and the DAES for chemistry are solved together. In [18] this was done by adding a chemistry related nonlinear first order term to the transport equation, which was then solved by a FE technique. Linearization was done by PICARD iteration. In a three-dimensional application [10], the same authors simplified the approach by explicit treatment of the reaction terms. In [5] a first order and a zeroth order term for the chemistry were added to the transport equation, which was then solved by a localized adjoint FE method. The reactive terms were calculated from a projection of nodal concentrations, based on the previous two timesteps.

Rather than adding chemistry related terms to a transport scheme, we will follow a classical method of lines (MOL) approach for the coupled system. This method was used in a code for reactive transport problems in general aquatic environments [24].

The transport equation is discretized in space by an EULERIAN scheme, so that the $n_{compounds}$ PDE's for transport are transformed to a system of $n_{compounds} \cdot n_{nodes}$ ODE's. These are then combined with the n_{nodes} DAES's for chemistry, leading to a large DAES of order $n_{compounds} \cdot n_{nodes}$. The DAES is integrated by the solver DASPK [3].

This scheme allows consistent linearization of the entire problem. Properties of the DAES solver, such as adaptive time stepping and adaptive choice of order, are applied to the coupled set of equations. Consistent initialization must be done only once for the entire simulation.

However, the computational effort of the approach is quite high, since the systems to be solved are large, nonlinear and stiff. In the decoupled approaches transport and chemistry schemes may be optimized independent of each other. Since the large system of equations generated for transport is linear, linearization concerns many small systems related to chemistry at every node. In contrast to that in the direct coupled approach one large system must be linearized. This will require more CPU time. However, less iterations are necessary, and therefore
the overall CPU time may be reduced. The resulting system of linear equations is highly nonsymmetric and requires computationally expensive solvers.

Spatial discretization of transport and chemistry must be identical, and the transport scheme must fit into the MOL approach, therefore only semi-discrete schemes may be used.

5. Transport Schemes

5.1. LANGRANGIAN and EULERIAN-LANGRANGIAN Schemes

The hyperbolic term in the transport equation (1) is the main cause for instabilities in numerical simulation of transport, using EULERIAN methods. These problems can be overcome if the transport equation (1) is transformed to is LANGRANGIAN form (11):

$$\frac{dc(\vec{x}(t))}{dt} = \vec{\nabla}\cdot\left(\mathbf{D}\vec{\nabla}c_i\right) + \phi r_i \qquad \textbf{(11)}$$

where the characteristic $\vec{x}(t)$ fulfills (12):

$$\frac{\partial\vec{x}(t)}{\partial t} = \vec{v}\,. \qquad \textbf{(12)}$$

Rather than balancing mass in a fixed control volume as in EULERIAN methods, LANGRANGIAN methods balance mass in control volumes moving in space over time.

5.1.1 Particle Tracking - Random Walk

In the particle tracking/random walk approach, virtual particles carrying mass are tracked along characteristics. Dispersion is described by stochastical deviation from the tracks [23]. Concentrations must be calculated by summing up the mass of all particles in each control volume.

Negative concentrations can not occur. If a sufficient number of particles is used, the results are very accurate. However, the applicability in reactive transport is limited. Source-sink terms independent of the flow field are difficult to implement. As concentrations are obtained only from post-processing, the method can be applied for reactive transport only in an operator-split coupling scheme. This was done in [9].

After each transport step, concentrations in the cells of a structured grid were calculated from summation of mass from the particles in the cells. Geochemistry was calculated in each cell and the mass was redistributed to the particles. This scheme is illustrated in figure 2.

Note that spatial redistribution after each timestep introduces artificial diffusion, thus destroying the advantage of the particle trakking approach to be non-diffusive.

5.1.2 Eulerian-Langrangian methods

The shortcomings of the particle tracking approach for diffusive/ dispersive transport has led to combinations of LANGRANGIAN schemes for advection and EULERIAN schemes for diffusion and source-sink terms. The EULERIAN-LANGRANGIAN localized adjoint method (ELLAM) [6] may be an example for mass-conservate EULERIAN-LANGRANGIAN schemes.

ELLAM is a space-time FEM formulation, where the edges of the elements follow characteristics in the space-time domain. The weighting function fulfills the adjoint PDE of the original problem. The framework allows direct access to the treatment of boundary conditions, spatially distributed source-sink terms and first-order decay terms.

Figure 3 shows two variants of spa ce-time discretization. In the moving grid approach, all observation points are tracked along characteristics, wheras in the original ELLAM implementation concentrations are redistributed. The latter approach leads to artificial diffusion comparable to stabilized EULERIAN schemes.

If compounds of different mobility, and hence following different characteristics, react

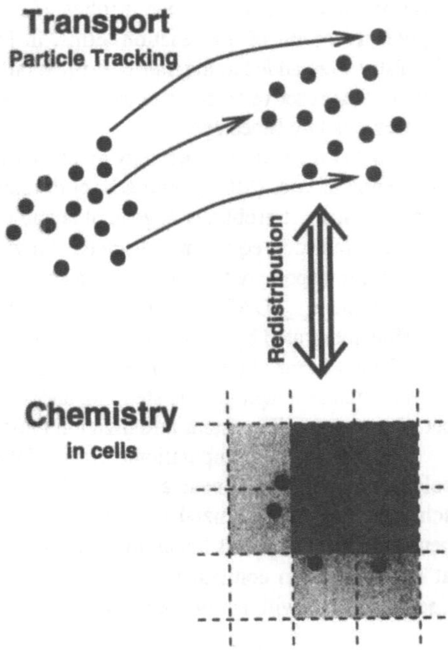

Fig. 2: Coupling of chemistry and transport in the framework of particle tracking / random walk as applied in [9].

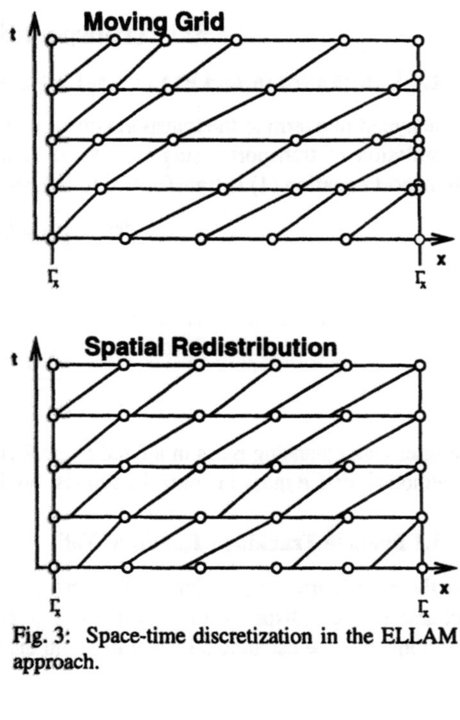

Fig. 3: Space-time discretization in the ELLAM approach.

with each other, a moving grid approach is difficult to implement. Also, implicit consideration of reactive source-sink terms would be complicated, if the space-time elements of reacting compounds did not coincide. Therefore practical application of ELLAM for reactive transport will be limited to operator-split coupling.

The ELLAM scheme requires no CFL condition for stabilization, therefore large timesteps may be applied for conservative transport. Unfortunately this advantage is destroyed by applying operator-split coupling.

5.2 High Resolution EULERIAN Schemes

As mentioned in section 3.1, linear EULERIAN techniques suffer either from excessive artificial diffusion (e.g. upwind differentiation and inconsistent PETROV-GALERKIN methods) or from oscillative behaviour (e.g. central differentiation, LAX-WENDROFF techniques, consistent PETROV-GALERKIN methods). Note that linear EULERIAN schemes are based on TAYLOR extension, which leads to oscillations near discontinuities if a finite number of higher order terms are taken into account.

In order to obtain monotonicity preserving solutions including the highest possible proportion of higher order terms, the choice of differentiation scheme must be dependent on the behaviour of the solution. Near discontinuities and local extrema the scheme will be closer to a low order solution such as obitained by a first order upwind scheme, wheras in smooth areas the scheme will be closer to a high order solution such as obitained by central differentiation. This leads to nonlinear schemes for the solution of an initially linear problem.

Two high resolution schemes are presented: The Flux-Corrected Transport (FCT) algorithm as predictor-corrector type of scheme, and the slope-limiter scheme as nonlinear MOL approach. It should be mentioned that nonlinear "discontinuity capturing" schemes also exist in space-time FEM [13], they are not addressed here.

5.2.1 Flux-Corrected Transport (FCT)

The Flux-Corrected Transport (FCT) algorithm was introduced in [1], extended to multidimensional application in [31] and transfered to FEM in [21].

The main idea of the scheme is to solve a timestep twice: Once using a monotone low-order scheme and once using an oscillatory high-order scheme. Antidiffusive fluxes (or element contributions to a node in FEM) are defined by subtracting low order fluxes from high order fluxes. These antidiffusive fluxes are limited in such a way that no new extrema, with respect to the solution of the previous timestep and the low order solution, can occur. Finally the corrected antidiffusive fluxes are added to obtain the solution of the timestep.

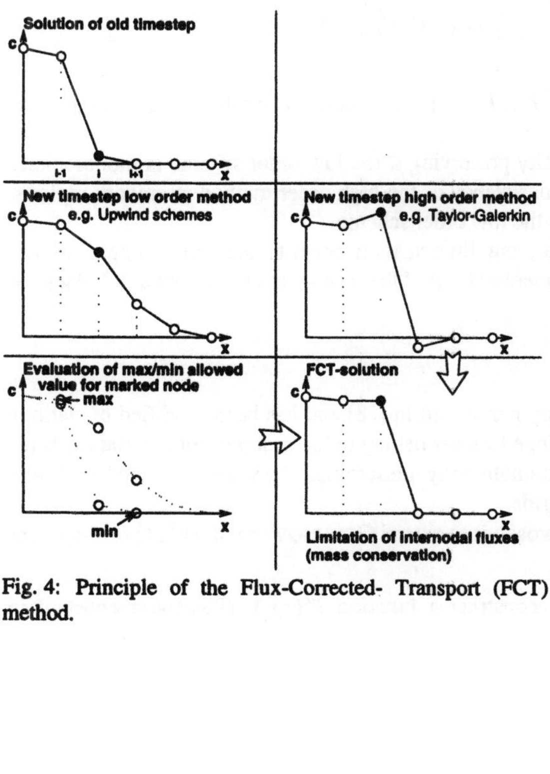

Fig. 4: Principle of the Flux-Corrected- Transport (FCT) method.

The principle is illustrated in figure 4.

Given the time integrated fluxes from cell *j-1* to cell *j* for a low order scheme $F_{j-1,j}^{l}$ and a high order scheme $F_{j-1,j}^{h}$, the antidiffusive flux is defined by $F_{j-1,j}^{ad} = F_{j-1,j}^{h} - F_{j-1,j}^{l}$. This antidiffusive flux is corrected in the following four steps:

1. Summarize all positive/negative antdiffusive fluxes at cell *j*:

$$P_j^+ = \max(0,F_{j-1,j}^{ad}) + \max(0,F_{j,j+1}^{ad}) \qquad P_j^- = \max(0,-F_{j-1,j}^{ad}) + \max(0,-F_{j,j+1}^{ad}).$$

2. Define the maximum/minimum allowable antidiffusive flux at cell *j*:

$$Q_j^+ = V_j(c_j^{\max}-c_j^l) \qquad Q_j^- = V_j(c_j^{\min}-c_j^l)$$

with the definition of c_j^{max} and c_j^{min}:

$$c_j^{\max} = \max(c_j^l,c_{j-1}^l,c_{j+1}^l,c_j(t_n),c_{j-1}(t_n),c_{j+1}(t_n))$$

$$c_j^{\min} = \min(c_j^l,c_{j-1}^l,c_{j+1}^l,c_j(t_n),c_{j-1}(t_n),c_{j+1}(t_n)).$$

3. Calculate the ratio of $Q_j^{\pm}$ and $P_j^{\pm}$:

$$R_j^+ = \min\left(1,\frac{Q_j^+}{P_j^+}\right) \quad if\, P_j^+ \neq 0 \qquad R_j^+ = 0 \quad if\, P_j^+ = 0$$

$$R_j^- = \min\left(1,\frac{Q_j^-}{P_j^-}\right) \quad if\, P_j^- \neq 0 \qquad R_j^- = 0 \quad if\, P_j^- = 0.$$

4. Define the correction factor $T_{j-1,j}$ for the antidiffusive flux $F_{j-1,j}^{ad}$:

$$T_{j-1,j} = \min\begin{pmatrix} R_{j-1}^-,R_j^+ & if & F_{j-1,j}^{ad} > 0 \\ R_{j-1}^+,R_j^- & if & F_{j-1,j}^{ad} \le 0 \end{pmatrix}.$$

The corrected antidiffuisve mass flux $T_{j-1,j}\, F_{j-1,j}^{ad}$ is then added to the low order flux $F_{j-1,j}^{l}$ to obtain the total flux.

The FCT algorithm is monotonicity preserving if the low order scheme is monotonicity preserving. It can reach the order of consistency of the high order method in smooth regions. Near discontinuities it will be closer to the low order scheme.

The FCT algorithm is nonlinear, but linearization consists only of a single predictor-limiter-corrector loop. Therefore implementation of the scheme is straightforward. It does not fit into the MOL approach.

5.2.2 Slope Limiter Methods

The slope limiter method was originally introduced in [28] and has been modified by various authors (see [12,17] for references). Slope limiters belong to the group of total variation diminishing (TVD) schemes, which are monotonicity preserving. They are restricted to Finite Volume discretizations on structured grids.

Main idea of the slope limiter approach is to extend GODUNOV's method [11] for hyperbolic conservation laws:

1. Given cell averaged data $\{c_j^n\}$, construct a function $\tilde{c}^n(x,t_n)$. (Piecewise constant in

GODUNOV's method.)

2. Solve the conservation law exactly with this data to obtain $\tilde{c}^n(x,t_{n+1})$.
3. Compute cell averages of the resulting solution to obtain $\{c_j^{n+1}\}$.

Rather than using piecewise constant functions, in the slope limiter approach piecewise linear reconstructions are applied. The principle of the scheme is illustrated in figure 5.

Defining a slope s_j inside of cell j the time integrated flux from cell j-1 to cell j is given in 1D by (13). For multidimensional and coupled problems semi-discretization is more appropriate than direct integration. This is done using the reconstructed upwind concentration at the interface of two cells.

$$F_{j-1,j} = v_{j-1,j}\,\Delta t\left(c_{j-1}^{n}+s_{j-1}^{n}\left(\frac{\Delta x_{j-1}-v_{j-1,j}\Delta t}{2}\right)\right). \quad (13)$$

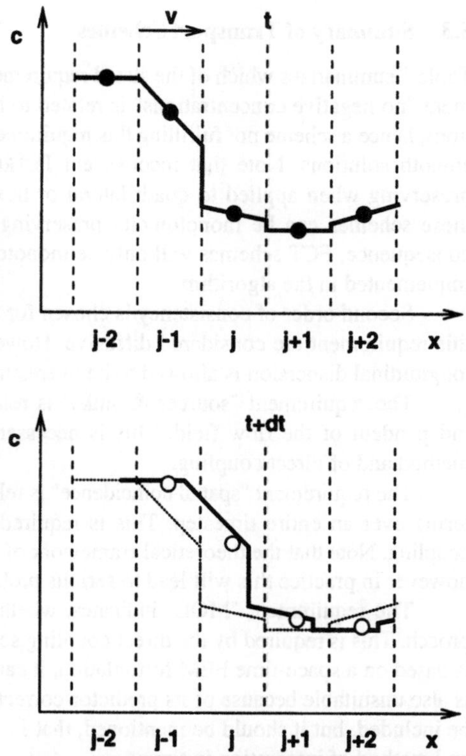

Fig. 5: Principle of the slope limiter procedure after [12]. Upper figure: Piecewise linear reconstruction. Lower figure: Solution of new timestep. Circles: Cell averaged concentrations

The crucial step of this method is the choice of the slopes. It can be shown, that the overall scheme is TVD, if the reconstruction is TVD.

With the definition of special slopes (14) one choice is ROE's *Superbee* limiter [25] (15), which is the sharpest limiter of second order accuracy.

$$s_{lin}^{+}=\frac{2(c_{j+1}-c_j)}{\Delta x_j+\Delta x_{j+1}} \qquad s_{lin}^{-}=\frac{2(c_j-c_{j-1})}{\Delta x_j+\Delta x_{j-1}} \qquad s_{max}^{+}=\frac{2(c_{j+1}-c_j)}{\Delta x_j} \qquad s_{max}^{-}=\frac{2(c_j-c_{j-1})}{\Delta x_j} \quad (14)$$

$$\begin{aligned}
&\textit{if} && s_{lin}^{+}\cdot s_{lin}^{-}<0 && \textit{then } s_j=0 && \textit{elseif} && |s_{max}^{+}|<|s_{lin}^{-}| && \textit{then } s_j=s_{max}^{+}\\
&\textit{elseif} && |s_{lin}^{-}|>|s_{lin}^{+}| && \textit{then } s_j=s_{lin}^{-} && \textit{elseif} && |s_{lin}^{+}|<|s_{max}^{-}| && \textit{then } s_j=s_{lin}^{+} \quad (15)\\
&\textit{elseif} && |s_{max}^{-}|<|s_{lin}^{+}| && \textit{then } s_j=s_{max}^{-}\,.
\end{aligned}$$

Note that the accuracy of the scheme is preserved only for high-order time integration. Since the limiter functions are nonlinear, implicit time integration requires an iterative solution. Therefore explicit RUNGE-KUTTA schemes are frequently applied to reduce computational costs. In the framework of direct coupling, as defined in section 4.3, the additional effort for linearization of the nonlinear transport scheme is tolerable because of existing nonlinearities resulting from the reactive DAES.

5.3 Summary of Transport Schemes

Table 1 summarizes which of the stated requirements are met by selected schemes. The requirement "no negative concentrations" is related to high PECLET numbers and discontinuous solutions, hence a scheme not fulfilling this requirement is restricted to low PECLET numbers and/or smooth solutions. Note that inconsistent PETROV-GALERKIN schemes are not monotonicity preserving when applied to quadrilateral or hexaedral elements. If mass-lumping is applied, these schemes can be monotonicity preserving for triangular and tetraedral elements. As a consequence, FCT schemes will only be monotonicity preserving if it is the low order scheme implemented in the algorithm.

Second order of consistency is chosen for the "accuracy" criterion. Schemes not meeting this requirement are considered diffusive. However, they may be applied in problems, where longitudinal dispersion is allowed to be overestimated.

The requirement "sources & sinks" is related to spatially distributed source/sink-terms independent of the flow field. This is necessary in the application of the iterative two-step method and of direct coupling.

The requirement "spatial coincidence" is related to the exchange of chemistry and transport terms over an entire timestep. This is required for the iterative two-step method and direct coupling. Note that the theoretical framework of ELLAM allows for complicated relationships, however in practice this will lead to serious problems.

The requirement "MOL" indicates, whether a scheme fits into the method of lines approach. This is required by the direct coupling scheme as defined in section 4.3. Since ELLAM is based on a space-time FEM formulation, it can not be used in that scheme. The FCT scheme is also unsuitable because of its predictor-corrector character. LAX-WENDROFF techniques may be included, but it should be mentioned, that LAX-WENDROFF terms are dependent on stepsize and method of integration in time.

Table 1: Restrictions of selected transport schemes in the framework of reactive transport

	No negative Concentrations	**Accuracy**	**Sources & Sinks**	**Spatial Coincidence**	**MOL**
Random Walk	+	+	-	-	-
ELLAM	+	+	+	-(+)	-
Standard FDM/FEM	-	+	+	+	+
Lax-Wendroff FDM/FEM	-	+	+	+	+(-)
Upwind FDM	+	-	+	+	+
Upwind FEM	+(-)	-	+	+	+
FCT FDM	+	+	+	+	-
FCT FEM	+(-)	+	+	+	-
Slope Limiter (FDM)	+	+	+	+	+

Table 2 summarizes the applicability of the selected transport schemes to the different coupling approaches.

Table 2: Applicability of selected transport schemes for selected coupling schemes.

	Operator Split	Iterative 2-Step Method	Direct Coupling
Random Walk	+	+	-
ELLAM	+	+	+
Standard FDM/FEM	-	+	+
Lax-Wendroff FDM/FEM	-	+	+
Upwind FDM	+	-	+
Upwind FEM	+(-)	-	+
FCT FDM	+	+	+
FCT FEM	+(-)	+	+
Slope Limiter (FDM)	+	+	+

6. Test Case for Comparison

Reductive dechlorination of tetrachloroethene in a 100m long one-dimensional streamtube is chosen as a test case. Six mobile compounds are observed: the four chlorinated ethenes, ethene itself and an unspecified electron donor. The sorbed phase of the mobile compounds and four immobile biomasses, each catalyzing one reduction step, are considered.

Kinetics of the reduction steps are described by double MICHAELIS-MENTEN terms (16).

$$v_{reac}^{i} = v_{\max}^{i} \frac{[CHC_i]}{[CHC_i]+K_m^{CHC_i}} \cdot \frac{[E_{don}]}{[E_{don}]+K_m^{E_{don}}} X^{i} \qquad \textbf{(16)}$$

where CHC_1 is Tetrachloroethene (PCE), CHC_2 is Trichloroethene (TCE), CHC_3 is *cis*-Dichloroethene (DCE), CHC_4 is Chloroethene (VC) and CHC_5 is Ethene (ETH). E_{don} is an unspecified electron donor. The biomass X_1 catalyzes the reduction from PCE to TCE, X_2 catalyzes the reduction from TCE to DCE, X_3 catalyzes the reduction from DCE to VC, and X_4 catalyzes the reduction from VC to ETH.

For the chlorinated ethenes, the reactive source-sink terms r_i in the transport equation (1) are defined by adding the microbial consumption and production terms to a kinetic mass transfer term (17).

$$r_{CHC_i} = -Y_{CHC_i} v_{reac}^{i} + Y_{CHC_i} v_{reac}^{i-1} - k_s \left([CHC_i] - \frac{[CHC_i^s]}{K_d^{CHC_i}} \right) \qquad \textbf{(17)}$$

where is the stoichiometric coefficient for CHC_i. The index s indicates the sorbed phase of the compound.

For the electron donor, additional competitive consumption is taken into account as first order decay term (18).

$$r_{E_{don}} = -\sum_{i=1}^{4} Y_{E_{don}} v_{reac}^{i} - k_{comp}[E_{don}] - k_s\left([E_{don}] - \frac{[E_{don}^{s}]}{K_d^{E_{don}}}\right). \tag{18}$$

For simplification, concentrations in the sorbed phase are defined by mass of sorbed compound per liquid volume rather than by mass of sorbed compound per mass of solid phase. Kinetic mass transfer is the only process considered for sorbed compounds (19 + 20):

$$\frac{\partial[CHC_i^{s}]}{\partial t} = k_s\left([CHC_i] - \frac{[CHC_i^{s}]}{K_d^{CHC_i}}\right) \tag{19}$$

$$\frac{\partial[E_{don}^{s}]}{\partial t} = k_s\left([E_{don}] - \frac{[E_{don}^{s}]}{K_d^{E_{don}}}\right). \tag{20}$$

The dechlorination process is considered to be the only energy source for biomass growth. No additional limitation of biomass growth is assumed. Decay of the biomass is described by a first order law.

$$\frac{\partial X^{i}}{\partial t} = Y_X^{i} v_{reac}^{i} - k_{dec}^{i} X^{i} \tag{21}$$

Table 3: Parameters for the test case model calculations.

Compound related parameters						
	E_{don}	**PCE**	**TCE**	**cis-DCE**	**VC**	**ETH**
K_m^{i} [µmol/L]	100	100	100	100	100	100
K_{OC}^{i} [L/kg]	2.2	364	126	59	8.2	6.0
$c_{initial}$ [µmol/L]	0	200	0	0	0	0
c_{inflow} [µmol/L]	10,000	0	0	0	0	0
Y_i [-]	1	1	1	1	1	1
k_{comp} [1/s]	10^{-6}					
Biomass related parameters						
	Step 1	**Step 2**	**Step 3**	**Step 4**		
v_{max}^{i} [µmol/s/mg]	$7.7\cdot10^{-2}$	$5.0\cdot10^{-2}$	$2.5\cdot10^{-2}$	$1.0\cdot10^{-2}$		
Y_x^{i} [mg/µmol]	0.001	0.001	0.001	0.001		
k_{dec} [1/s]	$1.0\cdot10^{-6}$	$1.0\cdot10^{-6}$	$1.0\cdot10^{-6}$	$1.0\cdot10^{-6}$		
$X_{initial}$ [mg/L]	0.01	0.01	0.01	0.01		
Soil properties						
$\rho = 2.6$ kg/L	$f_{OC} = 0.2\%$		$\phi = 0.3$	$\alpha_s = 1/d = 1.157\cdot10^{-5}/s$		
$v = 10^{-5}$ m/s	$D = 3.3\cdot10^{-7}$ m²/s					

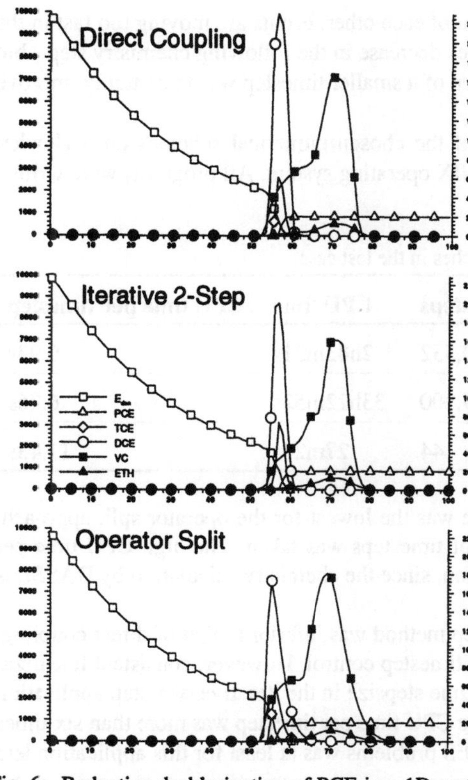

Fig. 6: Reductive dechlorination of PCE in a 1D system. Concentration length profiles of dissolved compounds 30 days after begin of electron donor injection. Comparison of three coupling methods for chemistry and transport.

where Y_X^i is the yield coefficient for biomass X^i in reduction step i. Equilibrium of sorption is described by the linear $f_{OC} K_{OC}$ concept (22).

$$K_d^i = f_{OC} K_{OC}^i \rho \frac{1-\phi}{\phi} \qquad (22)$$

where f_{OC} is the mass fraction of organic carbon in the soil, K_{OC}^i is the partitioning coefficient between organic carbon and water for compound i, and ρ is the soil density.

For the initial conditions of the test case, PCE is homogenously distributed over the entire domain. Sorption is at equilibrium, and no metabolites nor electron donor is present. Initial biomass concentrations are very low. At timepoint zero electron donor enriched water is injected into the domain, stimulating the dechlorinating biomass.

The model assumptions are explained in detail in [8]. Parameters of the model calculations are summarized in table 3. 50 days of injection have been simulated in the model runs. Figure 6 shows distributions of aqueous phase concentrations 30 days after begin of injection.

The following numerical schemes have been applied:

- Direct coupling as explained in section 4.3 with an integrated FDM semi- discretization of transport, using ROE's *Superbee* limiter. Time stepsize was adapted by the DASPK solver.
- Iterative 2-step coupling with a FEM semi-discretization of transport, applying flux-corrected transport and CRANK-NICHOLSON time integration for transport. Time stepsize was limited to 225s, in order to achieve convergence over the entire simulation. No adaptive timestepping scheme was applied.
- Operator split coupling with integrated FDM semi-discretization of transport, applying flux-corrected transport and explicit predictor-corrector time integration for transport. For time discretization the CFL number was set to unity, the resulting stepsize was 30,000s.

Physical explanations to the model problem are given in [8]. Comparison of the three coupling approaches show good qualitative agreement. The overall behaviour is the same for all simulations.

The operator split approach shows significant fronting. This is the result of reactive processes retarding the front velocities. This retardation is not addressed directly if transport and chemistry are calculated completely independent of each other. Fronts are moving too fast in the transport step, the related concentration changes decrease in the following chemistry step - but apparently not to a sufficient extend. The choice of a smaller timestep would certainly improve the behaviour of the operator split approach.

Table 4 summarizes the performance of the chosen numerical schemes on a Hewlett Packard 735-125 workstation under the HP-UNIX operating system. All programs were written in FORTRAN 77.

Table 4: Performance of the coupling approaches in the test case.

	Number of timesteps	CPU time	CPU time per timestep
Direct coupling	10,532	2h43m31s	0.93s
Iterative 2-step method	19,200	33h22m52s	6.26s
Operator-split approach	144	27m26s	11.43s

As expected, the computational expense was the lowest for the operator split approach. However, in this approach the least number of timesteps was taken. The high CPU time per timestep is related to the large time discretization, since the chemistry calculation by DASSL is longer for larger timesteps.

The performance of the iterative two-step method was inferior to that of direct coupling. This may be different when using an adaptive timestep control. However, consistent linearization shows its efficiency as well. The average time stepsize in the iterative two-step application was half that of the direct coupling, however the CPU time per timestep was more than six times higher. Therefore frequent restart of small DAES problems was at least for this application less efficient then integrating one large DAES.

Note that performance behaviour may be different for multidimensional applications, where there are more unknowns and the matrices of the linear problems are more ill-posed than in 1D problems.

7 Conclusions

It has been shown that integrating reactive processes into a transport model leads to additional constraints for the numerical transport scheme. The choice of the transport scheme is dependent on the choice of coupling scheme.

The presented coupling schemes may be characterized by the following summary:

- The operator split method is the coupling approach with the least number of additional requirements, allowing the application of any type of transport scheme. Since there is no control over the coupling error, discretization in time must be handle with care.
- Direct coupling is the most consistent method of coupling in terms of accuracy and stability. The approach is limited to EULERIAN transport schemes which fit into the method of lines approach. The resulting DAES may be very large in multidimensional applications and require highly stable solvers.
- The iterative two-step approach is somewhat of a compromise. It allows control over the coupling error as well as treating the reactive terms independent of the tansport terms. In order to guarantee convergence, small timesteps must be chosen.

In the test calculation, direct coupling performed better than the iterative two-step method. This may be partially a result of consistent linearization in the direct coupling approach. Adaptive time stepping of the solver used in the direct coupling approach contributes to the higher performance. The computational effort for direct coupling seems acceptable, at least for one-dimensional problems. In order to ensure accuracy and stability, direct coupling is preferable for this type of problems.

Although the operator split approach may be a poor choice from the view of numerical analysis of stiff systems, it has been applied successfully in many studies. However, for large problems in domains discretized by ten thousands of nodes, operator split may be the only practical solution for a multi-component reactive transport problem. The shortcomings of the scheme should be kept in mind. For each specific application the accuracy of operator splitting should be verified by comparison with implicit coupling schemes for 1D problems before application to multidimensional problems.

References

[1] BORIS, J.P. & D.L. BOOK (1973): Flux corrected transport I: Shasta, a fluid transport algorithm that works., J. Comput. Phys. 11: 38-69.

[2] BRENAN, K.E., S.L. CAMPBELL & L.R. PETZOLD (1989): Numerical solution of initial-value problems in differential-algebraic equations., Elsevier Science Publishing Co., New York.

[3] BROWN, P.N., A.C. HINDMARSH & L.R. PETZOLD (1994): Using Krylov methods in the solution of large-scale differential-algebraic systems.

[4] BRUN, A, P. ENGESGAARD & E.O. FRIND (1994): A coupled microbiology-geochemistry transport model for saturated groundwater flow., in: T.H. DRACOS & F. STAUFFER (eds.): Transport and reactive processes in aquifers., Balkema, Rotterdam: 457-462.

[5] CELIA, M.A., J.S. KINDRED & I. HERRERA (1989): Contaminant transport and biodegradation 1. A numerical model for reactive transport in porous media., Water Res. Res. 25(6): 1141-1148.

[6] CELIA, M.A., T.F. RUSSELL, I. HERRERA & R.E. EWING (1990): An Eulerian-Langrangian localized adjoint method for the advection-diffusion equation., Adv. Water Res. 13(4): 187-206.

[7] CIRPKA, O. & R. HELMIG (1994): Numerical simulation of contaminant transport and biodegradation in porous and fractured-porous media., in: A. PETERS ET AL. (eds.): Computational methods in water resources X, Kluwer Academic Publishers, Dordrecht: 605-612.

[8] CIRPKA, O. (1995): Influence of hydraulic aquifer properties on reductive dechlorination of Tetrachloroethene., in: R.E. HINCHEE, A. LEESON & L. SEMPRINI (eds.): Bioremediation of chlorinated Ethenes., Battelle, Columbus: 25-34.

[9] FABRIOL, R., J.-P. SAUTY & G. OUZOUNIAN (1993): Coupling geochemistry with a particle tracking transport model., J. Contam. Hydrol. 13: 117-129.

[10] FRIND, E.O., E.A. SUDICKY & J.W. MOLSON (1989): Three-dimensional simulation of organic transport with aerobic biodegradation., in: Groundwater Contamination, IAHS Publ. no. 185.

[11] GODUNOV, S.K. (1959), Math. Sb. 47: 271.

[12] HIRSCH, C. (1990): Numerical computation of internal and external flows, Vol 2: Computational methods for inviscid and viscous flows., Wiley & Sons, New York.

[13] HUGHES, T.J.R., M. MALLET & A. MIZUKAMI (1986): A new Finite Element formulation for computational fluid dynamics: II. Beyond SUPG., Comput. Meth. Appl. Mech. Eng. 54: 341-355.

[14] KALUARACHCHI, J.J. & J. MORSHED (1995): Critical assessment of the operator-splitting technique in solving the advection-dispersion-reaction equation: 1. First-order reaction., Adv. Water Res. 18(2): 89-100.

[15] KINZELBACH, W., W. SCHÄFER & J. HERZER (1991): Numerical modeling of natural and enhanced denitrification processes in aquifers., Water Res. Res. 27(6):1123-1135.

[16] LENSING, H.H. (1995): Numerische Modellierung mikrobieller Abbauprozesse im Grundwasser., Thesis (in German), Mitteilungen des Instituts f\"ur Hydrologie und Wasserwirtschaft der Universität Karlsruhe 51.

[17] LEVEQUE, R.J. (1992): Numerical methods for conservation laws., Birkhäuser Verlag, Basel.

[18] MACQUARRIE, K.T.B., E.A. SUDICKY & E.O. FRIND (1990): Simulation of biodegradable organic contaminants in groundwater 1. Numerical formulation in principal directions., Water Res. Res. 26(2): 207-222.

[19] MORSHED, J. & J.J. KALUARACHCHI (1995): Critical assessment of the operator-splitting technique in solving the advection-dispersion-reaction equation: 1. Monod kinetics and coupled transport., Adv. Water Res. 18(2): 101-110.

[20] PARKHURST, D.L., D.C. THORNSTENSON & L.N. PLUMMER (1980): PHREEQE - A computer program for geochemical calculations., USGS Water Res. Inv., 80-96, Washington D.C..

[21] PARROTT, A.K. & M.A. CHRISTIE (1986): FCT applied to the 2-D Finite Element solution of tracer transport by single phase flow in a porous medium., in: K.W. MORTON & M.J. BAINES (eds.): Numerical methods for fluid dynamics, Oxford University Press: 609-620.

[22] PETZOLD, L.R. (1982): A description of DASSL: A differential/algebraic system solver., SANDIA report SAND82-8637.

[23] PRICKETT, T.A., T.G. NAYMIK & G. LONNQUIST (1981): A "random walk" solute transport model for selected groundwater quality evaluations., Illinois State Water Survey, Bulletin 65.

[24] REICHERT, P. (1994): Concepts underlying a computer programfor the identification and simulation of aquatic systems., Schriftenreihe der EAWAG Nr.7, Swiss Federal Institute for Environmental Science and Technology, Dübendorf.

[25] ROE, P.L. (1985): Some contributions to the modeling of discontinuous flows., Lect. Notes Appl. Math. 22: 163-193.

[26] SCHWARZENBACH, R.P., P.M. GSCHWEND & D.M. IMBODEM (1993): Environmental organic chemistry., Wiley & Sons, New York.

[27] STRANG, G. (1968): On the construction of and comparison of difference schemes., SIAM J. Numer. Anal. 5(3): 506-517.

[28] VAN LEER, B. (1973): Towards the ultimate conservaticve difference scheme I. The quest of monotonicity., Springer Lecture Notes in Physics, 18: 163-168.

[29] WALTER, A.L., E.O. FRIND, D.W. BLOWES, C.J. PTACEK & J.W. MOLSON (1994): Modeling of multicomponent reactive transport in groundwater 1. Model development and evaluation., Water Res. Res. 30(11): 3137-3148.

[30] YANENKO, N.N. (1971): The method of fractional steps., Springer-Verlag, New York.

[31] ZALESAK, S.T. (1979): Fully multidimensional Flux-Corrected Transport algorithm for fluids., J. Comput. Phys. 31:335-362.

THEORETICAL AND NUMERICAL ASPECTS OF ELASTO-PLASTIC POROUS MEDIA MODELS

W. Ehlers, S. Diebels and D. Mahnkopf
Institut für Mechanik (Bauwesen), Universität Stuttgart,
D-70550 Stuttgart, Germany

SUMMARY

Within the general framework of porous media theories (mixture theories extended by the concept of volume fractions), the paper presents an overview of theoretical and computational aspects of the consolidation problem. In particular, consolidation phenomena are described by the assumption of a two-phase continuum model consisting of an elasto-plastic porous soil matrix saturated by a viscous pore-fluid. In the elastic regime, the porous soil is modeled by a finite elasticity law of Simo-Pister type, while the plastic range (first yielding and isotropic hardening) is governed by a "single surface" yield function together with a non-associated flow rule. The fluid viscosity is included in the drag force. In the present paper, a numerical example for a plain strain problem is carried out within the geometrically non-linear approach by the finite element method, where, in the elasto-plastic regime, use is made of an elastic predictor/plastic corrector scheme.

INTRODUCTION

Consolidation problems, such as the settlement of buildings, are characterized by the physical phenomenon that any additional load applied onto the surface of a fluid-saturated porous solid material yields time-dependent visco-elastic or visco-elasto-plastic deformations accompanied by a draining process. The description of this type of material behaviour naturally falls into the category of porous media theories, which is defined as the classical theory of mixtures extended by the concept of volume fractions. The reader, who is interested in the various details of this approach, is referred, e. g., to Bowen [3, 4], Ehlers [5] or de Boer *et al.* [2]. For simplicity and convenience, thermal effects as well as mass exchanges between the constituents are excluded from the following treatment. Based on this concept, the objective of the paper is to re-formulate and to solve the mechanical balance equations of a binary model of incompressible constituents (elasto-plastic solid skeleton and viscous pore-fluid). Within the framework of frictional materials, the plastic model equations consist of a single surface yield function [9] for the description of first yielding and isotropic hardening together with a non-associated flow rule to represent the evolution equation for the plastic strain rate. The viscous properties of the pore fluid are modeled by the momentum supply terms or the interaction force acting between the soil skeleton and the fluid, respectively.

Based on the weak formulation of the general finite model equations, plane strain initial boundary-value problems including geometrical and material non-linearities are solved within the finite element method (FEM), thus delivering a successful alternative to the different numerical studies based on the well-known Biot approach or on the work by Prévost [11, 12]; concerning the Biot approach, compare [1, 8, 16].

Proceeding from the macroscopic formulation of general porous media theories, the fluid-saturated porous medium under study is treated as an immiscible mixture of constituents φ^i with particles X^i ($i = S$: solid phase; $i = F$: fluid phase), where at any time t each spatial point of the current configuration is simultaneously occupied by particles X^i of constituents φ^i (superimposed continua). These particles proceed from different reference positions $\mathbf{X}_i$ at time t_0. Thus, each constituent is assigned its own motion function, viz.:

$$\mathbf{x} = \boldsymbol{\chi}_i(\mathbf{X}_i, t). \tag{1}$$

The volume fractions

$$n^i = n^i(\mathbf{x}, t) \tag{2}$$

are defined as the local ratios of the constituent volumes v^i with respect to the bulk volume v. Associated with each φ^i is an effective (or material) density ρ^{iR} which is defined as the mass of φ^i per unit of v^i and a partial (or bulk) density ρ^i defined as the mass of φ^i per unit of v. The density functions are related by

$$\rho^i = n^i \rho^{iR}. \tag{3}$$

From equation (3), it is immediately concluded that constituent incompressibility of φ^i proceeding from $\rho^{iR} =$ const. does not lead to macroscopic incompressibility, because the bulk density functions can still change through changes in the volume fractions.

From equation (1), each constituent is assigned its own velocity and acceleration fields,

$$\overset{\prime}{\mathbf{x}}_i = \frac{\partial \boldsymbol{\chi}_i(\mathbf{X}_i, t)}{\partial t}, \qquad \overset{\prime\prime}{\mathbf{x}}_i = \frac{\partial^2 \boldsymbol{\chi}_i(\mathbf{X}_i, t)}{\partial t^2}. \tag{4}$$

By use of the inverse motion functions, an equivalent representation of the above equations is

$$\overset{\prime}{\mathbf{x}}_i = \overset{\prime}{\mathbf{x}}_i\,(\mathbf{x}, t), \qquad \overset{\prime\prime}{\mathbf{x}}_i = \overset{\prime\prime}{\mathbf{x}}_i\,(\mathbf{x}, t). \tag{5}$$

Furthermore, suppose that, e. g., Γ is any differentiable function of $(\mathbf{x}, t)$, then, with respect to equation $(5)_1$, the material time derivatives of Γ corresponding to the individual motion functions of the respective φ^i are defined by

$$(\Gamma)'_i = \frac{\partial \Gamma}{\partial t} + \operatorname{grad}\Gamma \cdot \overset{\prime}{\mathbf{x}}_i\,. \tag{6}$$

From equation (1), the deformation gradient of φ^i and its inverse are

$$\mathbf{F}_i = \operatorname{Grad}_i \mathbf{x}\,, \qquad \mathbf{F}_i^{-1} = \operatorname{grad}\mathbf{X}_i\,. \tag{7}$$

In the above equations, the symbols "grad" and "Grad$_i$" mean partial differentiation with respect to the spatial position $\mathbf{x}$ and the reference position $\mathbf{X}_i$ of φ^i, respectively.

Excluding mass exchanges between the solid and the fluid phases, the mechanical constituent balance equations read [5]:
Saturation constraint:

$$n^S + n^F = 1\,. \tag{8}$$

Balance of mass:

$$(\rho^i)'_i + \rho^i \operatorname{div} \overset{\prime}{\mathbf{x}}_i = 0\,. \tag{9}$$

Balance of momentum:

$$\begin{aligned} \rho^i \overset{\prime\prime}{\mathbf{x}}_i &= \operatorname{div} \mathbf{T}^i + \rho^i \mathbf{b} + \mathbf{p}^i\,, \\ \mathbf{p}^S + \mathbf{p}^F &= \mathbf{0}. \end{aligned} \tag{10}$$

In the above relations, $\mathbf{T}^i$ is the symmetric partial Cauchy stress tensor for non-polar constituents [2, 10], $\mathbf{b}$ is the body force density, and $\mathbf{p}^i$ is the momentum production term of φ^i or the interaction force per unit of bulk volume, respectively. Finally, "div" is the divergence operator corresponding to "grad". Due to the microscopic incompressibility, $\rho^{iR} = \text{const.}$, the mass balance (9) is reduced by use of (3) to a balance of volume fractions:

$$(n^i)'_i + n^i \operatorname{div} \overset{\prime}{\mathbf{x}}_i = 0\,. \tag{11}$$

Integration yields

$$n^i = n^i_{0i} \det \mathbf{F}_i^{-1}\,, \tag{12}$$

where $(\ldots)_{0i}$ is a given quantity with respect to the reference configuration of φ^i.

In addition to the equations (4) and (5), the derivatives of the solid and the fluid motion can be expressed by

$$\mathbf{v}_S = \overset{\prime}{\mathbf{x}}_S = (\mathbf{u}_S)'_S\,, \qquad \mathbf{w}_F = \overset{\prime}{\mathbf{x}}_F - \overset{\prime}{\mathbf{x}}_S\,, \tag{13}$$

where $\mathbf{v}_S$ and $\mathbf{u}_S$ are the solid velocity and displacement functions and $\mathbf{w}_F$ is the seepage velocity describing the fluid motion relative to the deformed skeleton.

From the incompressibility constraint of both constituents ($\rho^{iR} = \text{const.}$), equations (8) and (11) combine to the volume balance of the whole aggregate:

$$\operatorname{div}(n^F \mathbf{w}_F + \mathbf{v}_S) = 0. \tag{14}$$

Finally, the incompressible model under study has to be completed by a set of constitutive assumptions for the solid and the fluid stress tensors $\mathbf{T}^i$ ($i \in \{S, F\}$) and for the interaction force $\mathbf{p}^F = -\mathbf{p}^S$. Each of these constitutive equations consists of two terms, where the former arise from the incompressibility constraint via the effective fluid pressure p, while the latter represent the so-called extra quantities, index $(\ldots)_E$, which are a result of the solid deformation and the fluid viscosity [5]:

$$\begin{aligned} \mathbf{T}^i &= -n^i p\, \mathbf{I} + \mathbf{T}^i_E\,, \\ \mathbf{p}^F &= p \operatorname{grad} n^F + \mathbf{p}^F_E\,. \end{aligned} \tag{15}$$

In the finite elastic range, the solid extra stress tensor can be expressed by an elasticity law firstly proposed in a different context by Simo and Pister [14]:

$$\mathbf{T}^S_E = \det \mathbf{F}_S^{-1} \left[2\mu^S \mathbf{K}_{Se} + \lambda^S \ln(\det \mathbf{F}_{Se})\, \mathbf{I} \right]. \tag{16}$$

Therein, μ^S and λ^S are the usual Lamé constants of the macroscopic solid material. Due to the microscopic incompressibility λ^S contains the structural compressibility, compare Ehlers [7]. As usual in finite plasticity theories, the deformation gradient is multiplicatively decomposed into elastic and plastic parts:

$$\mathbf{F}_S = \mathbf{F}_{Se}\mathbf{F}_{Sp}, \tag{17}$$

leading to the introduction of the plastic intermediate configuration, as shown in Figure 1. Then, the Lagrangian strain tensor of the reference configuration and its plastic part are given by

$$\mathbf{E}_S = \tfrac{1}{2}(\mathbf{F}_S^T\mathbf{F}_S - \mathbf{I}), \qquad \mathbf{E}_{Sp} = \tfrac{1}{2}(\mathbf{F}_{Sp}^T\mathbf{F}_{Sp} - \mathbf{I}), \tag{18}$$

while the elastic part of the strain tensor is defined as difference of the total and the plastic strains:

$$\mathbf{E}_{Se} := \mathbf{E}_S - \mathbf{E}_{Sp}. \tag{19}$$

Push forward rotation of the Lagrangian strain $\mathbf{E}_S$ by the orthogonal tensor $\mathbf{R}_S$, resulting from the polar decomposition of $\mathbf{F}_S$, leads to the Karni-Reiner strain

$$\mathbf{K}_S = \tfrac{1}{2}(\mathbf{F}_S\mathbf{F}_S^T - \mathbf{I}), \tag{20}$$

which is additively decomposed into elastic an plastic parts:

$$\mathbf{K}_{Se} = \tfrac{1}{2}(\mathbf{F}_{Se}\mathbf{F}_{Se}^T - \mathbf{I}), \quad \mathbf{K}_{Sp} = \mathbf{K}_S - \mathbf{K}_{Se}. \tag{21}$$

Further details of the multiplicative decomposition of the deformation gradient are given e. g. by Ehlers [5, 6].

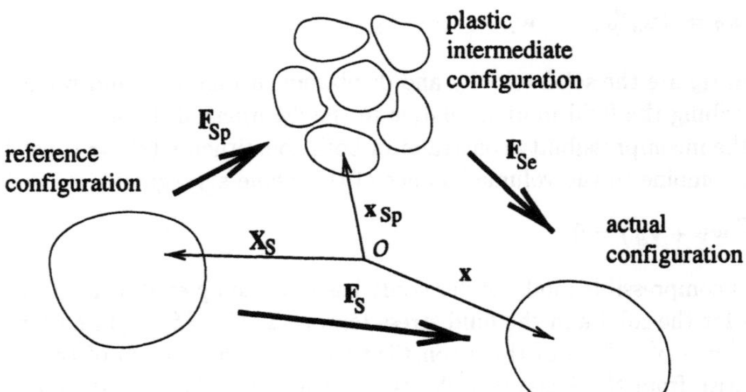

Figure 1. Reference configuration, plastic intermediate configuration and actual configuration according to the multiplicative split of the solid deformation gradient.

For convenience, even if viscous effects dominate the pore-fluid motion, $\mathbf{T}_E^F$ is *a priori* simplified toward $\mathbf{T}_E^F \approx \mathbf{0}$, i. e., the internal friction between the solid and the fluid constituents is only modeled by the extra momentum supply term

$$\mathbf{p}_E^F = -\frac{(n^F)^2\,\gamma^{FR}}{k^F}\,\mathbf{w}_F. \tag{22}$$

Therein, $\gamma^{FR} = \rho^{FR}|\,\mathbf{b}\,|$ is the effective (true) weight of the fluid, and k^F is the so-called D'Arcy permeability parameter.

PLASTIC SOIL PROPERTIES

In addition to the constitutive equations (15) – (22), the description of the elasto-plastic soil properties must be completed by the assumption of a suitable yield and hardening function together with a set of evolution equations for the plastic strain rate (the flow rule), and the material parameters governing the effect of isotropic hardening.

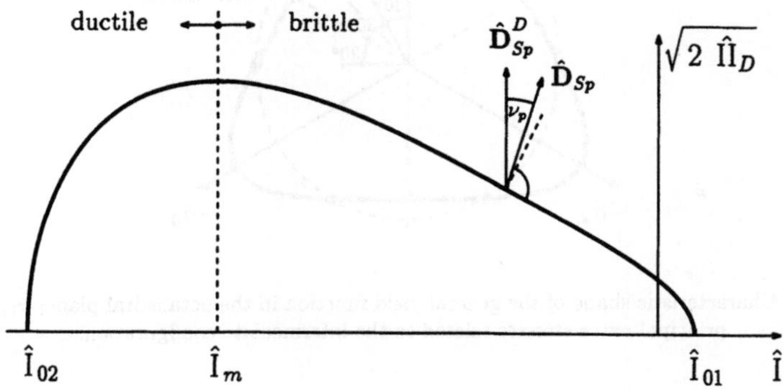

Figure 2. Characteristic shape of the general yield function in the hydrostatic plane. $\hat{\text{I}}$, $\hat{\text{II}}_D$: first and negativ deviatoric second principal invariants of the stress state on the intermediate configuration (positive sign means tension).

As was discussed in [9], the following yield function, compare Figures 2 and 3, proves to be very convenient for the description of both first yielding and isotropic hardening of frictional materials:

$$
\begin{aligned}
F &= \hat{\Phi}^{1/2} + \beta\,\hat{\text{I}} + \epsilon\,\hat{\text{I}}^2 - \kappa = 0\,,\\
\hat{\Phi} &= \hat{\text{II}}_D\,(1+\gamma\,\hat{\vartheta})^m + \tfrac{1}{2}\alpha\,\hat{\text{I}}^2 + \delta^2\,\hat{\text{I}}^4\,,\\
\hat{\vartheta} &= \hat{\text{III}}_D/\hat{\text{II}}_D{}^{3/2}.
\end{aligned}
\qquad (23)
$$

In this representation, $\mathcal{P} = \{\alpha,\,\beta,\,\gamma,\,\delta,\,\epsilon,\,m,\,\kappa\}$ is a set of seven material parameters, and $\hat{\text{I}}, \hat{\text{II}}_D$ and $\hat{\text{III}}_D$ are the first invariant and the negativ deviatoric second and the deviatoric third invariants of the Kirchhoff extra stress tensor $\hat{\boldsymbol{\tau}}^S_E$ of the plastic intermediate configuration

$$
\hat{\text{I}} = \hat{\boldsymbol{\tau}}^S_E \cdot \mathbf{I}\,, \qquad \hat{\text{II}}_D = \frac{1}{2}\,\hat{\boldsymbol{\tau}}^{SD}_E \cdot \hat{\boldsymbol{\tau}}^{SD}_E\,, \qquad \hat{\text{III}}_D = \frac{1}{3}\,\hat{\boldsymbol{\tau}}^{SD}_E \cdot (\hat{\boldsymbol{\tau}}^{SD}_E\,\hat{\boldsymbol{\tau}}^{SD}_E)\,, \qquad (24)
$$

where

$$\hat{\tau}_E^S = (\det \mathbf{F}_{Se})\ \mathbf{F}_{Se}^{-1}\ \mathbf{T}_E^S\ \mathbf{F}_{Se}^{T-1}. \tag{25}$$

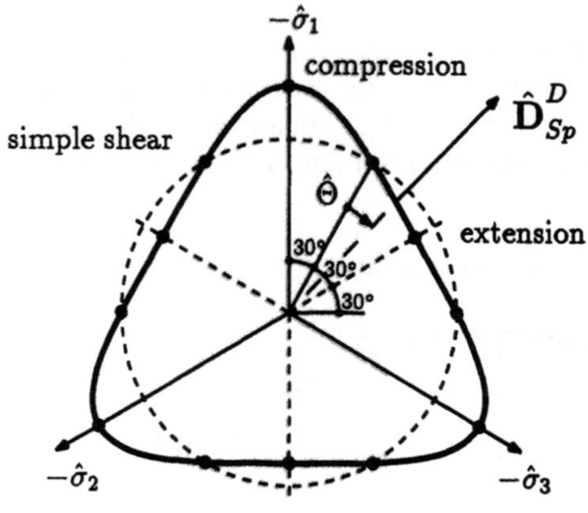

Figure 3. Characteristic shape of the general yield function in the octahedral plane; $\hat{\sigma}_1$, $\hat{\sigma}_2$, $\hat{\sigma}_3$: principal extra stresses related to the intermediate configuration.

The additional index $(\dots)^D$ characterizes the deviatoric part of the respective tensorial object. The definition of the Reuss variables

$$\begin{aligned} r &= \sqrt{2\,\hat{\mathrm{II}}_D}\,, \\ \Theta &= \tfrac{1}{3}\arcsin\left(\frac{\sqrt{27}}{2}\,\frac{\hat{\mathrm{III}}_D}{\hat{\mathrm{II}}_D{}^{3/2}}\right) \end{aligned} \tag{26}$$

leads to the geometric interpretation of the yield function exhibiting a closed and smooth shape in the principal stress space:

$$\begin{aligned} r\,(\Theta,\hat{\mathrm{I}}) &= \mathcal{F}_d(\Theta)\ r_s(\hat{\mathrm{I}})\,, \\ \mathcal{F}_d(\Theta) &= [1+\frac{2}{\sqrt{27}}\,\gamma\,\sin(3\Theta)]^{-m/2}\,, \\ r_s(\hat{\mathrm{I}}) &= \sqrt{2}\,[(\epsilon^2-\delta^2)\,\hat{\mathrm{I}}^4+2\beta\epsilon\,\hat{\mathrm{I}}^3+(\beta^2-\tfrac{1}{2}\,\alpha-2\epsilon\kappa)\,\hat{\mathrm{I}}^2-2\beta\kappa\,\hat{\mathrm{I}}+\kappa^2\,]^{1/2}\,. \end{aligned} \tag{27}$$

Therein, $r\,(\Theta,\,\hat{\mathrm{I}})$ is the yield radius given as the product of two independent functions, $r_s(\hat{\mathrm{I}})$ and $\mathcal{F}_d(\Theta)$, where $r_s(\hat{\mathrm{I}})$ represents the hydrostatic shape function in the simple shear regime, while $\mathcal{F}_d(\Theta)$ describes, in the deviatoric plane, the typical deviation from a circle with the shear radius r_s toward the compression and the extension radii, r_c and r_t, compare Figure 3. The reader, who is interested in an extensive discussion of the above yield function is referred to [9].

It is seen from equations (27) that the general set of material parameters $\mathcal{P}$ can be split into a set $\mathcal{P}_d = \{\gamma, m\}$ governing the deviatoric plane and into a set $\mathcal{P}_s = \{\alpha, \beta, \delta, \epsilon, \kappa\}$ governing the hydrostatic plane, where, in the case of isotropic hardening, only $\mathcal{P}_s$ depends on the plastic strain pathes, namely, through the deviatoric and the volumetric pathes, p^D and p^V, respectively. Thus,

$$\mathcal{P}_s = \mathcal{P}_s(\mathrm{p}^D, \mathrm{p}^V), \tag{28}$$

where

$$\mathrm{p}^D = \int_{(t)} \sqrt{\frac{2}{3} \hat{\mathbf{D}}^D_{Sp} \cdot \hat{\mathbf{D}}^D_{Sp}}\, Dt, \quad \mathrm{p}^V = \int_{(t)} \left| \frac{1}{3} (\hat{\mathbf{D}}_{Sp}) \cdot \mathbf{I} \right| Dt. \tag{29}$$

The plastic strain rate $\hat{\mathbf{D}}_{Sp} = \mathbf{F}^{-1}_{Se}\, \mathbf{K}^{\nabla}_{Sp}\, \mathbf{F}^{T-1}_{Se}$ on the intermediate configuration is equivalent to the pull back of the lower Lie derivative of the plastic Karni-Rainer tensor $\mathbf{K}_{Sp}$. The lower Lie derivative is given by $(\ldots)^{\nabla} = (\ldots)'_S - \mathbf{L}^T_S(\ldots) - (\ldots)\mathbf{L}_S$, where $\mathbf{L}_S$ is the velocity gradient.

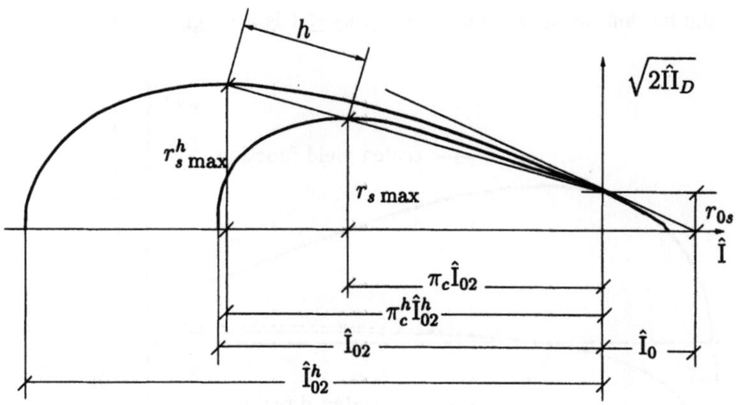

Figure 4. First yielding and isotropic hardening of the general yield function; simple shear meridians in the hydrostatic plane at $\Theta = 0°$.

As was shown in [9], the transition of the general yield surface toward the Drucker-Prager and the Mohr-Coulomb theories is easily guaranteed, if

$$\begin{aligned} \kappa &= c \cos\phi \;\longrightarrow\; r_{os} = \sqrt{2}\,\kappa, \\ \beta &= \tfrac{1}{3}\sin\phi \;\longrightarrow\; \varphi_{os} = \arctan\left(-\frac{\sqrt{2}}{3}\sin\phi\right); \end{aligned} \tag{30}$$

ϕ is the angle of internal friction, and c is the cohesion. The remainder of $\mathcal{P}_s$, $\mathcal{P}^h_s = \{\alpha, \delta, \epsilon\}$, must be determined from experimental data by use of a numerical fitting process including an overall convexity control. Usually, experimental data is available on the consolidation pressure, $p_c = -1/3\,\hat{\mathrm{I}}_{o2}$, the maximum yield radius $r_{s\,\max}$, and its location given by π_c, compare Figure 4. Thus, both at the onset of plastic deformations and in the isotropic hardening regime,

$$\mathcal{P}_s^h = f^h\,(\hat{\mathrm{I}}_{o2}^h\,,\, r_{s\,\max}^h\,,\, \pi_c^h)\,, \tag{31}$$

where the additional index $(\dots)^h$ characterizes the evolution of the initial values $\hat{\mathrm{I}}_{o2}$, $r_{s\,\max}$ and π_c during the hardening process:

$$\begin{aligned} \varphi^h &= \arctan \frac{r_{s\,\max} - r_{os}}{\pi_c\,|\,\hat{\mathrm{I}}_{o2}|}\,, & \hat{\mathrm{I}}_{o2}^h &= \hat{\mathrm{I}}_{o2} - \frac{\hat{\mathrm{I}}_o - \hat{\mathrm{I}}_{o2}}{\hat{\mathrm{I}}_o - \pi_c\,\hat{\mathrm{I}}_{o2}}\,h\cos\varphi^h\,, \\ r_{s\,\max}^h &= r_{s\,\max} + h\,\sin\varphi^h\,, & \pi_c^h &= \frac{1}{\hat{\mathrm{I}}_{o2}^h}\,(\pi_c\,\hat{\mathrm{I}}_{o2} - h\cos\varphi^h)\,. \end{aligned} \tag{32}$$

Therein, the evolution of the hardening length h, compare Figure 4, is given by

$$h = \mathrm{k}_1\,[\,\mathrm{k}_2\,\mathrm{p}^V + (1-\mathrm{k}_2)\,\mathrm{p}^D]\,, \tag{33}$$

where k_1 is the hardening parameter, and $0 \le \mathrm{k}_2 \le 1$ is a weighting factor.

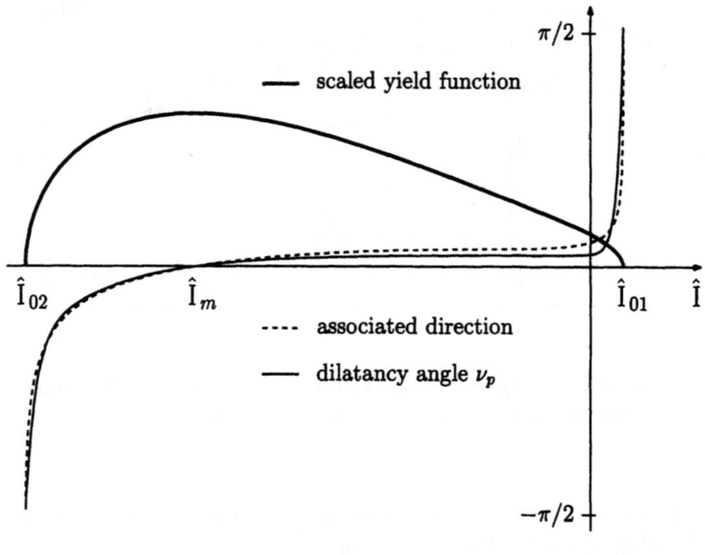

Figure 5. Plastic strain rate and dilatation angle in the hydrostatic plane.

In addition to the above results, the description of the plastic material response must be completed by a flow rule governing the evolution of the plastic strain $\hat{\mathbf{D}}_{Sp}$. As is known from experiments, the usual type of associated flow rules leads, by application to frictional materials, to extremely unrealistic results, especially, with respect to the description of plastic volumetric strains, i. e., dilatation and compression. Thus, one must apply a non-associated flow rule, which, in the present article, is given with respect to [6]:

$$\hat{\mathbf{D}}_{Sp} = \Lambda \left[\tfrac{1}{2}(\hat{\mathrm{II}}_D)^{-1/2} \hat{\boldsymbol{\tau}}_E^{SD} + \sqrt{\tfrac{1}{6}} \tan \nu_p \mathbf{I} \right]. \tag{34}$$

Close to experimental data, the above flow rule prescribes co-axial behaviour of the deviatoric part of $\hat{\mathbf{D}}_{Sp}$, while the volumetric part depends on an additional material function, the dilatation angle ν_p, expressed by three exponential functions

$$\nu_p = \nu_1 \exp\left[a_1(\hat{\mathrm{I}} - \hat{\mathrm{I}}_{01})\right] + \nu_2 \exp\left[a_2(\hat{\mathrm{I}} - \hat{\mathrm{I}}_m)\right] + \nu_3 \exp\left[a_3(\hat{\mathrm{I}} - \hat{\mathrm{I}}_{02})\right] + \nu_4 . \tag{35}$$

In (30) the parameters a_1, a_2, a_3, and ν_4 have to be determined by experiments, while the values of ν_1, ν_2, and ν_3 follow from the solution of a system of linear equations, $\nu_p(\hat{\mathrm{I}}_{01}) = \pi/2$, $\nu_p(\hat{\mathrm{I}}_m) = 0$, $\nu_p(\hat{\mathrm{I}}_{02}) = -\pi/2$, compare Figure 5.

Furthermore, the usual Λ-factor of plasticity theories can either be determined from the consistency condition or, and this is done in the following numerical examples, it can directly be computed from the yield surface using an elastic predictor/plastic corrector scheme.

WEAK FORMULATION

Based on consideration of three independent fields, the solid displacement $\mathbf{u}_S$, the seepage velocity $\mathbf{w}_F$ and the effective fluid pressure p, the corresponding three equations of the weak formulation can be obtained from the kinematics, the balance relations and the constitutive equations of the preceding sections, where, in the present article, acceleration terms are excluded from the formulation.

By combination of the solid and the fluid momentum balance equations, one obtains the first relation representing the momentum balance of the the whole aggregate to compute the solid displacement $\mathbf{u}_S$:

$$\begin{aligned} &\int_V \left(\mathbf{T}_E^S - p\,\mathbf{I}\right) \cdot \operatorname{grad} \delta\mathbf{u}_S \, dv - \\ &- \int_V \left(n^F \rho^{FR} + n^S \rho^{SR}\right) \mathbf{b} \cdot \delta\mathbf{u}_S \, dv = \int_{\partial V} \mathbf{t} \cdot \delta\mathbf{u}_S \, da . \end{aligned} \tag{36}$$

Therein, $\mathbf{t}$ is the total surface load vector acting upon both constituents. The advantage of using (31) instead of the weak form of the momentum balance of the solid is that the total surface load $\mathbf{t}$ is applied, while the load carried by the skeleton depends on the deformation.

Secondly, the weak formulation of the fluid momentum balance reads

$$\begin{aligned} &\int_V \left(\frac{\gamma^{FR}}{k^F} n^F \mathbf{w}_F - \rho^{FR} \mathbf{b}\right) \cdot \delta\mathbf{w}_F \, dv - \\ &- \int_V p \operatorname{div} \delta\mathbf{w}_F \, dv = - \int_{\partial V} p\,(\delta\mathbf{w}_F \cdot \mathbf{n}) \, da ; \end{aligned} \tag{37}$$

$\mathbf{n}$ is the outward oriented unit normal. The third equation,

$$\int_V \operatorname{div}(n^F \mathbf{w}_F + \mathbf{v}_S)\, \delta p \, dv = 0\,, \tag{38}$$

represents the volume balance of both constituents acting as an incompressibility constraint on the two-phase model under consideration. $\delta \mathbf{u}_S$, $\delta \mathbf{w}_F$ and δp are weighting functions corresponding to $\mathbf{u}_S$, $\mathbf{w}_F$ and p.

EXAMPLE

Within the finite element method, the following numerical example proceeds from quadratic shape functions for the solid displacement, while the seepage velocity and the fluid pressure are approximated by linear functions. By use of the standard Galerkin procedure, the shape functions and the weighting functions coincide.

Proceeding from the geometrically non-linear approach to the above weak formulation, the first example exhibits a fluid-saturated elasto-plastic soil block of 10 m width and 10 m height, compare Figure 5, treated as a two-dimensional plain strain problem. Half of the surface is perfectly drained, while the other part is undrained and loaded (permanently vertical) by a ramp load with an amplitude of $F = 60\,\mathrm{kN/m^2}$. The remainder of the outer boundary is rigid and undrained. The material parameters of the problem are summarized in Table 1. The choosen plastic parameters fit the yield condition (18) to the data given by Thamm [15], which were also used by Schad [13].

Table 1. Material properties of the solid and fluid constituents.
The description of first yielding is taken from [9].

soil matrix properties		
elastic domain	first yielding	hardening
$\mu^S = 5\,583\,\mathrm{kN/m^2}$	$\gamma = 1.555\,04$	
$\lambda^S = 8\,375\,\mathrm{kN/m^2}$	$m = 0.593\,5$	$k_1 = 10^4\,\mathrm{kN/m^2}$
	$c = 11\,\mathrm{kN/m^2}$	$k_2 = 0.5$
	$\phi = 21°$	
$\rho^{SR} = 2\,600\,\mathrm{kg/m^3}$	$r_{s\,\max} = 66.82\,\mathrm{kN/m^2}$	
$n_{0S}^S = 0.67$	$p_c = 258\,\mathrm{kN/m^2}$	
	$\pi_c = 0.66$	
pore-fluid properties		
$\rho^{FR} = 1\,000\,\mathrm{kg/m^3}$	$\gamma^{FR} = 10\,000\,\mathrm{kN/m^3}$	$k^F = 5 \cdot 10^{-5}\mathrm{m/s}$

On the drained part of the surface, the boundary condition prescribes zero pore pressure, while the maximum pressure $p_{\max}$ appears under the undrained boundary. During the consolidation process, $p_{\max}$ decreases with time until $p = 0$ in the whole do-

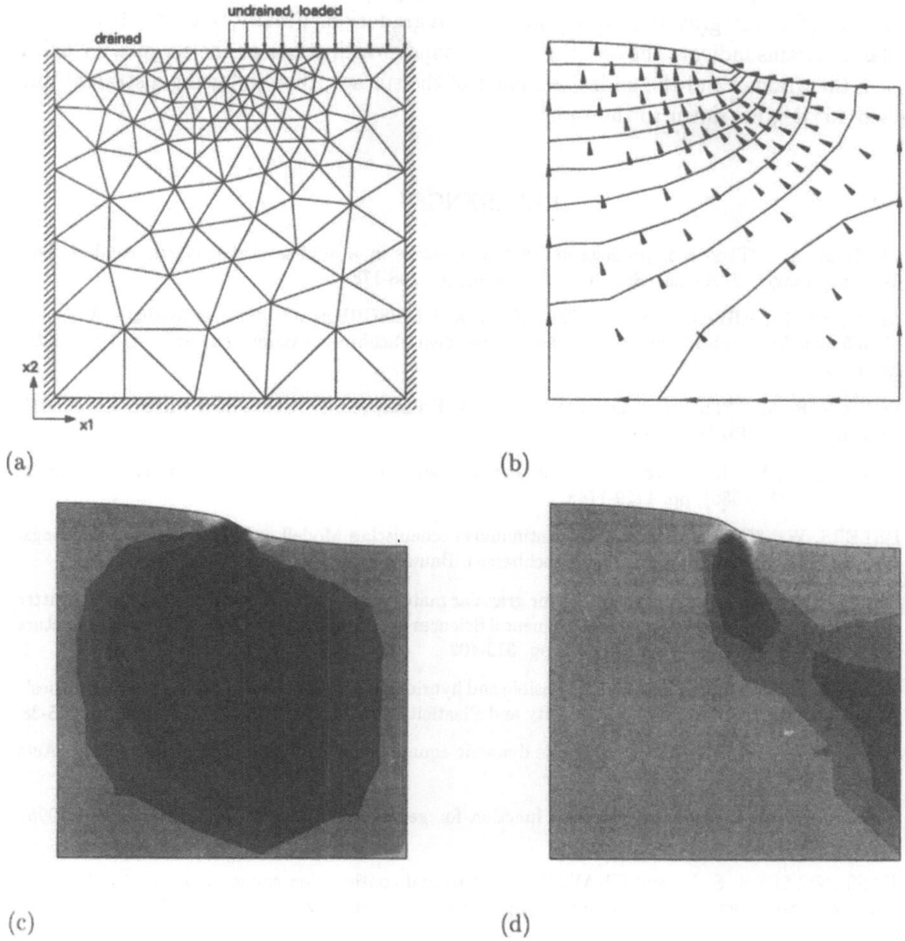

Figure 6. Undeformed grid and snap shots of the deformed grid at $t = 70\,s$. The solid deformations are scaled by the factor 50. (a) Undeformed grid and boundary conditons, (b) pore pressure isolines (distance: $4.06\,kN/m^2$) and directions of the seepage velocity, (c) shear extra stress distribution, and (d) horizontal plastic strains.

main, thus indicating both the maximum solid deformation and the end of the consolidation process. Figure 6 shows the unstructured finite element mesh and the deformed soil skeleton $t = 50\,s$ after having applied the external load. At this time, we are at the beginning of the consolidation process, what is easily seen from the maximum value of the fluid pressure ($71.3\,kN/m^2$ under the right part of the loaded surface), almost representing approximately 1/3 of the external load. The extra shear stress component yields a maximum of $8.3\,kN/m^2$ (light) and a minimum of $-15.9\,kN/m^2$ (dark), while the horizontal plastic strain component lies between $-7.1 \cdot 10^{-4}$ (light) and $10.4 \cdot 10^{-4}$ (dark). Both figures proceed from 10 gray steps using an identical gradation. Finally, note in passing that the plastic strains indicate the onset of a shear band, which starts at the singularity point between the drained and the undrained parts of the surface, then following a straight line directed 45° degrees down to the right.

REFERENCES

[1] BIOT, M. A.: "Theory of propagation of elastic waves in a fluid-saturated elastic solid, I. Low frequency range", J. Acoust. Soc. Am., **28** (1956), pp. 168-178.

[2] DE BOER, R., EHLERS, W., KOWALSKI. S. and PLISCHKA, J.: "Porous Media – A Survey of Different Approaches", Forschungsberichte aus dem Fachbereich Bauwesen, Heft 54, Universität Essen 1991.

[3] BOWEN, R. M.: "Theory of mixtures" in A. C. Eringen (ed.): "Continuum Physics", Vol. III, Academic Press (1976), pp. 1-127.

[4] BOWEN, R. M.: "Incompressible porous media models by use of the theory of mixtures", Int. J. Engng. Sci., **18** (1980), pp. 1129-1148.

[5] EHLERS, W.: "Poröse Medien – ein kontinuumsmechanisches Modell auf der Basis der Mischungstheorie", Forschungsberichte aus dem Fachbereich Bauwesen, Heft 47, Universität Essen 1989.

[6] EHLERS, W.: "Constitutive equations for granular materials in geomechanical context" in K. Hutter (ed): "Continuum Mechanics in Environmental Sciences and Geophysics", CISM Courses and Lecture Notes No. 337, Springer-Verlag (1993), pp. 313-402.

[7] EHLERS, W.: "Compressible, incompressible and hybrid two-phase models in porous media theories", Anisotropy and Inhomgeneity in Elasticity and Plasticity, **AMD**-Vol. **158**, ASME (1993), pp. 25-38.

[8] EHLERS, W. and KUBIK, J.: "On finite dynamic equations for fluid-saturated porous solids", Acta Mech., **105** (1994), pp. 101-117.

[9] EHLERS, W.: "A single surface yield function for geomaterials", Arch. Appl. Mech., **65** (1995), pp. 246-259.

[10] HASSANIZADEH, S. M. and GRAY, W. G.: "General conservation equations for multi-phase systems: 2. Mass, momenta, energy, and entropy equations", Adv. Water Resources, **2** (1979), pp. 191-203.

[11] PRÉVOST, J. H.: "Consolidation of anelastic porous media", ASCE: J. Engrg. Mech. Div., **107** (1981), pp. 69-186.

[12] PRÉVOST, J. H.: "Constitutive equations for pressure-sensitive soils: Theory, numerical implementation, and examples" in R. Dungar and J. A. Studer (ed.): "Geomechanical Modelling in Engineering Practice", Balkema (1986), pp. 331-350.

[13] SCHAD, H.: "Nichtlineare Stoffgleichungen für Böden und ihre Verwendung bei der numerischen Analyse von Grundbauaufgaben", Mitt. 10, Baugrundinst. Stuttgart 1979.

[14] SIMO, J. C., and PISTER, K. S.: "Remarks on rate constitutive equations for finite deformation problems: computational implications", Comput. Methods Appl. Mech. Eng., **46** (1984), pp. 201-215.

[15] THAMM, B. R., "Berechnug der Anfangssetzungen und der Anfangsporenwasserüberdrücke eines wassergesättigten normalverdichteten Tones", Mitt. 1, Baugrundinst. Stuttgart 1974.

[16] ZIENKIEWICZ, O. C., CHAN, A. H. C., PASTOR, M., PAUL, D. K., and SHIOMI, T.: "Static and dynamic behaviour of soils: A rational approach to quantitative solutions: I. Fully saturated problems", Proc. Royal Soc. Lond. A, **429** (1990), pp. 285-309.

NUMERICAL SIMULATION OF THE MULTIPHASE FLOW OF CONTAMINANTS IN POROUS MEDIA

Richard E. Ewing
Institute for Scientific Computation
Texas A&M University
College Station, Texas 77843–3404

SUMMARY

Three basic problem areas have dominated much of the recent research in reservoir simulation. First, one must obtain an effective model to describe the complex fluid/fluid and fluid/rock interactions that control recovery processes. This includes the problem of obtaining accurate reservoir descriptions at various length scales and including the effects of this heterogeneity in the reservoir simulators. Next, one must develop accurate discretization techniques that retain the important physical properties of the continuous models. Finally, one should develop efficient numerical solution algorithms that utilize the potential of the emerging computing architectures. We will discuss recent advances in each of these three areas.

1. INTRODUCTION

The understanding and prediction of the behavior of the flow of multiphase or multicomponent fluids through porous media are often strongly influenced by heterogeneities, either large-scale lithological discontinuities or quite localized phenomena. Considerable information can be gained about the physics of multiphase flow of fluids through porous media via laboratory experiments and pore-scale models; however, the length scales of these data are quite different from those required from field-scale understanding. The coupled fluid-fluid interactions are highly nonlinear and quite complex. The presence of heterogeneities in the medium greatly complicates this flow. We must understand the effects of heterogeneities coupled with nonlinear parameters and functions on different length scales. We can then use the simulators as "experimental tools" in the laboratory of high performance computing to simulate the process on increasingly larger length scales to develop intuition on how to model the effects of heterogeneities at various levels.

The simulation of processes such as bioremediation and chemically reacting flows involves the use of nonlinear reaction terms. The form of the reaction kinetics must be determined from laboratory tests at laboratory length scales. Scaling these nonlinear reaction terms may require a new form of reaction kinetics to model the process at the field scale.

The partial differential equation models used in the simulators are convection-dominated. Mixed finite element methods are described to treat the strong variation in coefficients arising from heterogeneities. An operator-splitting technique is then used to address the disparate temporal scales of convection, diffusion, and reaction terms. Convection is treated by time stepping along the characteristics of the associated pure convection problem, and diffusion is modeled via a Galerkin method for single phase flow and a Petrov-Galerkin technique for multiphase regimes. Eulerian-Lagrangian techniques, MMOC (modified method of characteristics) described by Douglas and Russell [22] or Ewing *et al.* [37] or ELLAM (Eulerian-Lagrangian Localized Adjoint Methods) introduced by Celia *et al.*[10] effectively treat the advection-dominated processes. Extensions of ELLAM to the multiphase regime appear in [31]. Accurate approximations of the fluid velocities needed in the Eulerian-Lagrangian time-stepping procedure are obtained by mixed finite element methods.

In order to scale the highly localized behavior of fine-scale fingering generated by heterogeneous media up to computational and field scales, we must develop techniques to obtain effective parameters for coarse-grid models which match fine-grid simulations. Ewing, Russell, and Young [39] presented a coarse-grid dispersion model of heterogeneity and viscous fingering to match fine-grid simulation of miscible displacement processes. They adjusted longitudinal and transverse dispersivities in a dispersion tensor to match recovery curves for simulations of viscous fingering on fine grids. Espedal *et al.* [28]. considered similar dispersion models to describe fingering processes in immiscible, two-phase flow. In this paper we combine these ideas for multiphase and multicomponent flow, using dispersion models coupled with accurate treatment of first-order transport effects for both types of flow. This coupling is very important for saturated-unsaturated models, which possess aspects of each process. Then accurate high-resolution numerical simulators are introduced and used as our experimental tools. Numerical results have illustrated the success of dispersion models for all of these problems.

Even at the finest grid level available using effectively scaled models, many phenomena cannot be resolved. Understanding and modeling of these localized phenomena require the use of adaptive or local grid refinement. Usual implementation of local grid refinement techniques destroys the efficiency of large-scale simulation codes. Techniques which involve a relatively coarse macro-mesh are the basis for domain decomposition methods [4,30] and associated parallel solution algorithms. Accuracy, efficiency of implementation, and adaptivity of these techniques are discussed [29,34,35]. Experiences in using these ideas in three-dimensional multiphase industrial petroleum simulators are briefly described [3,32].

In Section 2 we present multiphase flow model equations for field-scale simulations. In Section 3 we discuss mixed finite element discretization techniques to accurately approximate fluid velocities in the presence of heterogeneities. In Section 4 we present efficient iterative methods. Section 5 contains a description of operator-splitting techniques.

2. MODEL EQUATIONS FOR POROUS MEDIA FLOW

The simplest and the most popular is the model of fully saturated incompressible porous media. In this case the water (or the liquid) phase occupies the whole pore space and the flow is due to the nonuniform pressure distribution. The mathematical formulation is based on the mass balance equation and Darcy's law (see, e.g., [2,53]):

$$\nabla \cdot (\rho \mathbf{u}) = F \quad \text{and} \quad \mathbf{u} = -\frac{\mathbf{K}}{\mu}(\nabla p - \rho \mathbf{g}), \qquad \text{in } \Omega, \tag{1}$$

where $\mathbf{u}$ is the volumetric flux of water, F is the source or sink of fluid, ρ is the fluid density, $\mathbf{K}$ is the absolute permeability tensor, μ is the dynamic fluid viscosity, p is the fluid pressure, and $\mathbf{g}$ is the acceleration vector due to gravity.

Darcy's law provides a relation between the volumetric flux in the mass conservation equation and the pressure in the fluid. This relation is valid for viscous dominated flows which occur at relatively low velocities.

The transport of a contaminant that is dissolved in the water is described by the following equation:

$$\frac{\partial(\theta c)}{\partial t} + \nabla \cdot (\rho \mathbf{u} c) - \nabla \cdot (\theta \mathbf{D} \nabla c) + \beta \theta c = G(c), \qquad \text{in } \Omega, \quad t > 0. \tag{2}$$

Here c is the concentration of the contaminant, $\mathbf{D}$ is the dispersion tensor, β is the reaction rate, $\theta = \phi\rho$, ϕ is the porosity, and G is the source/sink term. In addition to Equations (1) and (2), initial and boundary conditions are specified.

When either an air or vapor phase or a nonaqueous phase liquid contaminant (NAPL) is present, the equations describing two phase, immiscible flow in a horizontal porous medium are given by

$$\frac{\partial(\phi\rho_w S_w)}{\partial t} - \nabla \cdot \left(\mathbf{K} \frac{\rho_w k_{rw}}{\mu_w} \nabla (p_w - \rho_w \mathbf{g}) \right) = q_w \rho_w, \quad x \in \Omega, \quad t \in J, \tag{3}$$

$$\frac{\partial(\phi\rho_a S_a)}{\partial t} - \nabla \cdot \left(\mathbf{K} \frac{\rho_a k_{ra}}{\mu_a} \nabla (p_a - \rho_a \mathbf{g}) \right) = q_a \rho_a, \quad x \in \Omega, \quad t \in J, \tag{4}$$

where the subscripts w and a refer to the water and air phases respectively, S_i is the saturation, p_i is the pressure, ρ_i is the density, k_{ri} is the relative permeability, μ_i is the viscosity, and q_i is the external flow rate, each with respect to the ith phase.

One of the saturations can be eliminated, since $S_w + S_a = 1$. The pressure between the two phases is described by the capillary pressure

$$p_c(S_w) = p_a - p_w. \tag{5}$$

Although formally, Equations (1) and (2) seem quite different from those in Equations (3) and (4), the latter system may be rearranged in a form which very closely resembles the former system. In order to use the same basic simulation techniques in our sample computations to treat both miscible and immiscible displacement, we will follow the ideas of Chavent [11].

Let Ω in $\mathbf{R}^3$ represent a porous medium. The global pressure p and total velocity $\mathbf{v}$ formulation of a two-phase water (w) and air (a) flow model in Ω is given by the following equations [13]:

$$S_a c_a \frac{dp}{dt} + \nabla \cdot \mathbf{v} = -\frac{\partial \phi(p)}{\partial t} + q(x, S_w), \qquad x \in \Omega, \quad t > 0, \tag{6}$$

$$\mathbf{v} = -\mathbf{K}\lambda(\nabla p - \mathbf{G}_\lambda), \qquad x \in \Omega, \quad t > 0, \tag{7}$$

$$\phi \frac{\partial S_w}{\partial t} + \nabla \cdot (f_w \mathbf{v} - \mathbf{K}\lambda_a q_w \delta\rho \mathbf{g} - \mathbf{D}(S_w) \cdot \nabla S_w) = -S_w \frac{\partial \phi(p)}{\partial t} + q_w, \quad x \in \Omega, \quad t > 0. \tag{8}$$

The global pressure and total velocity are defined by

$$p = \frac{1}{2}(p_w + p_a) + \frac{1}{2} \int_{S_c}^{S} \frac{\lambda_a - \lambda_w}{\lambda} \frac{dp_c}{d\xi} d\xi \qquad \text{and} \qquad \mathbf{v} = \mathbf{v}_w + \mathbf{v}_a. \tag{9}$$

where $p_c(S_c) = 0$. Further, $d/dt \equiv \phi(\partial/\partial t) + \mathbf{v}_a/S_a \cdot \nabla$, $\lambda = \lambda_w + \lambda_a$ is the total mobility, $\lambda_i = k_{ri}/\mu_i$, $i = w, a$, is the mobility for water and air, and $\mathbf{K}$ is the absolute permeability tensor.

The gravity forces G_λ and capillary diffusion term $D(S)$ are expressed as

$$\mathbf{G}_\lambda = \frac{\lambda_w \rho_w + \lambda_a \rho_a}{\lambda}\mathbf{g} \quad \text{and} \quad \mathbf{D}(S_w) = -\mathbf{K}\lambda_a f_w \frac{dp_c}{dS_w} \tag{10}$$

and the compressibility c_a and fractional flow of water f_w are defined by

$$c_a = \frac{1}{\rho_a}\frac{d\rho_a}{dp_a} \quad \text{and} \quad f_w = \frac{\lambda_w}{\lambda}. \tag{11}$$

We note that in this formulation, the only diffusion/dispersion term is capillary mixing described by Equation (10).

The phase velocities for water and air, which are needed in transport calculations, are given by:

$$\begin{aligned} \mathbf{v}_w &= f_w \mathbf{v} + \mathbf{K}\lambda_a f_w \nabla p_c - \mathbf{K}\lambda_a f_w \delta\rho \mathbf{g}, \\ \mathbf{v}_a &= f_a \mathbf{v} - \mathbf{K}\lambda_w f_a \nabla p_c + \mathbf{K}\lambda_w f_a \delta\rho \mathbf{g}, \end{aligned} \tag{12}$$

where $f_\alpha = \lambda_\alpha/\lambda$, $\alpha = w, a$, and $\delta\rho = \rho_a - \rho_w$. Within the groundwater literature, the pressure normally is scaled by the gravity potential function. Equation (6) would then be given in terms of the pressure head. We should also note that if the Richards approximation, infinite mobility of air or $p_a = 0$, is valid, Equation (5) can be replaced by: $p_c(S_w) = -p_w$. We may note that the phase velocity for air is given by Equation (12) even if the Richards approximation is used.

If Γ is the boundary of Ω, general boundary conditions for Equations (6)–(8) can be given by a combination of expressions as described in [13]. Normally the boundary conditions will be nonlinear functions of the physical boundary conditions for the original two-pressure formulation [13]. This means that we may have to iterate on the boundary conditions as a part of the solution process. Our experience is that this does not cause problems.

Both single and multiphase codes used in our simulation utilize a physical dispersion tensor with different longitudinal and transverse terms. Although this is clearly natural for single phase contaminant modeling, the local physics of multiphase flows does not normally involve a dispersion phenomena. However, via perturbation analysis, Espedal and Langlo have developed a natural dispersion tensor arising from heterogeneous flow at larger length scales following the single phase work of Dagan [20]. Description of these concepts appear in Espedal *et al.* [26,27,54]. Furtado *et al.* [49] have stochastically arrived at a dispersion phenomenon with effects somewhere between transport and diffusion in origin. We feel that this corresponds to the need to match the gross permeability effects with first-order transport concepts and the finer-scale effects with dispersion models.

Even if complete information is known about the reservoir properties in a highly heterogeneous medium, the problem of how to represent this medium on coarse-grid blocks of different length scales still remains. The power of supercomputers must be brought to bear for simulation studies using homogenization and statistical averaging to represent fine-scale phenomena on coarser grids.

3. MIXED METHODS FOR ACCURATE VELOCITY APPROXIMATIONS

In this section, we describe mixed finite element methods for the accurate approximation of the total velocity $\mathbf{v}_t$. Among the disadvantages of the conforming discretizations are the lack of local mass conservation of the numerical model and some difficulties in computing the phase velocities needed in the transport and saturation equations. The straightforward numerical differentiation is far from being justifiable in problems formulated in a highly heterogeneous medium with complex geometry. On the other hand, the mixed finite element method [8] offers an attractive alternative. In fact, this method conserves mass cell by cell and produces a direct approximation of the two variables of interest—pressure and velocity. Below we explain briefly the mixed finite element method for the pressure equation.

To describe the mixed method we introduce two Hilbert spaces. Let

$$W = L^2(\Omega), \qquad \boldsymbol{V} = \left\{\varphi \in L^2(\Omega)^3,\ \nabla\cdot\varphi \in L^2(\Omega)\right\}.$$

The inner product in $L^2(\Omega)$ is denoted by $(\cdot,\cdot)$. For the sake of simplicity, $(\cdot,\cdot)$ is also used as the inner product in the product space $L^2(\Omega)^3$.

The pressure equation is written in the following mixed weak form: find $(p,\mathbf{v}) \in W\times \boldsymbol{V}$ such that

$$\begin{aligned}
(A\mathbf{v},\varphi) - (p,\nabla\cdot\varphi) &= (\mathbf{G}_\lambda,\varphi), && \forall\,\varphi\in \boldsymbol{V},\ t>0,\\
(C(p,S_w)p_t,\psi) + (\nabla\cdot\mathbf{v},\psi) &= (f(p,S_w),\psi), && \forall\,\psi\in W,\ t>0,\\
p(0)\in L^2(\Omega) &\text{ is the given initial pressure.}
\end{aligned} \tag{13}$$

Here $C(p,S_w) = S_a c_a$, $p_t = \partial p/\partial t$ and $A = (\mathbf{K}\lambda)^{-1}$ as in Equations (6) and (7). We note that A is always symmetric and positive definite which leads to a well defined problem. This is in contrast to Equation (3) or (4) where the relative permeability $k_{r\alpha}$ vanishes when the phase α is absent in some subregion of Ω. We note that if there were nonhomogeneous boundary conditions on $\partial\Omega$ they should be added to the right hand side $(f(p,S_w),\psi)$. Corresponding changes in the bilinear forms in the left hand side should be introduced in the case of Robin boundary conditions. Obviously Equation (13) forms a nonlinear problem. To solve it one can use Picard linearization (see, e.g., [12]) or any other feasible approach.

We triangulate the domain Ω in tetrahedra with characteristic diameter h. Next we introduce the finite element spaces $W_h\subset W$ and $\boldsymbol{V}_h\subset \boldsymbol{V}$ of piecewise polynomials with respect to the triangulation and time discretization $t_n = n\Delta t$, $n = 0,1,\ldots$. The mixed finite element approximation $(P^n,\boldsymbol{V}^n)\in W_h\times\boldsymbol{V}_h$ of $(p(t_n),\mathbf{u}(t_n))\in W\times\boldsymbol{V}$ is the solution of the following problem:

$$\begin{aligned}
(A^n\boldsymbol{V}^n,\varphi_h) - (\nabla\cdot\varphi_h,P^n) &= (\mathbf{G}^n_\lambda,\varphi_h), && \forall\,\varphi_h\in\mathbf{V_h},\\
\frac{1}{\Delta t}(C^n(P^n-P^{n-1}),\psi_h) + (\nabla\cdot\boldsymbol{V}^n,\psi_h) &= (f^n,\psi), && \forall\,\psi_h\in W_h,\\
P^0\in W_h &\text{ is expressed through given initial data.}
\end{aligned} \tag{14}$$

This is an implicit Euler approximation of a nonlinear problem which can be solved by Picard or Newton iterations.

The resulting system of linear equations has the form of a saddle point problem defined on a pair of finite dimensional spaces W_h and $\boldsymbol{V}_h$:

$$\begin{pmatrix}\mathbf{A} & \mathbf{B}^T\\ \mathbf{B} & -\mathbf{D}\end{pmatrix}\begin{pmatrix}V^n\\ P^n\end{pmatrix} = \begin{pmatrix}F\\ G\end{pmatrix}, \tag{15}$$

where $F \in \boldsymbol{V}_h$ and $G \in W_h$ are given and $P^n \in W_h$ and $\boldsymbol{V}^n \in \boldsymbol{V}_h$ represent the unknown approximate solution on the time level t_n. Here $\mathbf{A} : \boldsymbol{V}_h \mapsto \boldsymbol{V}_h$ is a linear symmetric, and positive definite operator. In addition, the linear map $\mathbf{B}^T : W_h \mapsto \boldsymbol{V}_h$ is the adjoint of $\mathbf{B} : \boldsymbol{V}_h \mapsto W_h$. $\mathbf{D} : W_h \mapsto W_h$ is either $(1/\Delta t)\mathbf{M}$ with $\mathbf{M}$ similar to the mass matrix in W_h for time dependent problems or $\mathbf{0}$, for steady state problems. The existence and uniqueness of a solution is guaranteed by the fact that the pair of spaces $(W_h, \boldsymbol{V}_h)$ satisfies the *inf-sup* condition of Babuška-Brezzi [8].

Clearly, the accuracy of the fluid velocities that govern the advection of the fluids is a limiting factor in large-scale simulations. Ewing, Lazarov, and Wang [36] have shown that along certain lines, the fluid velocities are considerably more accurate than in general, and have quantified this superconvergence phenomenon. Since these loci of higher accuracy correspond to Gauss points which are used for quadrature points in the simulation codes, we can take advantage of this greater accuracy without any extra computation from post processing. This observation allows considerably greater accuracy on coarse grids.

Since many flow phenomena are highly localized, the coarse grids dictated by the storage requirements of the supercomputer are often not capable of resolving the physics of flow, and local grid refinement techniques must be utilized. Special methods are required to maintain the efficiency of the code while incorporating the local refinement. Ewing, Lazarov, Russell, and Vassilevski [33] and Ewing and Wang [47] have analyzed the accuracy and efficiency of the use of local grid refinement in the context of mixed methods. Some superconvergence is retained (theoretically and computationally), again along predictable lines associated with Gauss points.

4. EFFICIENT ITERATIVE METHODS

In this section we give an overview of some relevant iterative methods for solving the systems of linear equations arising from finite element discretizations of the differential problems discussed above.

Because of significant progress made in developing iterative techniques for systems coming from Galerkin discretizations, we shall not concentrate on this issue. We only mention that domain decomposition preconditioners are particularly useful when the models are implemented on distributed parallel computers [7].

The indefinite systems resulting from mixed finite elements are much more difficult to solve iteratively. In general, the properties of the discrete operators involved in the definition of the system must be well understood in order to design an efficient iterative scheme for solving it. For example, if Raviart-Thomas pairs of finite dimensional spaces are used (see [8]) then the operator $\mathbf{A}$ is well conditioned, whereas the Schur complement operator $\mathbf{B}\mathbf{A}^{-1}\mathbf{B}^T$ exhibits a condition number growth like h^{-2}, where h is the discretization parameter. It is well known that in our setting $\mathbf{B}\mathbf{A}^{-1}\mathbf{B}^T$ behaves like a discretization of a second-order operator.

A classical approach [48] to the solution of (15) is to perform Gaussian block elimination and obtain the reduced system

$$\mathbf{B}\mathbf{A}^{-1}\mathbf{B}^T P^n = \mathbf{B}\mathbf{A}^{-1}F - G. \qquad (16)$$

The classical Uzawa algorithm is a linear iteration for solving (16). Alternatively, one can apply preconditioned conjugate gradient to (16). A disadvantage of these approaches is that they require the action of $\mathbf{A}^{-1}$ which substantially reduces their efficiency. Recent convergence results for inexact variants of the Uzawa algorithm [6], where the evaluation of $\mathbf{A}^{-1}$ is avoided, make such algorithms still attractive. The approach here uses appropriately scaled preconditioners. It is worth mentioning that there are efficient preconditioners that are scaled appropriately by default. For example, multigrid is automatically scaled properly according to the theory of the inexact Uzawa algorithms whereas a scaling constant must be computed for the incomplete Cholesky factorization.

The positive definite reformulation of (15) suggested in [5] is an attractive method for solving the indefinite problem. This approach utilizes preconditioners for $\mathbf{A}$ and $\mathbf{BA}^{-1}\mathbf{B}^T$ and results in a positve definite system which can be solved efficiently by conjugate gradient. However, the preconditioner for $\mathbf{A}$ must be properly scaled. This scaling factor is either known *a priori* or can be obtained by employing computationally cheap procedures for estimating the largest eigenvalue, such as the power method. In general, the linear iteration corresponding to the inexact Uzawa methods converges slower than the conjugate gradient method just described. However, the Uzawa method may be advantageous when implementated on distributed memory parallel computers since inner products are not required.

The hybrid mixed formulation [8] provides another way for computing the solution to (15). The idea behind this method is to impose continuity of the normal components of the velocity at the interelement interfaces by Lagrange multipliers. From a computational point of view, the main benefit in this formulation is in the fact that the resulting system of linear equation can be reduced to a system for the multipliers which is symmetric and positive definite. The condition number of the latter system is like $O(h^{-2})$ and should be preconditioned. Using equivalence arguments, a domain decomposition preconditioner for this problem is constructed in [15]. The preconditioner suggested in [14] is based on algebraic substructuring techniques and also uses equivalence arguments. Both approaches give rise to well conditioned linear systems.

5. OPERATOR-SPLITTING TECHNIQUES

In multiphase or multicomponent flow models, the convective, hyperbolic part is a linear function of the velocity. An operator-splitting technique has been developed to solve the purely hyperbolic part by time stepping along the associated characteristics [22,37,38,55]. We first obtain the non-divergence form of Equation (2) with $\theta = 1$ by using the product rule for differentiation on the $\nabla \cdot \mathbf{u}c$ term and applying Equation (1) to obtain

$$\phi\frac{\partial c}{\partial t} + \mathbf{u}\cdot\nabla c - \nabla\cdot\mathbf{D}\nabla c = q(\tilde{c} - c) . \qquad (17)$$

Next, the first and second terms in Equation (17) are combined to form a directional derivative along what would be the characteristics for the equation if the tensor $\mathbf{D}$ were zero. The resulting equation is

$$\nabla\cdot(\mathbf{D}\nabla c) + q(\tilde{c} - c) = \phi\frac{\partial c}{\partial t} + \mathbf{u}\cdot\nabla c \equiv \phi\frac{\partial c}{\partial \tau} . \qquad (18)$$

The system obtained by modifying Equations (1) and (2) in this way is solved sequentially. An approximation for $\mathbf{u}$ is first obtained at time level $t = t^n$ from a solution of Equations (1) and (2) with the fluid viscosity μ evaluated via some mixing rule at time level t^{n-1}. Equations (1) and (2) can be solved as a mixed finite element method for a more accurate fluid velocity as in the last section. Let $C^n(x)$ and $\mathbf{U}^n(x)$ denote the approximations of

$c(x,t)$ and $\mathbf{u}(x,t)$, respectively, at time level $t = t^n$. The directional derivative is then discretized along the "characteristic" mentioned above as

$$\phi \frac{\partial c}{\partial \tau}(x, t^n) \approx \phi \frac{C^n(c) - C^{n-1}(\bar{x}^{n-1})}{\Delta t} , \tag{19}$$

where $\bar{x}^{n-1}$ is defined for an x as

$$\bar{x}^{n-1} = x - \frac{\mathbf{U}^n(x)\Delta t}{\phi}. \tag{20}$$

This technique is a discretization back along the "characteristic" generated by the first-order derivatives from Equation (18). Although the advection-dominance in the original Equation (18) makes it non-self-adjoint, the form with directional derivatives is self-adjoint and discretization techniques for self-adjoint equations can be utilized. This modified method of characteristics can be combined with either finite difference or finite element spatial discretizations.

In multiphase flow, the convective part is nonlinear. A similar operator-splitting technique to solve Equation (8) needs reduced time steps because the pure hyperbolic part may develop shocks. An operator-splitting technique has been developed for multiphase flows [16,17,18,23,24,25] which retains the long time steps in the characteristic solution without introducing serious discretization errors.

Ignoring gravity, the operator splitting gives the following set of equations:

$$\phi \frac{\partial \bar{S}}{\partial t} + \frac{d}{dS}\mathbf{f}^m(\bar{S}) \cdot \nabla \bar{S} \equiv \phi \frac{d}{d\tau}\bar{S} = 0 , \tag{21}$$

$$\phi \frac{\partial S}{\partial \tau} + \nabla \cdot (\mathbf{b}^m(S)S) - \nabla \cdot (D(S)\nabla S) = \mathbf{q}(\mathbf{x}, t) , \tag{22}$$

$t_m \le t \le t_{m+1}$, together with proper initial and boundary conditions. As noted earlier, the saturation S is coupled to the pressure/velocity equations, which will be solved by mixed finite element methods described in the last section.

The splitting of the fractional flow function into two parts: $\mathbf{f}^m(S) + \mathbf{b}(S)S$, is constructed [24] such that $\mathbf{f}^m(S)$ is linear in the shock region, $0 \le S \le S_1 < 1$, and $\mathbf{b}(S) \equiv 0$ for $S_1 \le S \le 1$. Further, Equation (21) produces the same unique physical solution after a shock has been fully developed as

$$\frac{\partial S}{\partial t} + \nabla \cdot (\mathbf{f}^m(S) + \mathbf{b}(S)S) = 0 \tag{23}$$

with an entropy condition imposed. This means that, for a fully developed shock, the characteristic solution of Equation (21) always will produce a unique solution and, as in the single-phase case, we may use long time steps Δt without loss of accuracy.

Unfortunately, the modified method of characteristics techniques described above generally do not conserve mass. Also, the proper method for treating boundary conditions in a conservative and accurate manner using these techniques is not obvious. Recently, M.A. Celia, T.F. Russell, I. Herrera, and the author have devised Eulerian-Lagrangian localized adjoint methods (ELLAM) [10,52], a set of schemes that are defined expressly for conservation of mass properties.

The ELLAM formulation was motivated by localized adjoint methods [9,50], which are one form of the optimal test function methods discussed above [1,18,21,24]. We briefly describe these methods. Let

$$Lu = f, \qquad x \in \Omega \text{ or } (x,t) \in \Omega, \tag{24}$$

denote a partial differential equation in space or space-time. Integrating against a test function ϕ, we obtain the weak form

$$\int_\Omega Lu\phi d\omega = \int_\Omega f\phi d\omega. \tag{25}$$

If we choose test functions ϕ to satisfy the formal adjoint equation $L^*\phi = 0$ and $\phi = 0$ on the boundary, except at certain nodes or edges denoted by ℓ_i, then integration by parts (the divergence theorem in higher dimensions) yields

$$\sum_i \int_{\ell_i} uL^*\phi d\omega = \int_\Omega f\phi d\omega. \tag{26}$$

Various different test functions can be used to focus upon different types of information. Herrera has built an extensive theory around this concept; see [50] for references. The theory is quite general and can deal with situations where distributions do not apply, such as when both u and ϕ are discontinuous.

As in the work of Demkowitz and Oden [21], we want to localize these test functions to maintain sparse matrices. Certain choices of space-time test functions which are useful for linear equations of the form (2) have been described in [21,56].

We next extend the ELLAM techniques to the nonlinear multiphase flow equations (see e.g., [16,17,18,19,31,51,56,57]). We consider the divergence form of the multiphase flow equation given by Equation (8) with ϕ assumed constant in time and ignoring the gravity term for simplicity:

$$LS \equiv \phi\frac{\partial S}{\partial t} + \nabla\cdot(f_w\mathbf{v}) - \nabla\cdot D\nabla S = q_w, \quad x \in \Omega,\ t \in J, \tag{27}$$

$$(f_w\mathbf{v} - D\nabla S)\cdot\nu = h, \quad x \in \partial\Omega,\ t \in J, \tag{28}$$

where ν is the outward unit normal to the boundary $\partial\Omega$. Let $\Sigma = \Omega \times J$ denote the space-time domain. Then we obtain a weak formulation of Equation (27) by integrating against a test function $w = w(x,t)$. This yields a weak form, $\int_\Sigma (LS)w\,dxdt = \int_\Sigma q_w\,dxdt$. We obtain the specific equation

$$\begin{aligned}\int_\Omega\int_J \phi(Sw)_t dtdx + \int_J\int_\Omega \nabla\cdot(f_w\mathbf{v} - D\nabla S)wdtdx + \int_\Sigma D\nabla S\cdot\nabla w\,dxdt \\ - \int_\Sigma(\phi Sw_t + f_w\mathbf{v}\cdot\nabla w)dxdt = \int_\Sigma q_w w\,dxdt.\end{aligned} \tag{29}$$

Then, as in [57], we begin to study the time dependence of the potentially useful test functions by looking at a semidiscrete scheme on the time interval $J^{n+1} = [t^n, t^{n+1}]$ or over the space time region $\sum^{n+1} = \Omega \times J^{n+1}$. By applying the divergence theorem to (29), we obtain

$$\begin{aligned}\int_\Omega \phi S(x,t^{n+1})w(x,t^{n+1})dx + \int_{\Sigma^{n+1}} D\nabla S\cdot\nabla w\,dxdt + \int_{J^{n+1}}\int_{\partial\Omega}(f_w\mathbf{v} - D\nabla S)\cdot\nu\,wd\sigma dt \\ - \int_{\Sigma^{n+1}}(\phi Sw_t + f_w\mathbf{v}\cdot\nabla w)dxdt = \int_\Omega \phi S(x,t^n)w(x,t^n)dx + \int_{\Sigma^{n+1}} q_w w\,dxdt.\end{aligned} \tag{30}$$

In order to consider the ELLAM formulation from [10] directly, we should look for solutions of the adjoint to treat the term of the form

$$\int_{\Sigma^{n+1}} SL^* w\, dxdt = 0\,. \tag{31}$$

Since L is not a linear operator, we must perform some linearizations before we apply the analogue of Equation (31) to treat the fourth term in Equation (30).

Motivated by [23], we split f_w to obtain

$$f_w = \bar{f}(S)S + b(S)S. \tag{32}$$

For $0 \le S \le S^1, b(S)S$ is an antidiffusive term causing the fronts to tend to sharpen. For $S^1 \le S \le 1, b(S)S$ is a diffusive term. Using these definitions, the fourth term in Equation (30) can be written as

$$\begin{aligned}\int_{\Sigma^{n+1}} & S\left(\phi w_t + \left\{\bar{f}(S) + b(S)\right\} \mathbf{v} \cdot \nabla w\right) dxdt \\ &= \int_{\Sigma^{n+1}} S\left(\phi w_t + \bar{f}\mathbf{v} \cdot \nabla w\right) dxdt + \int_{\Sigma^{n+1}} Sb\mathbf{v} \cdot \nabla w\, dxdt.\end{aligned} \tag{33}$$

We cannot, in general, determine a test function w that satisfies $\phi w_t + \bar{f}\mathbf{v} \cdot \nabla w = 0$, even locally within each small space-time element. However, we will make a choice of test functions that will make this term small. Analysis of the size of this term will be presented elsewhere.

By choosing a test function $w(x, t)$ that is constant in time along the characteristics that define the moving Lagrangian frame of reference, we can make the first term in Equation (33) small. If the test function were a standard chapeau basis function in the x-direction it would also make second term in Equation (30) small. This would be an effective test function if the second term on the right side of Equation (33) were zero or were small. However, in many multiphase flow problems, the $b(S)\mathbf{v}$ term is not small and the use of characteristics has not symmetrized the form which is analagous to the form in Equation (30). As above, the use of an upwinded form of the test function for constant x will efficiently treat the b term from (33) together with the D term from Equation (30).

We thus arrive at a choice of $w(x, t)$ which is constant along the characteristics determined by the directional derivative along τ with $\bar{f}$ defined from the convex hull of f_w, as in Equation (32) (see also [23]). Using these test functions, our approximation scheme can be defined in the interior of the region on prisms as in [57].

Therefore, if Σ_{ij}^{n+1} does not meet the boundary of the total space-time computational region, the approximation procedure can be defined in Σ_{ij}^{n+1} as follows:

$$\begin{aligned}\int_{\Omega_{ij}} \phi S(x, t^{n+1}) w(x, t^{n+1}) dx &+ \int_{\Sigma_{ij}^{n+1}} (D\nabla S - b(\tilde{S}^{n+1})\mathbf{v}S) \cdot \nabla w(x, t^{n+1}) dxdt \\ = \int_{\Omega_{ij}^*} \phi S(x, t^n) w(x, t^n) dx &+ \int_{\Sigma_{ij}^{n+1}} q_w w\, dxdt,\end{aligned} \tag{34}$$

where Ω_{ij}^* is the region at t^n spanned by the feet of the characteristics whose heads are in Ω_{ij} and $\tilde{S}^{n+1}$ is some approximation of S^{n+1} obtained by extrapolating values from earlier time levels. As discussed in [56,57], if the time integrals along the characteristics are approximated by a one-point backward Euler rule at t^{n+1}, with $\Delta t = t^{n+1} - t^n$, we obtain the Modified Method of Characteristics (MMOC) formulation presented in Equation (21).

Differences between ELLAM and MMOC for linear partial differential operators have been discussed in [56,57].

These comparisons also apply in the nonlinear problems considered here. Reference [57] also contains excellent discussions of the errors involved in numerical integration along the characteristics via various tracking algorithms when the coefficients are spatially dependent, and for the terms arising when the adjoint equation is not completely satisfied.

Since one motivation for considering ELLAM instead of MMOC techniques was to obtain more accurate treatment of the boundary conditions, we note that we can extend our previous treatment of these terms for constant coefficients to the nonlinear case (see [19,31]).

Recently ELLAM techniques have been extended to a wide variety of applications [19,40,41,42,43,44,45,46,58]. Optimal order error estimates have been developed for advection [41], advection-diffusion [44], advection-reaction [40,41,42,43,44,45], and advection-diffusion-reaction [42,58] systems.

REFERENCES

1. J.W. Barrett and K.W. Morton, Approximate symmetrization and Petrov-Galerkin methods for diffusion-convection problems, *Comp. Meth. in Appl. Mech. and Eng.* **45** (1984), 97–122.
2. J. Bear, *Dynamics of Fluids in Porous Media*, Dover Publications, Inc., 1988.
3. B.A. Boyett, M.S. El-Mandouh, and R.E. Ewing, R.E., Local grid refinement for reservoir simulation, *Computational Methods in Geosciences, Frontiers in Applied Mathematics* (W.E. Fitzgibbon and M.F. Wheeler, eds.), SIAM, Philadelphia, PA, 1992, 15–28.
4. J.H. Bramble, R.E. Ewing, J.E. Pasciak, and A.H. Schatz, A preconditioning technique for the efficient solution of problems with local grid refinement, *Computer Methods in Applied Mechanics and Engineering* **67** (1988), 149–159.
5. J.H. Bramble and J.E. Pasciak, A Preconditioning Technique for Indefinite Systems Resulting from Mixed Approximations of Elliptic Problems, *Math. Comp.* **50** (1988), 1–18.
6. J.H. Bramble, J.E. Pasciak, and A.T. Vassilev, Analysis of the Inexact Uzawa Algorithm for Saddle Point Problems, *SIAM J. Numer. Anal.*, (to appear).
7. J.H. Bramble, J.E. Pasciak, J. Wang, and J. Xu, Convergence Estimates for Product Iterative Methods with Applications to Domain Decomposition, *Math. Comp.* **57**, (1991), 1–21.
8. F. Brezzi and M. Fortin, *Mixed and Hybrid Finite Methods*, Springer-Verlag, New York, 1991.
9. M.A. Celia, I. Herrera, E. Bouloutas, and J.S. Kindred, A new numerical approach for the advection-diffusive transport equation, *Numerical Methods for PDE's* **5** (1989), 203–226.
10. M.A. Celia, T.F. Russell, I. Herrera, and R.E. Ewing, An Eulerian-Lagrangian localized adjoint method for the advection-diffusion equation, *Advances in Water Resources*, **13** (1990), 187–206.
11. G. Chavent, A new formulation of diphasic incompressible flows in porous media, *Lecture Notes in Mathematics* **503**, Springer-Verlag, 1976.
12. G. Chavent and J. Jaffre, *Mathematical Models and Finite Elements for Reservoir Simulation: Single Phase, Multiphase and Multicomponent Flows Through Porous Media*, North-Holland, Amsterdam, 1986.
13. Z. Chen, R.E. Ewing, and M.S. Espedal, *Multiphase flow simulation with various boundary conditions*, *Computational Methods in Water Resources* (A. Peters, G. Wittum, B. Herrling, U. Meissner, C.A. Brebbia, W.G. Gray, and G.F. Pinder, eds.), Kluwer Academic Publishers, Netherlands, 1994, 925–932.
14. Z. Chen, R. Ewing, Y. Kuznetsov, R. Lazarov, and S. Maliassov, Multilevel Preconditioners for Mixed Methods for Second Order Elliptic Problems, *Technical Report ISC-94-14-MATH*, Institute for Scientific Computation, Texas A& MUniversity, 1994.
15. L. Cowsar, J. Mandel, and M. Wheeler, Balancing Domain Decomposition for Mixed Finite Elements, *Math. Comp.*, (to appear).
16. H.K. Dahle, Adaptive characteristic operator splitting techniques for convection-dominated diffusion problems in one and two space dimensions, *IMA Volumes in Mathematics and Its Applications* **II**, Springer Verlag, 1988, 77–88.
17. H.K. Dahle, M.S. Espedal, and R.E. Ewing, Characteristic Petrov-Galerkin subdomain methods for convection diffusion problems, *IMA Volume 11, Numerical Simulation in Oil Recovery* (M.F. Wheeler, ed.), Springer-Verlag, Berlin, 1988, 77–88.

18. H.K. Dahle, M.S. Espedal, R.E. Ewing, and O. Sævareid, Characteristic adaptive sub-domain methods for reservoir flow problems, *Numerical Methods for Partial Differential Equations* **6** (1990), 279–309.
19. H. Dahle, R.E. Ewing, and T.F. Russell, Eulerian-Lagrangian localized adjoint methods for a nonlinear advection-diffusion equation, *Comput. Meth. Appl. Mech. Eng.*, (to appear).
20. G. Dagan, *Flow and Transport in Porous Formations*, Springer-Verlag, Berlin-Heidelberg, 1989.
21. L. Demkowitz and J.T. Oden, An adpative characteristic Petrov-Galerkin finite element method for convection-dominated linear and nonlinear parabolic problems in two space variables, *Comp. Meth. in Appl. Mech. and Eng.* **55** (1986), 63–87.
22. J. Douglas, Jr. and T.F. Russell, Numerical methods for convection dominated diffusion problems based on combining the method of Characteristics with Finite Element or Finite Difference Procedures, *SIAM J. Numer. Anal.* **19** (1982), 871–885.
23. M.S. Espedal and R.E. Ewing, Petrov-Galerkin subdomain methods for two-phase immiscible flow, *Comp. Meth. Appl. Mech. and Eng.* **64** (1987), 113–135.
24. M.S. Espedal, R.E. Ewing, and T.F. Russell, Mixed methods, operator splitting, and local refinement techniques for simulation on irregular grids, *Proceedings 2nd European Conference on the Mathematics of Oil Recovery* (D. Guerillot and O. Guillon, eds.), Editors Technip, Paris, 1990, 237–245.
25. M.S. Espedal, R.E. Ewing, T.F. Russell, and O. Sævareid, Reservoir simulation using mixed methods, a modified method of characteristics, and local grid refinement, *Proceedings of Joint IMA/SPE European Conference on the Mathematics of Oil Recovery*, Cambridge University, July 25–27 1989.
26. M.S. Espedal, R. Hansen, P. Langlo, O. Sævareid, and R.E. Ewing, Heterogeneous porous media and domain decomposition methods, *Proceedings 2nd European Conference on the Mathematics of Oil Recovery* (D. Guerillot and O. Guillon, eds.), Paris, Editors Technip, 1990, 157–163.
27. M.S. Espedal, P. Langlo, O. Sævareid, E. Geslifosa, and R. Hansen, Heterogeneous reservoir models, local refinement, and effective parameters, *Proceedings of Eleventh SPS Symposium on Reservoir Simulation, SPE 21231*, 1991, 307–316.
28. M.S. Espedal, P. Langlo, O. Sævareid, E. Gislefoss, and R. Hansen, Heterogeneous reservoir models, local refinements, and effective parameters, *SPE 21231, Proceedings of Eleventh SPE Symposium on Reservoir Simulation*, Anaheim, CA, 1991, 307–316.
29. R.E. Ewing, Adaptive grid refinements for transient flow problems, *Adaptive Methods for Partial Differential Equations* (J.E. Flaherty, P.J. Paslow, M.S. Shephard, and J.D. Vasilakis, eds.), SIAM, Philadelphia, PA, Chapter 14, 1989, 194–205.
30. R.E. Ewing, Domain decomposition techniques for efficient adaptive local grid refinement, *Domain Decomposition Methods* (T.F. Chan, R. Glowinski, J. Periaux, and O.B. Widlund), SIAM, Philadelphia, PA, 1989, 192–206.
31. R.E. Ewing, Operator splitting and Eulerian-Lagrangian localized adjoint methods for multiphase flow, *The Mathematics of Finite Elements and Application* (J. Whiteman, ed.), *MAFELAP* **199**, Academic Press, Inc., San Diego, CA, 1991, 215–232.
32. R.E. Ewing, B.A. Boyett, D.K. Babu, and R.F. Heinemann, Efficient use of locally refined grids for multiphase reservoir simulations, *SPE 18413, Proceedings Tenth SPE Symposium on Reservoir Simulation*, Houston, TX, February 6–8, 1989, 55–70.
33. R.E. Ewing, R.D. Lazarov, T.F. Russell, and P.S. Vassilevki, Local refinement via domain decomposition techniques for mixed finite element methods with rectangular Raviart-Thomas elements, *Domain Decompositions for Partial Differential Equation* (T. Chan, R. Glowinski, J. Periaux, and O. Widlund, eds.), SIAM, Philadelphia, 1990, 98–114.
34. R.E. Ewing, R.D. Lazarov, and P.S. Vassilevski, Local refinement techniques for elliptic problems on cell-centered grids, I: Error analysis, *Math. Comp.* **56(194)** (1991), 437–462.
35. R.E. Ewing, R.D. Lazarov, and P.S. Vassilevski, P.S., Local refinement techniques for elliptic problems on cell-centered grids, II: Optimal order two-grid iterative methods, *Numer. Linear Algebra Appl.*, **1(4)** (1994).
36. R.E. Ewing, R.D. Lazarov, and J. Wang, Superconvergence of the velocities along the Gaussian lines in the mixed finite element methods, *SIAM J. Numer. Anal.* **28(4)** (1991), 1015–1029.
37. R.E. Ewing, T.F. Russell, and M.F. Wheeler, Simulation of miscible displacement using mixed methods and a modified method of characteristics, *SPE 12241, Proceedings Seventh SPE Symposium on Reservoir Simulation*, San Francisco, CA, 1983, 71–82.
38. R.E. Ewing, T.F. Russell, and M.F. Wheeler, Convergence analysis of an approximation of miscible displacement in porous media by mixed finite elements and a modified method of characteristics,

Comp. Meth. Appl. Mech. Eng. **47** (1984), 73–92.
39. R.E. Ewing, T.F. Russell, and L.C. Young, An anisotropic Coarse-grid dispersion model of heterogeneity and viscous fingering in five-spot miscible displacement that matches experiments and fine-grid simulations, *SPE 18441, Proceedings Tenth SPE Symposium on Reservoir Simulation*, Houston, Texas, Texas, February 6–8, 1989, 447–466.
40. R.E. Ewing and H. Wang, An Eulerian-Lagrangian localized adjoint method for variable-coefficient advection-reactionproblems, *Advances in Hydro-Science and Engineering* (S. Wang, ed.) **1, Part B**, University of Mississippi Press, 1993, 2010–2015.
41. R.E. Ewing and H. Wang, Eulerian-Lagrangian localized adjoint methods for linear advection or advection-reaction equations and their convergence analysis, *Computational Mechanics* **12** (1993), 97–121.
42. R.E. Ewing and H. Wang, Eulerian-Lagrangian localized adjoint methods for variable coefficient advection-diffusive-reactive equations in groundwater contaminant transport, *Advances in Optimization and Numerical Analysis* (S. Goméz and J.P. Hennart, eds.) **275**, Kluwer Academic Publishers, Netherlands, 1994, 185–205.
43. R.E. Ewing and H. Wang, Eulerian-Lagrangian localized adjoint methods for reactive transport in groundwater, *Environmental Studies: Mathematical Computational, and Statistical Analysis, IMA Volume in Mathematics and its Application* (M.F. Wheeler, ed.), **79**, Springer-Verlag, Berlin, 1995, 149–170.
44. R.E. Ewing and H. Wang, Optimal-order convergence rate for Eulerian-Lagrangian localized adjoint method for reactive transport and contamination in groundwater, *Numerical Methods in PDE's* **11(1)** (1995), 1–31.
45. R.E. Ewing and H. Wang, An optimal-order estimate for Eulerian-Lagrangian localized adjoint methods for variable-coefficient advection-reaction problems, *SIAM J. Numer. Anal.* (to appear).
46. R.E. Ewing, H. Wang, and T.F. Russell, Eulerian-Lagrangian localized adjoint methods for convection-diffusion equations and their convergence analysis, *IMA J. Numerical Analysis*, (to appear).
47. R.E. Ewing and J. Wang, Analysis of mixed finite element methods on locally refined grids, *Numerische Mathematik* **63** (1992), 183–194.
48. R.E. Ewing and M.F. Wheeler, Computational aspects of mixed finite element methods, *Numerical Methods for Scientific Computing* (R. Steelman, ed.), North-Holland, Amsterdam, 1983, 163–172.
49. J. Furtado, J. Glimm, W.B. Lindquist, and L.P. Pereira, Characterization of mixing length growth for flow in heterogeneous porous media, *Proceedings of Eleventh SPE Symposium on Reservoir Simulation*, Anaheim, CA, 1991, 317–322.
50. I. Herrera, Unified formulation of numerical methods I. Green's formula for operators in discontinuous fields, *Numerical Methods for PDE's* **1** (1985), 25–44.
51. I. Herrera and R.E. Ewing, Localized adjoint methods: Application of multiphase flow problems, *Proceedings of the Fifth Wyoming Enhanced Oil Recovery Symposium*, Casper, Wyoming, May 10–11 1989, 147–166.
52. I. Herrera, R.E. Ewing, M.A. Celia, and T.F. Russell, Eulerian-Lagrangian localized adjoint method: The theoretical framework, *Numerical Methods for PDE's* **9** (1993), 431–457.
53. D. Hittel, *Fundamentals of Soil Physics*, Academic Press, 1980.
54. O. Langlo and M. Espedal, Heterogeneous reservoir models, two-phase immiscible flow in 2-d, *Mathematical Modeling in Water Resources, Computational Methods in Water Resources* (T.F. Russell, R.E. Ewing, C.A. Brebbia, W.G. Gray, and G.F. Pinder, eds.) **IX(2)**, Elsevier Applied Science, London, 1992, 71–80.
55. T.F. Russell, The time-stepping along characteristics with incomplete iteration for Galerkin approximation of miscible displacement in porous media, *SIAM J. Numer. Anal.* **22** (1985), 970–1013.
56. T.F. Russell, Eulerian-Lagrangian localized adjoint methods for advection-dominated problems, *Proceedings of 13th Biennial Conference on Numerical Analysis*, Dundee, Scotland, June 27–30 1989. Pitmann Publishing Company.
57. T.F. Russell and R.V. Trujillo, Eulerian-Lagrangian localized adjoint methods with variable coefficients in multiple divergences, *Proceedings 7th International Conference on Computational Methods in Water Resources*, Venice, Italy, (to appear).
58. H. Wang, R.E. Ewing, and M.A. Celia, Eulerian-Lagrangian localized adjoint methods for reactive transport with biodegradation, *Numerical Methods for PDE's* **11(3)** (1995), 229–254.

DIRECT AND INVERSE MODELING OF MULTIPHASE FLOW SYSTEMS

S. Finsterle

Lawrence Berkeley National Laboratory
University of California
Earth Sciences Division, MS 50E
Berkeley, CA, 94720
(510) 486-5205

SUMMARY

A modeling study is presented which demonstrates how the combination of simulation and optimization techniques can be used to improve the design of a multi-component remediation system. A series of computer codes has been developed at the Lawrence Berkeley National Laboratory to solve forward and inverse problems in groundwater hydrology. Simulations of non-isothermal, three-phase flow of volatile organic compounds in three-dimensional heterogeneous media were performed. Inverse modeling capabilities have been developed which can be used for both automatic model calibration and optimization of remediation schemes. In this study, we discuss a sequence of simulations to demonstrate the potential use of numerical models to design and analyze cleanup of a contaminated aquifer.

INTRODUCTION

Industrial processes involving metallurgical operations frequently cause contamination of soil and groundwater due to the release of solvents such as chlorinated hydrocarbons and other volatile organic compounds (VOCs). In the subsurface, these chemicals tend to remain as nonaqueous phase liquids (NAPL), and are transported under multiphase flow conditions due to gravitational, viscous and capillary forces. Other physiochemical processes also taking place include evaporation into the soil gas phase and dissolution into the groundwater. In the gaseous phase, VOCs are transported by multi-component diffusion in addition to density and pressure driven advection. In the liquid phase, dissolved contaminants may cause density effects which affect advective transport. Simulation of contaminant behavior in the subsurface requires accounting for all these mechanisms. Accurate process description and robust implementation of multiphase flow in porous media including phase changes are essential for the modeling of air sparging, steam flooding, and vapor extraction as some of the most effective remediation technologies such as air sparging, steam flooding, etc.

The design of a cleanup operation for a contaminated aquifer poses problems of a hydrological, technical, environmental, and economic nature. Decisions have to be made regarding aquifer testing for site characterization, the design of a containment system, the choice of an appropriate remediation technology, which usually requires specifying an injection and pumping schedule, and the layout of a monitoring system. First, it is crucial to characterize the relevant hydrogeological features at a given site. Information about the system state under natural or test conditions have to be collected to calibrate the numerical model. Secondly, a remediation technology is chosen, and cleanup operations are simulated to improve the understanding of the fate of the contaminants in the subsurface. A comparison between alternative approaches is performed to design and optimize the remediation strategy.

Most of the tasks described above require solving some kind of optimization problem, i.e. the maximization or minimization of a performance measure. An incomplete list is given in Table 1. The testing and data collection program can be optimized to obtain as much information about the plume location and hydrogeologic properties of the aquifer as possible. The numerical model to be used for the subsequent studies has to be calibrated against the available data. Model calibration involves minimizing the difference between the observed and the calculated system state. It is important to realize that the result of the subsequent modeling studies as well as the conclusions drawn regarding the optimum remediation system design depend on the accuracy with which the relevant features at the site are represented in the numerical model. Once an appropriate remediation technology has been selected, operational details such as the layout of the containment system, temperature and rate of air or steam injection, the pumping schedule, etc. can be optimized to increase the effectiveness and efficiency of cleanup. This step requires the definition of a cost function which comprises actual operational costs as well as hypothetical costs for residual contamination. It is obvious that selection of the remediation technology and optimization of the operational parameters is an iterative process. The overall performance of the optimized remediation alternatives are compared to determine the most appropriate strategy. Finally, the design of a monitoring system is improved by evaluating the amount of information that can be obtained from a limited number of sensors.

Table 1 Optimization Problems in Aquifer Remediation

Objective	Task	Simulation Problem	Optimization Problem
data collection for plume mapping and site characterization	test design, data collection	simulate plume spreading and hydraulic tests	maximize sensitivity of data with respect to parameters of interest
site characterization	model calibration	simulate system state under natural or test conditions	minimize difference between measured and calculated system state
effectiveness, efficiency	technology selection, determination of operational parameters	simulate cleanup operation	minimize cost function comprising actual and hypothetical costs
control	design of monitoring system	simulate migration of residual contamination	maximize obtainable information

The modeling study presented in this paper aims at demonstrating how the combination of simulation and optimization techniques can be used to improve the design of a multi-component remediation system. Computer programs have been developed at the Lawrence Berkeley National Laboratory to solve forward and inverse problems in groundwater hydrology. Simulations of non-isothermal, three-phase flow of volatile organic compounds in three-dimensional heterogeneous porous media were performed. Inverse modeling capabilities have been developed which can be used for both automatic model calibration and optimization of remediation schemes. We first discuss the modeling approach used to simulate multiphase contaminant transport, before we provide some information about the optimization algorithms. A sequence of simulations of a hypothetical contaminated site demonstrates the potential use of numerical models to design and evaluate cleanup operations.

MODELING MULTIPHASE CONTAMINANT TRANSPORT

Sophisticated numerical simulators have been developed at the Lawrence Berkeley National Laboratory to model multiphase, multi-component flow and transport in the subsurface. The main computer code used in this study is T2VOC [1] for three-phase (gas, water, NAPL), non-isothermal flow of water, air, and a volatile organic compound (VOC) in three-dimensional heterogeneous porous media. T2VOC is an extension of the TOUGH2 general-purpose simulation program [2] which uses a general integral finite difference formulation for mass and energy balance equations. The balance equations for three mass components κ (κ= w-water, a-air, c-chemical) and heat (κ= h-heat) are written in integral form for some flow region, V_n, having a surface area Γ_n, as follows:

$$\frac{\mathrm{d}}{\mathrm{d}t}\int_{V_n} M^{\kappa}\,\mathrm{d}V_n = \int_{\Gamma_n} \mathbf{F}^{\kappa} \bullet \mathbf{n}\,\mathrm{d}\Gamma_n + \int_{V_n} q^{\kappa}\,\mathrm{d}V_n \quad . \tag{1}$$

Here M^{κ} is the amount of component κ per unit porous medium volume, $\mathbf{F}^{\kappa}$ is the total flux of component κ into V_n, $\mathbf{n}$ is the inward unit normal vector on Γ_n, and q^{κ} is the generation rate per unit volume. The three fluid components air, water, and VOC may be present in different proportions in any of the three phases gas, aqueous, and NAPL. Thus, the mass accumulation terms (κ=w, a, c) contain a sum over the three phases β (β = g-gas, w-aqueous, n-NAPL):

$$M^{\kappa} = \phi \sum_{\beta} S_{\beta}\,\rho_{\beta}\,X_{\beta}^{\kappa} + \delta^{\kappa c}\,\rho_b\,\rho_w\,X_w^c\,K_D \quad . \tag{2}$$

Here ϕ is the porosity, S_{β} is the saturation occupied by phase β, ρ_{β} is β-phase density, and X_{β}^{κ} is the mass fraction of component κ in phase β. The organic chemical accumulation term (κ=c, $\delta^{\kappa c}$ = 1) includes the effect of linear equilibrium adsorption onto the solid phase, where ρ_b is the dry bulk density of the soil, and K_D is the solid-aqueous distribution coefficient for the organic chemical [3]. The heat accumulation term (κ=h) includes contributions from both the solid and the fluid phases:

$$M^h = (1 - \phi)\,\rho_R\,C_R\,T + \phi \sum_{\beta} S_{\beta}\,\rho_{\beta}\,u_{\beta} \; , \tag{3}$$

where ρ_R is the soil grain density, C_R is the heat capacity of the soil grains, T is the temperature, and u_β is the specific internal energy of phase β.

Each phase flows in response to pressure and gravitational forces according to the multiphase version of Darcy's law which includes the effects of relative permeability and capillary pressure between the phases. The mass flux terms of component κ (κ=w, a, c) include a sum of contributions from the three phases β (β = g, w, n):

$$\mathbf{F}^\kappa = \sum_\beta \mathbf{F}_\beta^\kappa \, . \tag{4}$$

For aqueous and NAPL phases (β = w, n) we have $\mathbf{F}_\beta^\kappa = \mathrm{X}_\beta^\kappa \mathbf{F}_\beta$, where the phase fluxes $\mathbf{F}_\beta$ are given by

$$\mathbf{F}_\beta = -\mathrm{k}\frac{k_{r\beta}\,\rho_\beta}{\mu_\beta}\left(\nabla P_\beta - \rho_\beta\, \mathbf{g}\right) \, . \tag{5}$$

Here k is the absolute permeability, $k_{r\beta}$ is the relative permeability of phase β, μ_β is the β-phase dynamic viscosity, P_β is the fluid pressure in phase β, and $\mathbf{g}$ is the gravitational acceleration vector. Mass flux in the gas phase includes advection and multicomponent diffusion:

$$\mathbf{F}_g^\kappa = -\mathrm{k}\left(1 + \frac{b}{P_g}\right)\frac{k_{rg}\,\rho_g}{\mu_g} X_g^\kappa \left(\nabla P_g - \rho_g\, \mathbf{g}\right) + \mathbf{J}_g^\kappa \, . \tag{6}$$

Here b is the Klinkenberg factor [4], which accounts for increased effective gas permeability at low pressures, and $\mathbf{J}_g^\kappa$ is the diffusive mass flux of component κ in the gas phase. The diffusive flux is driven by a mass fraction gradient, ∇X_g^κ, and is a function of saturation, tortuosity, and the multicomponent molecular diffusion coefficient (for details see [1]).

Aqueous and gas phase pressures are related by

$$P_w = P_g + P_{cgw} \, , \tag{7}$$

where $P_{cgw} < 0$ is the gas-water capillary pressure. The NAPL phase pressure is related to the gas phase pressure by

$$P_n = P_g + P_{cgn} \, , \tag{8}$$

where $P_{cgn} < 0$ is the gas-NAPL capillary pressure. Assuming the NAPL is the intermediately wettable phase, the NAPL-water capillary pressure, P_{cnw}, is given by

$$P_{cnw} = P_{cgw} + P_{cgn} = P_w + P_n \, . \tag{9}$$

The heat flux includes both conduction and convection

$$\mathbf{F}^h = - K \nabla T + \sum_{\beta} h_{\beta} \mathbf{F}_{\beta} \ , \tag{10}$$

where K is the overall porous medium thermal conductivity, h_{β} is the β-phase specific enthalpy, and $\mathbf{F}_{\beta}$ is the total mass flux of phase β.

Finally, the sink and source term in Eq. (1) includes external contributions, e.g. the release of contaminant or the injection of steam as well as the production of a three-phase fluid mixture during forced vacuum extraction. In addition, a sink term may be included that permits an approximate representation of biodegradation of VOCs (for details see [1]).

Water properties in the liquid and vapor state are calculated using steam table equations given by the International Formulation Committee [5]. Thermophysical properties of the NAPL phase such as saturated vapor pressure and viscosity are calculated as functions of temperature, while specific enthalpy and density are computed as functions of both temperature and pressure. Gas phase thermophysical properties including molecular diffusivities are considered to be functions of temperature, pressure, and gas phase composition. A general equation of state is provided to compute the necessary NAPL/VOC thermophysical and transport properties. Based on semi-empirical corresponding states methods, the chemical parameters are calculated as functions of critical properties such as the critical temperature and critical pressure, which are readily available for many substances (see for example [6]). This approach makes T2VOC applicable to a variety of contamination problems involving different NAPLs that are either denser or lighter as water, and that have different vapor pressures and solubilities.

For numerical solution, the balance equations (1) are discretized in space based on an integral finite difference formulation [7]. Time is discretized fully implicitly using first-order backward finite differences. Discretization results in a set of nonlinear coupled algebraic equations which are solved simultaneously by means of Newton-Raphson iterations. A conjugate gradient algorithm is used to solve the linear equations arising at each iteration step [8].

OPTIMIZATION ALGORITHM

As discussed in the introduction, we use the simulation capabilities in combination with optimization techniques to perform model calibration and to support the design of a remediation system. The first task is referred to as inverse modeling, whereas optimization of a remediation design is a prevalent groundwater management problem. In both applications a performance measure is either minimized or maximized by adjusting certain input parameters or design variables. For example, model calibration consists of reducing the differences between the simulation results and the measurements, such as water potentials or temperatures, by adjusting the model input parameters, such as the absolute permeability or thermal conductivity of the porous medium. A cleanup operation can be improved by minimizing, for example, the remediation time, which can be achieved by increasing pumping rates or steam temperature. Increasing pumping rates or steam temperature, however, leads to higher energy costs. Therefore, the objective function to be minimized should also reflect these costs in order to obtain an optimal remediation design.

Solving an optimization problem occurs in two steps. First, a performance measure has to be defined which is an arbitrary function of T2VOC output variables, and which depends on certain T2VOC input parameters. This objective function contains contributions from different sources which have to be appropriately weighted against each other. Secondly, a minimization algorithm is needed which is capable of updating T2VOC input parameters in order to reduce the value of the objective function. Note that the objective function is usually highly non-linear, and - in certain cases - even discontinuous. In the remainder of this section we present the general form of the objective function and discuss the bases of the minimization algorithm.

We first introduce the residual vector **r** which holds the differences between the measured and calculated state variables in the case of model calibration, and the differences between the desired and calculated costs in the case of design optimization:

$$\mathbf{r} = \mathbf{y}^* - \mathbf{y}(\mathbf{p}) \ . \tag{11}$$

In the case of model calibration, the vector **y*** contains the actual data. For test design, the elements of vector **y*** are synthetically generated data, and for management problems, they represent target quantities such as zero costs or the maximum possible amount of contaminant removed from the aquifer. Note that **y(p)** is a function of the input parameters which are summarized in vector **p**. The elements of the parameter vector **p** are either hydrogeologic properties, initial and boundary conditions, or pumping and injection rates. In general the residuals have different units and may cover many orders of magnitudes. This requires specifying a weighting matrix of dimensions $m \times m$, where m is the total number of residuals. In inverse modeling, the weighting matrix **W** can be considered to be the inverse of the covariance matrix **C** holding measurement errors:

$$\mathbf{W} = \mathbf{C}^{-1} \ . \tag{12}$$

If no correlations are present, the vector of the weighted residuals **z** contains elements of the form

$$z_i = \frac{r_i^* - r_i}{c_i} \ , \tag{13}$$

where c_i is the square root of the i-th diagonal element of matrix **C**. The objective function to be minimized can be defined as follows:

$$Z = \sum_{i=1}^{m} \rho(z_i) \ . \tag{14}$$

In maximum likelihood theory, the function ρ reflects the negative logarithm of the probability density function of z. For normally distributed residuals, we obtain

$$\rho_i = x_i^2 \tag{15}$$

which yields the standard least squares estimator. The so-called L_1-estimator is obtained by minimizing the sum of the absolute values of the weighted residuals, thus

$$\rho_i = |x_i| \ . \tag{16}$$

The choice of an appropriate estimator is somewhat arbitrary. For inverse modeling applications, it is common to use least squares or one of the robust estimators described in [9]. For management problems, i.e. if r_i represents a cost function, the L_1-estimator is most frequently applied.

Next an algorithm has to be developed that minimizes the objective function Z. Note that the functions $\mathbf{y}(\mathbf{p})$ are highly nonlinear which immediately renders the direct solution of the minimization problem infeasible. The iterative procedure involves computing a correction vector $\Delta\mathbf{p}_k$ such that the new parameter set

$$\mathbf{p}_{k+1} = \mathbf{p}_k + \Delta\mathbf{p}_k \tag{17}$$

leads to a reduced objective function, $Z_{k+1} < Z_k$, at each iteration k. A number of minimization algorithms are available to solve non-linear least-squares problems as defined by Eqs. (14) and (15) [10]. One of the most general and most robust methods is the Levenberg-Marquardt modification of the Gauss-Newton minimization algorithm [11, 12]. This procedure involves solving the following system of equations:

$$(\mathbf{J}_k^T \mathbf{W} \mathbf{J}_k + \lambda \cdot \mathbf{D}_k) \cdot \Delta\mathbf{p}_k = - \mathbf{J}_k^T \mathbf{W} \mathbf{r}_k \ . \tag{18}$$

Here, $\mathbf{J}$ is the sensitivity matrix with elements $J_{ij} = \partial y_i / \partial p_j$. $\mathbf{D}$ denotes a matrix of order n (n being the number of parameters to be estimated) with elements equivalent to the diagonal elements of matrix $(\mathbf{J}_k^T \mathbf{W} \mathbf{J}_k)$. The basic idea of this method is to move in the parameter space along the steepest descent direction far from the minimum, and switching continuously to the Gauss-Newton algorithm as the minimum is approached. This is achieved by decreasing the scalar λ, known as the Levenberg parameter, if an iteration is successful and increasing it if the objective function is increased. The minimum of the objective function, which may be only a local one, is detected iteratively by solving Eq. (18) for $\Delta\mathbf{p}_k$. The parameter set is then updated according to Eq. (17). Under the assumption of normality and linearity, a detailed error analysis of the final residuals and the estimated parameters can be conducted (for details see [13]).

Note that the Levenberg-Marquardt algorithm is also general enough to solve management problems usually formulated in terms of the L_1-estimator, i.e. using Eqs. (14) and (16). However, there are two shortcomings that may prevent one from using the Levenberg-Marquardt method for arbitrary cost functions. First, in the presence of multiple local minima it cannot be assured that the optimum solution is detected. While this difficulty can be alleviated, the uncertainty in the definition of the cost function and the qualitative nature of the design process in general does not make it absolutely necessary to identify the global minimum. The second problem, however, precludes the use of the Levenberg-Marquardt algorithm for certain types of cost functions. The calculation of matrix $\mathbf{J}$ requires that the functions $\mathbf{y}(\mathbf{p})$ are continuous. In remediation technology, the cost functions are often discontinuous or even discrete. For example, installation costs depend on the number of wells being drilled, making the total cost a discontinuous function if the number of wells is subjected to the optimization process. In these cases, a minimization method is required that does not rely on the calculation of derivatives. The method of simulated annealing is able to handle discontinuous objective functions that exhibit many local minima. The term simulated annealing is used by analogy to thermodynamics, specifically with the way metals slowly cool and anneal. The analogy and algorithm is described in [14]. The basic idea is to take a random step $\Delta\mathbf{p}$ which is always accepted if it leads to a reduction of the objective function. Furthermore, an uphill move is sometimes accepted with a certain, decreasing probability enabling the algorithm to escape from local minima. There is no need to calculate derivatives which makes simulated annealing applicable to several types of objective functions (discrete, discontinuous, or continuous). The main drawback of simulated annealing is its inefficiency, a result of the randomness of step $\Delta\mathbf{p}$ which almost always proposes an uphill move.

The minimization algorithms discussed in this section have been implemented in a computer program named ITOUGH2 [1]. ITOUGH2 performs automatic model calibration for the TOUGH2 family of codes [2], including T2VOC [1]. Furthermore, ITOUGH2 supports sensitivity analysis, the design of experiments, and provides uncertainties for model predictions using linear error propagation analysis or Monte Carlo simulations. It can also be used to solve groundwater management problems such as the design of a cleanup operation [15]. In the remainder of this paper we demonstrate that the combination of sophisticated process simulation and efficient and robust optimization techniques is a powerful tool to study subsurface flow and transport problems.

MODELING STUDY

The illustrative sample problem discussed in this modeling study deals with contaminant migration from a dense solvent trichloroethylene (TCE) spill into the unsaturated zone of a highly heterogeneous soil, followed by a remediation operation using air sparging and soil vapor extraction. The effect of subsurface barriers on cleanup efficiency is also studied.

We consider a shallow heterogeneous deposit containing both a saturated and a unsaturated zone. The water table is at depth of 12 m and exhibits a head gradient of about 0.03. The model domain has dimensions of 30 m × 30 m × 20 m in the x, y, and z directions, respectively, and is discretized in 15 × 15 × 20 = 4500 grid blocks. A spatially correlated permeability field is generated using simulated annealing techniques. A log-normal permeability distribution is assumed with a geometric mean of 10^{-10} m^2 and a standard deviation of one order of magnitude. During the annealing process, the permeability field is rearranged in a random manner, and the difference between the resulting variogram and a prescribed variogram is minimized using the algorithm described in the previous section. We prescribed a spherical variogram with a correlation length of 20 m in the horizontal direction. No vertical correlation is considered. The generated permeabilities were then ordered into 9 groups, covering values from 10^{-12} to 10^{-8} m^2. Porosities are in the range from 0.27 to 0.43, and are assumed to vary in proportion to the logarithm of permeability. The capillary pressure was scaled in inverse proportion to the logarithm of permeability [16]. The permeability distribution is shown in Figure 1.

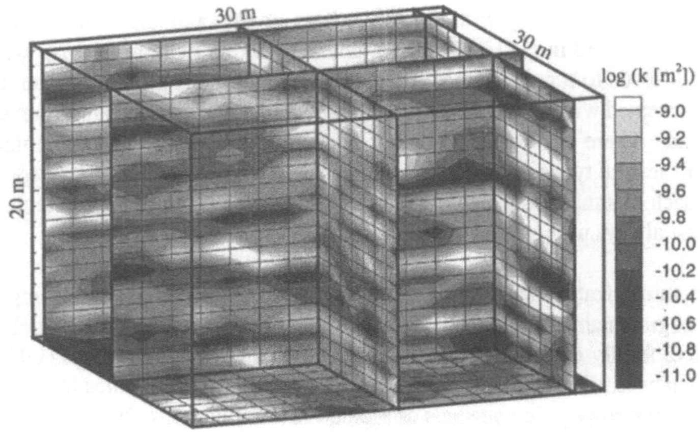

Fig. 1 Permeability distribution

The first part of the simulation aims at developing a gravity-capillary equilibrium in a saturated-unsaturated flow system that exhibits a pressure gradient in the saturated zone. An atmospheric boundary condition is applied at the surface, and a single-phase liquid Dirichlet boundary condition with a hydrostatic pressure distribution is specified at the lower part of the two opposite faces of the model. The depth to the water table is fixed at -12 m at the back face, and at -13 m at the front face. The left and right vertical faces as well as the bottom are modeled as no flow boundaries. After reaching steady-state flow conditions, the simulation continues by spilling a total of 6000 kg of TCE at the center of the model domain, 2 m below ground surface, for a period of 2 weeks.

The next part of the simulation allows for free redistribution of TCE over a 6 week period under the combined action of gravitational, capillary and pressure forces. The NAPL phase slowly slumps downward and also spreads due to capillary forces, leading to somewhat reduced NAPL contents compared to the situation immediately after the spill. Part of the NAPL plume evaporates, and contaminated soil gas of higher density flows downwards until it reaches the water table where it spreads horizontally. This density effect prevents the contaminant from being discharged naturally to the atmosphere. High aqueous TCE contents are observed near the water table, and the contaminated groundwater flows downstream and deeper into the saturated zone due to increased water density. Figure 2 shows the NAPL content (the product of NAPL saturation and porosity), TCE content in the gas phase (the product of VOC concentration in the gas phase [kg/m^3], gas saturation, and porosity), and the TCE content in the aqueous phase (the product of TCE concentration in the aqueous phase [kg/m^3], water saturation, and porosity) after the redistribution period. Note that the concentration of TCE in one phase is related to the concentration in another phase due to an assumption of local chemical equilibria. TCE vapors spreading into the system lead to a contamination of the pore water due to the solubility of the chemical in water. At this point in time, a preliminary cleanup operation is designed consisting of air sparging in combination with soil vapor extraction. In addition, four vertical cutoff walls are emplaced to contain the contaminant plume.

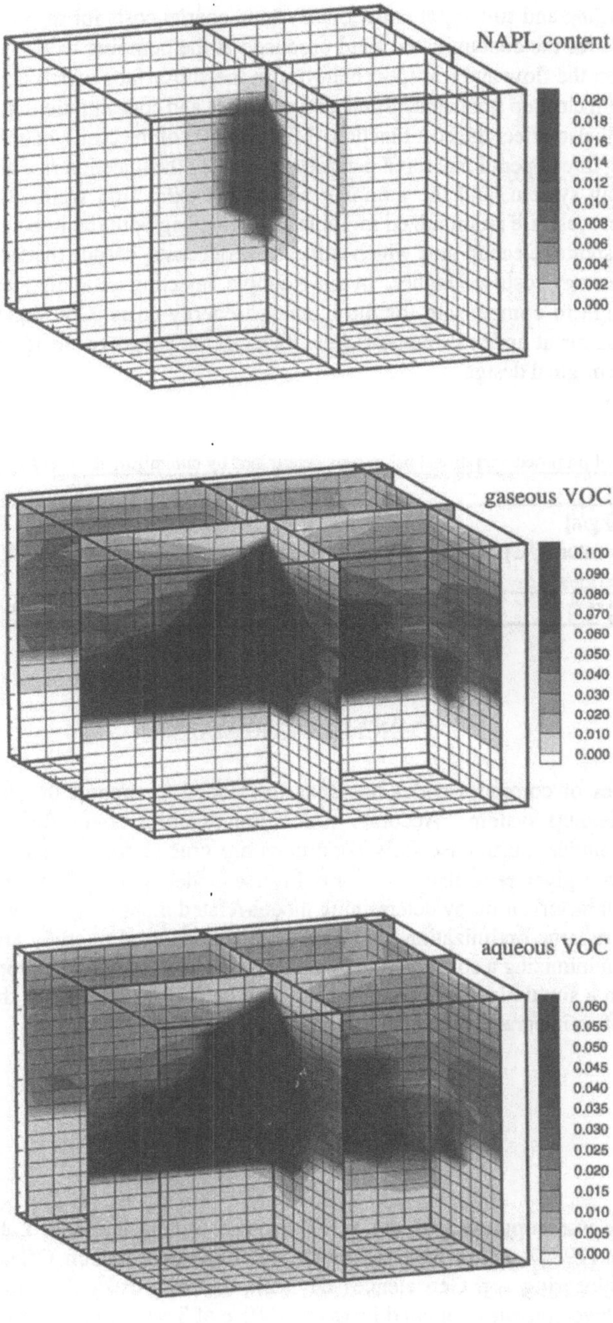

Fig. 2 **NAPL content (top), vaporized VOC content (center), and dissolved VOC content (bottom) after redistribution**

In the next step of the process, a cost function is defined which includes pumping costs for the air sparging and soil vapor extraction system, energy costs for the heating system, and treatment costs for the contaminated fluid produced at the pumping well. Note that all these costs depend on the flow and transport behavior in the subsurface which determines the time required to reach cleanup standards. Simulated pressures and concentrations at the injection and production well also affect the cost function. Minimization of the cost function is accomplished by determining three operational input parameters, namely the injection rate and temperature of the air sparging system, and the pumping rate at the extraction well. The results of the optimization process are summarized in Table 2. Increasing both the injection and extraction rate leads to a shorter cleanup time which reduces overall costs, despite the fact that the specific pumping and energy costs are higher. In this example, injecting air at high temperatures is not effective enough to compensate for the increased energy costs. Consequently, ITOUGH2 proposes to use air at ambient temperatures. Overall the proposed design is about twice as efficient as the original design.

Table 2 Set of proposed operational parameters determined by minimizing the cost function

Parameter	Original Design	Optimized Design
Injection rate [kg/s]	0.01	0.03
Injection temperature [°C]	90.00	40.00
Extraction rate [kg/s]	0.01	0.04
Cost function ["$"]	1090.00	550.00

CONCLUDING REMARKS

A series of computer codes has been developed to support the design of a multi-component cleanup system. Accurate simulation capabilities for flow and transport of contaminants under multiphase flow conditions are crucial for predicting the fate of the contaminant for a given remediation scenario. Inverse modeling techniques provide an essential tool for site characterization by determining model-related input parameters for the numerical simulator. The same optimization algorithms can also be used to study alternative cleanup operations by minimizing a cost function that reflects the efficiency of the proposed design. The results of such a simulation-optimization study can be used as a basis for the evaluation and optimization of an integrated remediation technology.

ACKNOWLEDGMENT

This work was supported by the WIPP Project, Sandia National Laboratories, under Document No. 129847, and by the Director, Office of Energy Research, Office of Basic Energy Sciences, Engineering and Geosciences Division, and in part by the In Situ Remediation Technology Development Integrated Program, Office of Technology Development, Office of Environmental Restoration and Waste Management, U.S. Department of Energy, under contract DE-AC03-76SF00098.

REFERENCES

[1] Falta, R.W., K. Pruess, S. Finsterle, and A. Battistelli, T2VOC User's Guide, Report No. LBL-36400, Lawrence Berkeley National Laboratory, Berkeley, Calif., March 1995.

[2] Pruess, K., TOUGH2 - A General Purpose Numerical Simulator for Multiphase Fluid and Heat Flow, Report No. LBL-29400, Lawrence Berkeley National Laboratory, Berkeley, Calif., May 1991.

[3] Freeze, R.A. and J.A. Cherry, Groundwater, Prentice-Hall, Englewood Cliffs, N.J., 1979.

[4] Klinkenberg, L.J., The Permeability of Porous Media to Liquids and Gases, in *API Drilling and Production Practice*, 200-213, 1941.

[5] International Formulation Committee, A Formulation of the Thermodynamic Properties of Ordinary Water Substance, IFC Secretariat, Düsseldorf, Germany, 1967.

[6] Reid, R.C., J.M. Prausnitz, and B.E. Poling, The Properties of Gases and Liquids, McGraw-Hill, New York, 1987.

[7] Narasimhan, T.N. and P.A. Witherspoon, An Integrated Finite Difference Method for Analyzing Fluid Flow in Porous Media, *Water Resour. Res.*, *12*(1), 57-64, 1976.

[8] Moridis, G.J., and K. Pruess, T2CG1: A Package of Preconditioned Conjugate Gradient Solvers for the TOUGH2 Family of Codes, Report LBL-36235, Lawrence Berkeley National Laboratory, Berkeley, Calif., 1995.

[9] Finsterle, S., ITOUGH2 User's Guide, Report No. LBL-34582, Lawrence Berkeley National Laboratory, Berkeley, Calif., August 1993.

[10] Scales, L.E., Introduction to Non-Linear Optimization, Springer, New York, 1985.

[11] Levenberg, K., A Method for the Solution of Certain Nonlinear Problems in Least Squares, *Q. Appl. Math.*, 2, 164-168, 1944.

[12] Marquardt, D.W., An Algorithm for Least-Squares Estimation of Nonlinear Parameters, *J. Soc. Ind. Appl. Math.*, *11*(2), 431-441, 1963.

[13] Finsterle, S., and K. Pruess, Solving the Estimation-Identification Problem in Two-Phase Flow Modeling, *Water Resour. Res.*, 31(4), 913-924, April 1995.

[14] Press, W.H., S.A. Teukolsky, W.T. Vetterling, B.P. Flannery, Numerical Recipes in FORTRAN: The Art of Scientific Computing, Second Edition, Cambridge University Press, 1992.

[15] Finsterle, S., and K. Pruess, Optimizing Multiphase Aquifer Remediation Using ITOUGH2, Report No. LBL-36088, Lawrence Berkeley National Laboratory, Berkeley, Calif., June 1994.

[16] Leverett, M.C., Capillary Behavior in Porous Solids, *Trans. Soc. Pet. Eng. AIME*, 142, 152-169, 1941.

A three-dimensional numerical model for the calculation of complex flow and transport phenomenas in reservoirs

C. Forkel, O. Bergen, J. Köngeter

Institut für Wasserbau und Wasserwirtschaft, Aachen University of Technology, Mies-van-der-Rohe Str. 1, 52056 Aachen, Germany

Summary

This paper presents a real three-dimensional numerical model for the calculation of complex flow and transport phenomenas in reservoirs and lakes. The turbulent processes are simulated with the large-eddy technique. Finite elements of arbitrary shape and with unstructured mesh topology are used to approximate spatial operators. The model is verified by the calculation of a straight channel flow and the comparison with data from the literature. Additionally a physical model experiment is recalculated for the verification of the model in complex geometries. Furthermore a field validation is presented where an observed inflow of turbid water into a reservoir is recalculated. After this validation of the model the reaction of the reservoir to different wind conditions is examined.

Introduction

The flow and transport phenomenas in large natural geometries such as lakes and reservoirs are very complex and in most cases totally three-dimensional. This is due to the irregular boundaries, the large volume of the water body and the variability of the external forces on the water body. A wind driven flow in a reservoir can not be modeled two-dimensionally, also density dependent flow and the influence of the Coriolis forces are difficult to be modeled properly with a two-dimensional model. Numerical models that claim to be capable of simulating flow and transport in such reservoirs correctly, must therefore at least be able to model the large scale processes in three dimensions. A numerical model is needed which is able to fit the irregular boundaries of natural reservoirs. The finite element method seems to be the best possibility to fulfill this requirement.

Another problem is the large spectrum of the turbulences that exists in large natural geometries. For the widely used standard statistical turbulence models based on the Reynolds-equations it is very difficult to model the whole spectrum, which is strongly dependent on the reservoir and its boundary conditions. Furthermore the statistical parameters are hardly transferable from one application to the other. Direct numerical simulations do not need any statistical parameters, but the necessity of directly simulating the complete eddy spectrum would require far more computer capacity than available.

A very promising approach in between the two above mentioned turbulence models is the large-eddy simulation technique. Table 1 shows the range of accuracy and computational costs between the three different general approaches. A large-eddy model simulates only the large scale quantities of the eddy spectrum directly on a relatively fine discretization level and models the small (subgrid) scale turbulences with a suitable turbulence model. This turbulence model can then be very simple, because the modeled energy spectrum of the eddies is limited. Only few statistical turbulence parameters are necessary which are transferable to other applications. In this paper we therefore present a large-eddy approach for the simulation of turbulent flow and transport phenomenas in complex geometries such as lakes and reservoirs with the finite element method.

Table 1: Turbulence modeling approaches, their accuracy and computational costs

averaging or filtering grade	simulation type	averaging procedure	smallest simulated eddies	accuracy of calculation	computational costs
high	use of the Reynolds-equations with statistical turbulence models	temporal average (spatial average not necessary)	non turbulent flow structures	small	small
medium		spatial	very large eddies	small	small
small	large-eddy simulation	and/or	medium large eddies	medium	medium
very small		temporal	smaller eddies	high	high
without averaging	direct numerical simulation	—	dissipative eddies	very high	out of range

Mathematical background

Flow

The spatially filtered Navier-Stokes equations are

$$\frac{\partial \overline{v_i}}{\partial x_i} = 0 \qquad (1)$$

and

$$\frac{\partial \overline{v_i}}{\partial t} + \overline{v_j}\frac{\partial \overline{v_i}}{\partial x_j} = -\frac{\partial \overline{P}}{\partial x_i} + \nu \frac{\partial^2 \overline{v_i}}{\partial {x_j}^2} - \frac{\partial \tau_{ij}}{\partial x_j} + \overline{F_i} \qquad (2)$$

with the subgrid scale stresses

$$\tau_{ij} = \overline{v_i v_j} - \overline{v_i}\,\overline{v_j}\,. \qquad (3)$$

In these equations v_i represents the three velocity components in x,y,z-direction, P is the kinematic pressure and ν is the kinematic viscosity. F_i includes all internal force terms like buoyancy and Coriolis forces. The overbar denotes a spatially filtered variable. The filtering functions used in equations 1 and 2 depend on the numerical approximation functions (see also chapter 'Numerical Approach').

All subgrid scale turbulences are modeled with a Smagorinsky [12] turbulence model

$$\tau_{ij} - \frac{2}{3} k \delta_{ij} = - 2 \nu_t \overline{S_{ij}} \tag{4}$$

with the eddy viscosity

$$\nu_t = (c_s \Delta)^2 \cdot (\overline{S_{ij}}\,\overline{S_{ij}})^{1/2}, \tag{5}$$

the filtered deformation tensor

$$\overline{S_{ij}} = \frac{1}{2} \cdot \left(\frac{\partial \overline{v_i}}{\partial x_j} + \frac{\partial \overline{v_j}}{\partial x_i} \right) \tag{6}$$

and the filter width

$$\Delta = \left(\frac{1}{n_k} \sum_{i=1}^{n_k} (\Delta_k)_i^2 \right)^{1/2}. \tag{7}$$

n_k is the number of the element edges at each finite element node, $(\Delta_k)_i$ is the length of an element edge. c_s is the statistical turbulence model parameter with $c_s \approx 0{,}1$. Germano et al. [6] and Lilly [10] have developed a method to calculate the value of c_s as a spatial and temporal variable by introducing a double filtering approach. This method is also implemented in the turbulence model, but is not described here in detail.

Transport

A similar approach is used for the conservative transport equation with the concentration of the transported solvent C, the molecular diffusion coefficient D_m, the subgrid scale dispersion E_i, the sink/source term M_c and the transport dispersion parameter c_t, which can be related to the corresponding flow turbulence parameter c_s.

$$\frac{\partial \overline{C}}{\partial t} + \frac{\partial}{\partial x_i}(\overline{v_i}\,\overline{C}) - D_m \frac{\partial^2 \overline{C}}{\partial x_i^2} + \frac{\partial}{\partial x_i} E_i = \overline{M_c} \tag{8}$$

with

$$E_i = -2 D_t \cdot \frac{\partial \overline{C}}{\partial x_i} \tag{9}$$

and

$$D_t = (c_t \Delta)^2 \cdot (\overline{S_{ij}}\,\overline{S_{ij}})^{1/2}. \tag{10}$$

Numerical approach

The numerical model is based on a finite element model by Daniels [2] for the solution of the transient incompressible Navier-Stokes equations and the transport equation. Time-accurate solutions are obtained by using the 'projection 2' algorithm which was developed and analyzed by Gresho [7]. Modifications to the original algorithm were made where it was appropriate. The model consists of different time integration schemes (explicit, semi-implicit, implicit) to solve the momentum equations and the transport equation.

The element type is a hexahedron with a trilinear approximation function for the velocities and the concentrations and a constant pressure in the element. The numerical approximation functions leads to a triangle filter for the velocities and to a box filter for the kinematic pressure (Forkel et al. [3] and Forkel [4]). Because of the arbitrary shape of the elements all integrations on the element level are obtained by full 2*2*2 Gauss point integration in the general case. For more details of the numerical model see Daniels [2] and Forkel [4].

Verification examples

Straight channel flow

As a first verification example the simulation of a turbulent straight channel flow was chosen. This example has been investigated thoroughly by experiments, direct numerical simulations and large-eddy simulations and an abundance of data is available for the comparison with the results of the presented model. Out of many simulation runs for the verification of the model only one example will be shown here. The modeled geometry is shown in figure 1, the discretization consists of 24*20*24 nodes in x,y,z-direction with constant spacing. The geometry is similar to a simulation by Piomelli et al. [11], but the discretization is coarser, as they used a grid of 32*63*32 nodes.

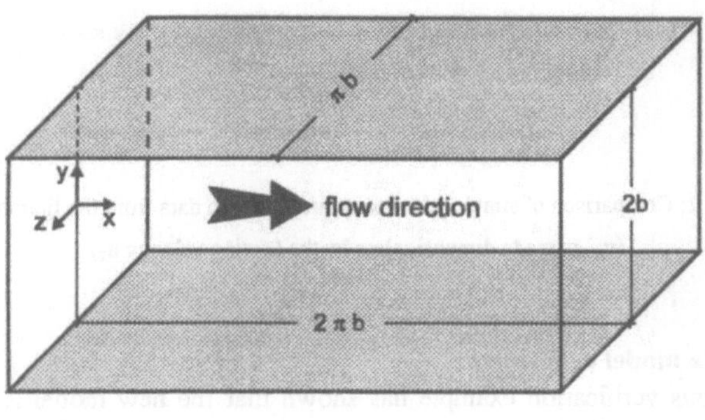

Fig. 1: Model geometry by Piomelli et al. [11], Re_τ=640

At the walls we used the logarithmic law of the wall as proposed by Unger and Friedrich [13], the outflow boundary is described by zero pressure/stress boundary conditions. At the inflow boundary we used periodic boundary conditions with

$$v_i(t, x = 0) = v_i(t - \Delta t, x = 2\pi b) . \qquad (11)$$

The simulated area is extended in x-direction (additional 5 nodes with same spacing) to avoid the damping influence of the outflow boundary condition. The turbulent initial condition was calculated in a previous channel flow simulation. The temporal discretization should be explizit. Other temporal discretizations normally lead to unstable or overdiffusive results [4].

Forkel [4] shows in his investigations on temporal discretizations that similar stable results without extensive numerical diffusion can be achieved with the semi-implicit scheme of Leismann and Frind [9] and small time steps.

Figure 2 shows the velocity profiles of our simulations with the Smagorinsky (c_S=0,1) and the Germano-Lilly model in comparison to the profile of Piomelli et al. [11] achieved with the finer spatial discretization. The velocities are averaged over time and over planes parallel to the walls. The comparison shows a rather good agreement between the three curves. The curve of Piomelli et al. [11] is smoother due to the fact that they use a much longer time interval for averaging. More results from other simulation runs and a more detailed evaluation of the presented verification is given by Forkel [4].

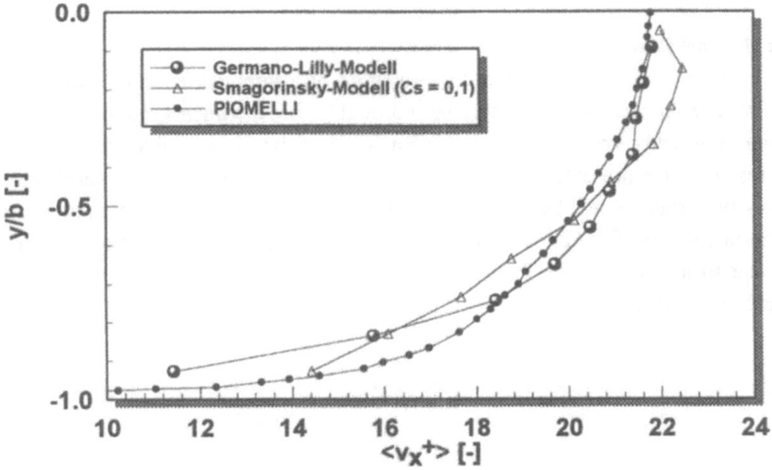

Fig. 2: Comparison of simulated velocity profiles with data from the literature (v_x^+ is made dimensionless by the friction velocity u_τ)

Recalculation of a model experiment

The previous verification example has shown that the new model is able to simulate turbulent flow in a relatively simple geometry correctly. As the model is designed for the simulation of laminar and turbulent flow and transport in complex geometries much more verification is needed. This is done by the recalculation of a physical model experiment. This experiment includes a turbulent tracer injection into an initially laminar flow through a very complex, three-dimensional geometry (figure 3).

The tracer movement is recorded by a video camera through a plexi glass plate which is fixed on top of the model. The information is converted by digital image processing technique into digital data, the measured concentrations are evaluated by reference measurements and can then be compared with the results of the numerical model. Figure 4 shows the very good agreement between the calculation and the measured data at different times after the beginning of the tracer injection.

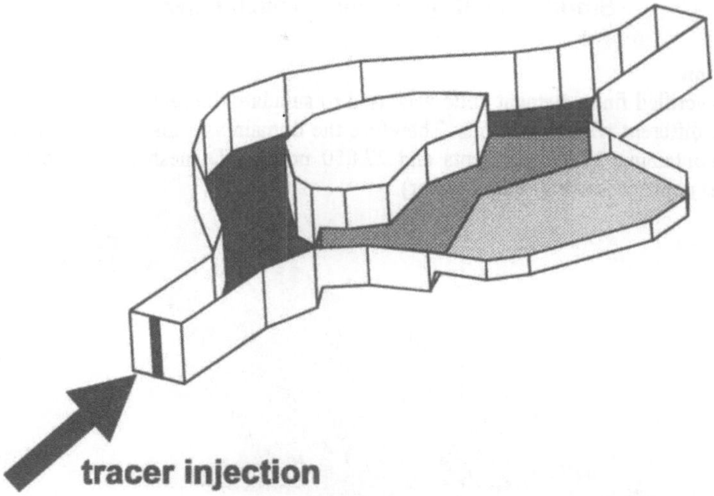

Fig. 3: Model geometry of the tracer experiment (5* vertically enlarged)

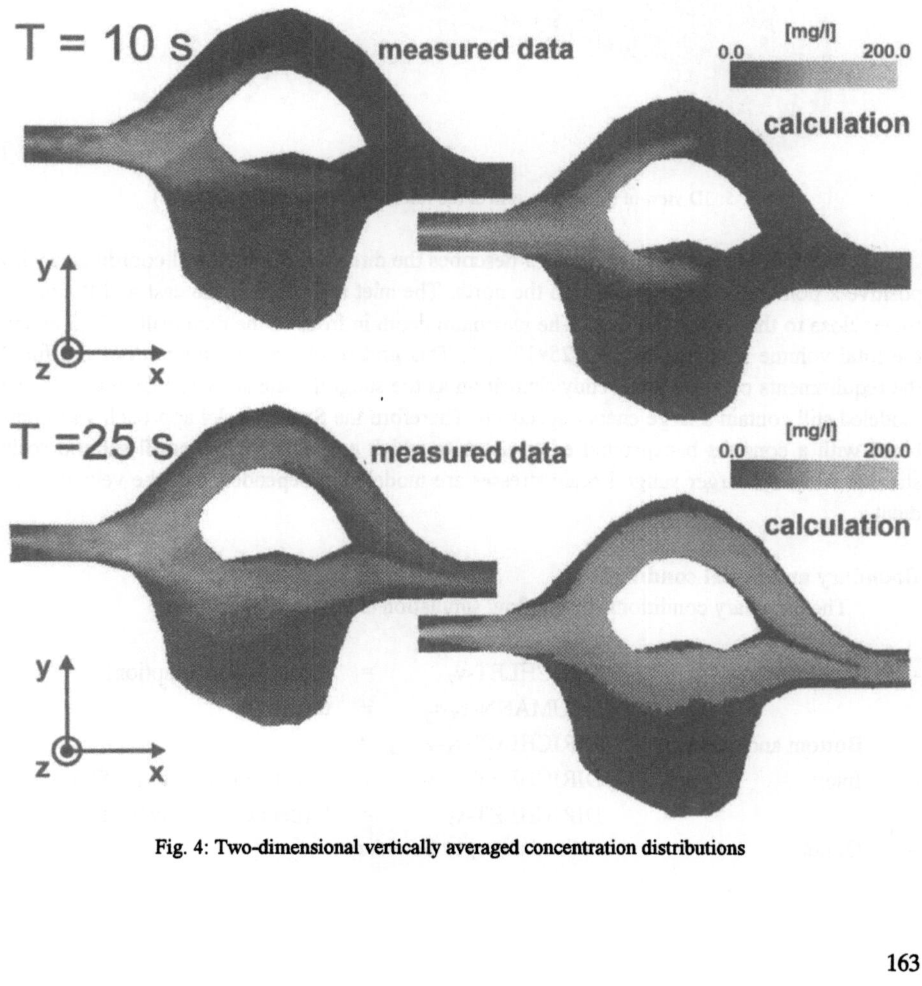

Fig. 4: Two-dimensional vertically averaged concentration distributions

Simulation of the Dreilägerbach reservoir

Discretization

The verified finite element code was used to simulate the reaction of the Dreilägerbach reservoir to different wind conditions. Therefore the domain was discretized with a finite element grid containing 23.972 elements and 27.850 nodes. 3D mesh refinements were made where it was appropriate (e.g. outlet tower).

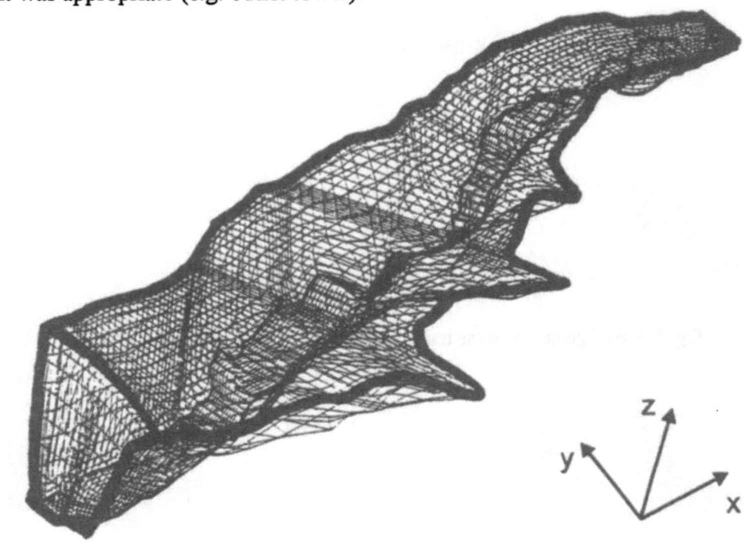

Fig. 5: 3D view of the surface grid of the reservoir (10* vertically enlarged)

The coordinate system in figure 5 describes the direction of the global coordinates with positive x pointing eastwards and y to the north. The inlet is located in the east and the outlet tower close to the wall in the west. The maximum depth in front of the dam wall is 25,5 m and the total volume is approximatly $3{,}25 * 10^6$ m^3. This grid resolution of course does not fulfill the requirements of a real large-eddy simulation as the subgrid scale stresses which have to be modeled still contain a large energy spectrum. Therefore the Smagorinsky approach was combined with a constant background eddy viscosity which accounts for the small subgrid scale stresses while the larger subgrid scale stresses are modeled in dependence of the velocity gradient.

Boundary and initial conditions

The boundary conditions for the flow simulation are defined as follows

–	Water surface:	DIRICHLET-v_z	=	0 (rigid lid assumption)
		NEUMANN-τ_x-τ_y	$\neq$	0 (wind)
–	Bottom and dam wall	DIRICHLET-v_x-v_y-v_z	=	0
–	Inlet:	DIRICHLET-v_x-v_y	$\neq$	0 (defined velocity profile)
		DIRICHLET-v_z	=	0 (defined velocity profile)
–	Outlet:	defined as „open boundary“		

The wind induced shear stress at the water surface can be determined by the empirical formula

$$\tau = \rho_a c_d u_w^2 \qquad (12)$$

ρ_a = density of the air = 1,23 kg/m^3
c_d = drag coefficient ≅ 1,5$*10^{-3}$
u_w = wind speed 10 m above the water surface.

The initial conditions for the flow field must be divergence free. Therefore the velocities in all three directions were defined to be equal to zero which made a first flow field calculation necessary to achieve realistic initial conditions.

The boundary conditions for the transport simulation were defined as DIRICHLET-concentrations at the inlet, zero-flux NEUMANN boundary conditions through the water surface and the dam wall and an open boundary at the outlet tower.

Simulation results

In the Dreilägerbach reservoir a strong inflow of turbid water was observed during times of heavy rain. One event with turbid inflow water was documented by taking pictures from an airplane and was used for a first verification run. At that time strong winds with wind speeds of 50 km/h blowing from west to east were observed. The early time of the year (spring) and the intense wind induced mixing made density effects negligible so that they were not considered in this simulation. For details of density dependent flow in lakes and reservoirs see [5].

Figure 6 shows the calculated velocity distributions after a simulated time of 20 hours in horizontal slices at different depths in the reservoir. At the watersurface (figure 6a) the wind pushes the water towards the inflow region. Only very close to the dam wall the coriolis deflection at the water surface can be seen, afterwards the basin morphology inhibits this movement to the right hand side.

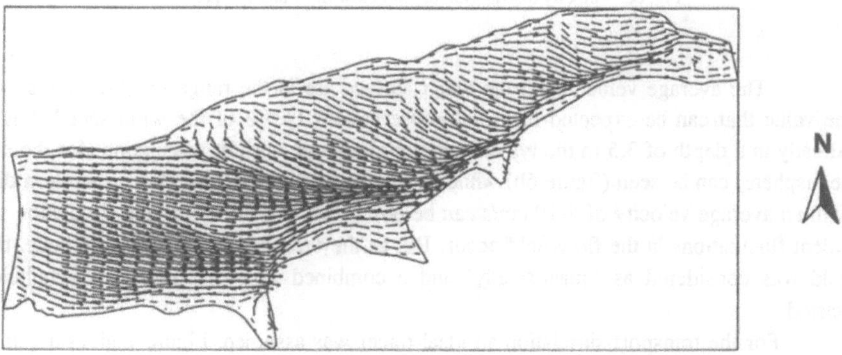

Fig 6a: Velocity distribution at the water surface

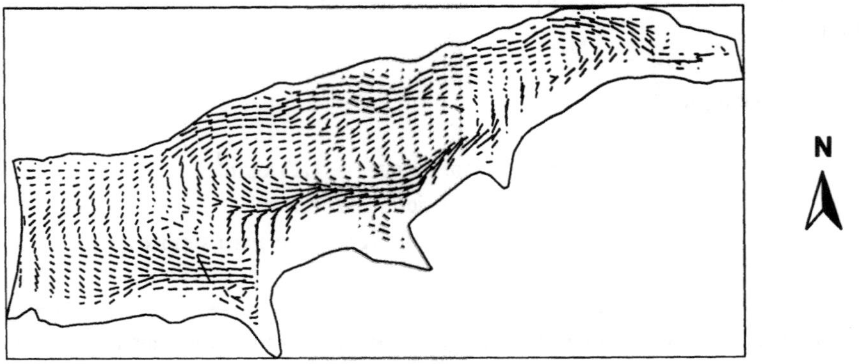

Fig. 6b: Velocity distribution 3,5 m below the water surface

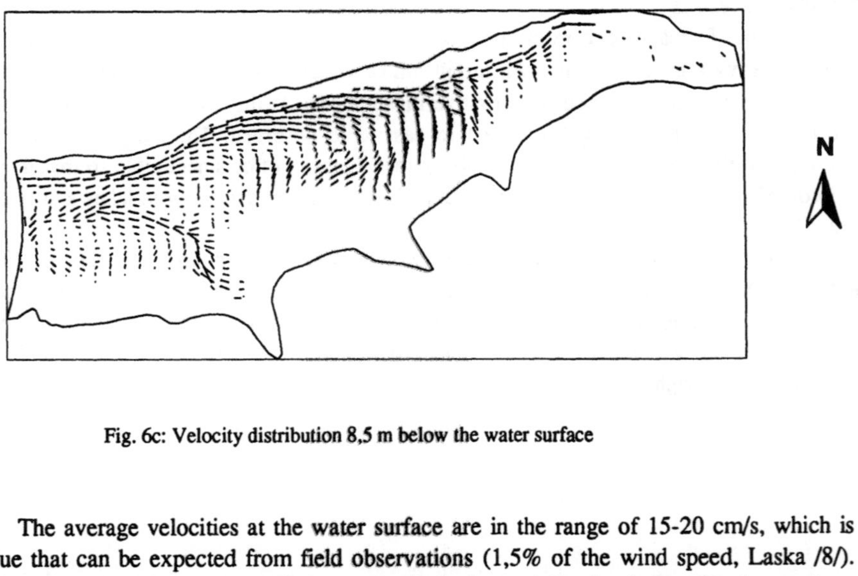

Fig. 6c: Velocity distribution 8,5 m below the water surface

The average velocities at the water surface are in the range of 15-20 cm/s, which is the value that can be expected from field observations (1,5% of the wind speed, Laska /8/). Already in a depth of 3,5 m the typical counter clockwise coriolis circulation (for the northern hemisphere) can be seen (figure 6b) while in a depth of 8,5 m a main flow in western direction with an average velocity of 5-10 cm/s can be observed. At this time of the simulation still turbulent fluctuations in the flow field occur. But as they reached a constant amplitude, the flow field was considered as 'quasi-steady' and a combined flow and transport simulation was started.

For the transport simulation an ideal tracer was assumed. Figure 7 gives a comparison between the calculated tracer distribution and a picture of the field observation. The results are in excellent agreement.

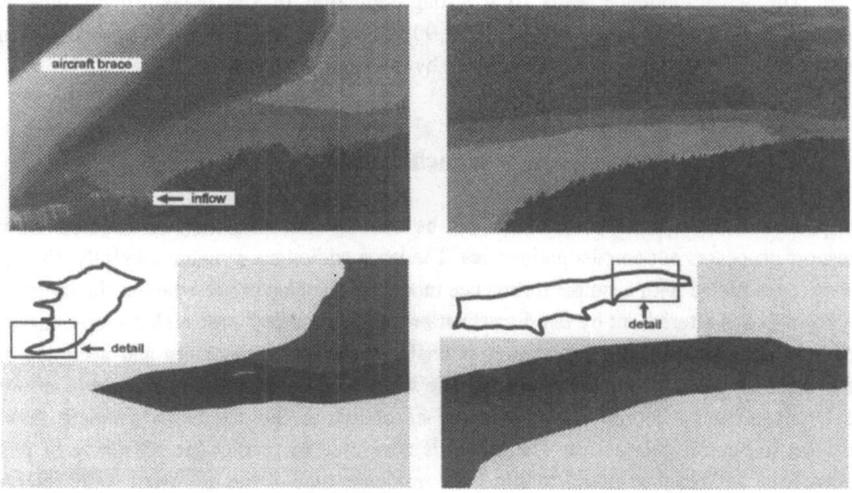

Fig. 7: Comparison between observed (top - edited at the computer) and calculated (bottom) tracer distribution

Application

After this verification run numerous flow and transport simulations were carried out to estimate the influence of the wind speed to the specific residence time. The duration of the inflow of turbid water was defined to 80 hours, afterwards clean water enters the reservoir. Figure 8 gives a comparison of the calculated tracer breakthrough curves for no wind, 50 km/h westwind and 50 km/h eastwind. The concentration is made dimensionless by division through the concentration of the inflowing water.

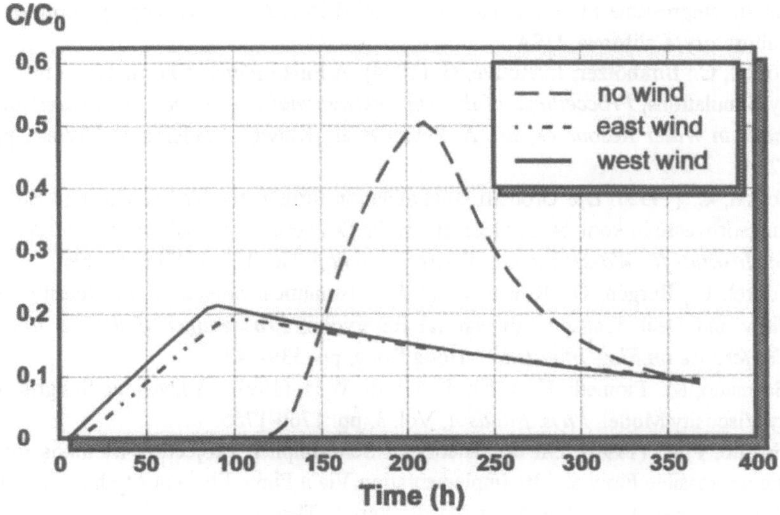

Fig. 8: Breakthrough curves at the outlet tower for different wind conditions

The wind influence leads to a strong reduction of the breakthrough time which decreases from approximately 210 hours to 90 (westwind) and 100 hours (eastwind). In this case the breakthrough time is hardly affected by the wind direction.

Conclusions

The presented three-dimensional model can be used for the simulation of turbulent flow and transport processes in complex geometries. The large eddies are simulated directly, the smaller eddies are modeled with a simple turbulence model. A question is still whether the large energy spectrum that is filtered out by the discretization can be modeled with such a simple turbulence model or not. Anyhow the verification examples show the applicability and accuracy of the method for simple and complex problems. The successful recalculation of the turbid inflow into the Dreilägerbach reservoir documents the capability of the model to simulate flow and transport in natural geometries. The model is now used to predict the efficiency of possible constructive devices that intend to affect the residence time of the inflowing water (Bergen et al. [1]). The large-eddy turbulence modelling is a very promising approach to describe the turbulent processes in large scale turbulent flows. Whether the single statistical parameter is applicable to different lakes and wind conditions should be investigated in detail.

Literature

1. Bergen, O.; Forkel, C.; Köngeter, J. (1995): Berechnung der Trübstoffausbreitung in der Dreilägerbachtalsperre, final report, *Institut für Wasserbau und Wasserwirtschaft*, RWTH Aachen.
2. Daniels, H. (1992): PASTIS-3D, Finite-Element Projection Algorithm Solver for Transient Incompressible Flow Simulations, *UCRL-MA-111833*, Lawrence Livermore National Laboratory, California, USA.
3. Forkel, C.; Birkhölzer, J.; Rouvé, G. (1994): A Finite Element Formulation for Large Eddy Simulations, *Proceedings of the Xth International Conference on Computational Methods in Water Resources*, Ed. A. Peters et al., Kluwer Academic Publishers, pp. 1217-1224.
4. Forkel, C. (1995): Die Grobstruktursimulation turbulenter Strömungs- und Stoffausbreitungsprozesse in komplexen Geometrien, Ph.D Thesis, to be published in *Mitteilungshefte des Instituts für Wasserbau und Wasserwirtschaft*, No. 102, RWTH Aachen.
5. Forkel, C.; Bergen, O.; Rouvé, G. (1994): Numerical Simulation of Density Dependent Flow and Heat Transport in Natural Reservoirs, *Proceedings of the 1st International Conference on Flow Interaction*, Hong Kong, pp. 539-542.
6. Germano, M.; Piomelli, U.; Moin, P.; Cabot, W.H. (1991): A Dynamic Subgrid-Scale Eddy Viscosity Model, *Phys. Fluids A*, Vol. 3, pp. 1760-1765.
7. Gresho, P.M. (1990): On the Theory of Semi-Implicit Projection Methods for Viscous Incompressible Flow and Its Implementation Via a Finite Element Method that also Introduces a Nearly-Consistent Mass Matrix, Part 1: Theory, *International Journal for Numerical Methods in Fluids*, Vol. 11, J. Wiley & Sons, Inc., pp. 587-620.

8. Laska, M. (1981): Characteristics and modelling of physical limnology processes; *Mitteilungsheft Nr. 54 der Versuchsanstalt für Wasserbau, Hydrologie und Glaziologie, ETH Zürich*, Hrsg.: Prof. Dr. D. Vischer.
9. Leismann, H.M.; Frind, E.O. (1989): A Symmetric-Matrix Time Integration Scheme for the Efficient Solution of Advection-Dispersion Problems, *Water Resources Research*, Vol. 25, No. 6, pp. 1133-1139.
10. Lilly, D.K. (1992): A Proposed Modification of the Germano Subgrid-Scale Closure Method, *Phys. Fluids A*, Vol. 4, No. 3, pp. 633-635.
11. Piomelli, U.; Ferziger, J.H.; Moin, P. (1987): Models for Large Eddy Simulations of Turbulent Channel Flows Including Transpiration, *Report TF-32*, Department of Mechanical Engineering, Stanford University, California.
12. Smagorinsky, J. (1963): General Circulation Experiments with the Primitive Equations, I. The Basic Experiment, *Monthly Weather Review*, Vol. 91, pp. 99-104.
13. Unger, F.; Friedrich, R. (1993): Numerical Simulation of Fully-Developed Turbulent Pipe Flow, *Notes on Numerical Fluid Mechanics*, Vol. 38, Vieweg Verlag, Braunschweig, pp. 201-215.

ON NUMERICAL SOLUTION METHODS FOR NONLINEAR PARABOLIC PROBLEMS

JÜRGEN FUHRMANN *
WEIERSTRASS INSTITUTE
FOR APPLIED ANALYSIS AND STOCHASTICS
MOHRENSTR. 39
10117 BERLIN
fuhrmann@wias-berlin.de

SUMMARY

A framework for the numerical solution of nonlinear parabolic equations, e.g. Richard's equation or the equation describing the movement of a viscous compressible fluid in a porous medium, are discussed. Special attention is given to problems with coefficients varying in space. The solution scheme is based on an implicit time discretization combined with Newton's method to solve the time step problems. For the solution of the linear problems, iterative methods are used. The usage of algebraic multilevel preconditioners on structured and unstructured meshes is discussed.

1 INTRODUCTION

We consider the following equation of saturated/unsaturated flow of a viscous, compressible fluid through a porous medium:

$$\frac{\partial (\rho\theta\theta_r)}{\partial t} - \nabla \cdot \frac{kk_r\rho}{\mu}(\nabla p - \rho\gamma) = q\,. \tag{1.1}$$

The meaning of the unknowns and their constitutive relationships are as follows:

p	pressure / capillary pressure
θ	pore volume
$\theta_r(p)$	saturation
$\rho(p)$	fluid density
$\mu(\lvert\nabla(p-\rho\gamma)\rvert)$	viscosity
k	permeability of the porous medium
$k_r(p)$	relative permeability
γ	gravity force
q	sources/drains

*This work has been supported by the SFB 359 of the Deutsche Forschungsgemeinschaft.

We consider this equation in a space-time domain $\Omega \times [0,T]$ with initial condition $p(x,0) = p_0(x)$ and appropriate boundary conditions of Neumann, Dirichlet and Cauchy type. Here, Ω is a one-, two- or three-dimensional polygonal domain. To be able to describe heterogeneities in the porous medium, we assume further that all parameters may depend on the space variable x.

Special cases of this equation describe

- the porous medium equation
- saturated/unsaturated groundwater flow (Richard's equation)
- saturated flow of a compressible Bingham fluid (in the context of oil reservoir simulation)

The main focus of this paper is the description of a numerical method for the solution of (1.1) in two and three space dimensions on unstructured meshes. The assumption of space dependency of the coefficients makes the use of the Kirchoff transformation difficult. So we use a rather straightforward approach consisting in

- Rothe's method for the time discretization
- space discretization by a finite volume method
- solution of the discrete nonlinear problems using Newton's method with line search
- solution of the linear problems by preconditioned Krylov subspace methods.

2 Time Discretization

First, write equation 1.1 as

$$\partial_t B(p) + A(p) = q \tag{2.1}$$

where

$$\begin{aligned} B(p) &= \rho\theta\theta_r, \\ -A(p) &= \nabla \cdot \frac{kk_r\rho}{\mu}(\nabla p - \rho\gamma). \end{aligned}$$

For given

$$0 = t_0 < t_1 < \ldots < t_N = T$$

an implicit Euler scheme using Rothe's method is defined by

$$\frac{B(p^n) - B(p^{n-1})}{t_n - t_{n-1}} + A(p^n) = q, \quad n = 1 \ldots N. \tag{2.2}$$

Note that we do not differentiate B. This scheme together with a careful solution of the time step problem will allow an exact discrete mass balance. The price to pay in comparison with schemes as in [JK91] is that in every time step we have to deal with the full nonlinearity of the problem.

To determine t_i, we use a simple adaptive time step control scheme. The basic idea consists in holding the value

$$\delta_p^n = ||p^n - p^{n-1}||_\infty \tag{2.3}$$

within a given range around an optimal δ_p^{opt}. This scheme is intended to be able to fit to fast processes in the beginning of an evolution and to be aware of the exponential decay towards a stationary solution — if such a solution exists — as time goes to infinity. The L^∞- norm in (2.3) could be replaced e.g. by the L^2-norm or the change of the total mass, but we can hope that the choice of the L^∞ norm enables the scheme better to adapt to local phenomena. The sensitivity of the approximation error to the parameter δ_p^{opt} has been tested numerically in section 5. Here is a short sketch of the algorithm:

procedure time_step_control($\delta_p^{opt}, \xi, \delta_t^0$)
begin
    $t_0 = 0$
    $\delta_t = \delta_t^0$
    $n = 1$
    while $t < T$ do
        $t_n = \max(T, t + \delta_t)$
        solve (2.2)
        if ***solution failed*** then goto reject
        if $\delta_p^n > \xi\delta_p^{opt}$ then goto reject
        $t = t_n$
        $n = n + 1$
        $\delta_t = \delta_t \frac{\delta_p^{opt}}{\delta_p^{n-1}}$
        continue
    reject:
        $\delta_t = \frac{\delta_t}{\xi}$
    done
end

This is only the skeleton of the algorithm implemented. So, measures have to be taken to accept the first time step after we changed a Dirichlet boundary condition. Further, a maximal and a minimal time step seem to be useful to ensure robustness.

3 Space discretization

For the space discretization, we use a finite volume method on two- and three-dimensional simplex meshes. In the implementation, the meshes can be generated by the mesh generator IBG [Sch93] or by the kernel of KASKADE [BER94]. The discretization should be able to utilize the ability of these mesh generators to handle domains with different materials.

At the other hand, the convection type term which arises in the incompressible case may demand upwinding, which can easily be defined in a finite volume framework. Further, the finite volume type schemes used in [Fuh94b, Fuh94a] performed well.

As a consequence, we use a scheme which actually could be seen as an element based finite volume approach.

So let

$$\bar{\Omega} = \bigcup_{\sigma \in \Sigma} \bar{\sigma} \subset \mathbb{R}^d$$

be a partition of the d-dimensional polygonal domain Ω into simplicial finite elements σ.
For a simplex σ let $X_\sigma = \{x_\sigma^i\}_{i=1\ldots d+1}$ denote the set of corners of σ.
Let $X_\Sigma = \bigcup_{\sigma \in \Sigma} X_\sigma$ be the set of discretization points.
Let V_σ denote the space of linear functions on σ, identified with $\mathbb{R}^{d+1}$. Note that under these circumstances, for $u \in V_\sigma$, the gradient ∇u is well defined.
Let

$$V_\Sigma = \{f \in H^1(\Omega) | \quad f|_\sigma \in V_\sigma \quad \forall \sigma \in \Sigma\}$$

be the space of piecewise linear functions. We identify V_Σ with the discrete space $\mathbb{R}^{\dim V_\Sigma}$ in the usual way.

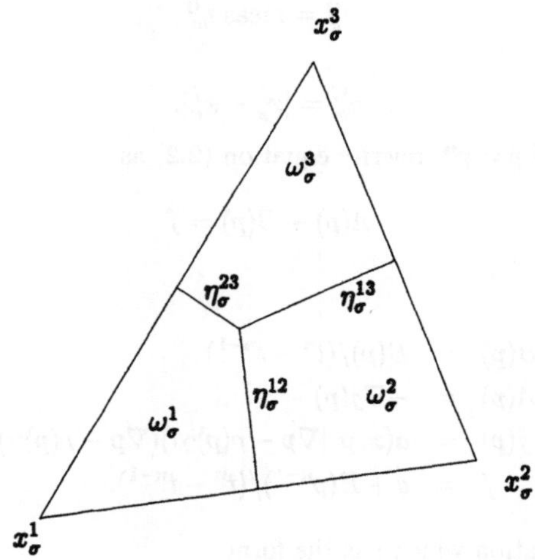

Figure 1: *A two-dimensional simplex σ*

Now we can define the restriction of a function to a simplex.

$$\begin{array}{rcl} \pi_\sigma : V_\Sigma & \rightarrow & V_\sigma \\ u & \mapsto & u_\sigma = \left(u(x_\sigma^1) \quad \ldots \quad u(x_\sigma^{d+1}) \right)^T \\ \pi_\sigma^* : V_\sigma & \rightarrow & V_\Sigma \\ u_\sigma & \mapsto & u_\sigma^* \end{array}$$

where for any $x \in X_\Sigma$, $u_\sigma^*(x)$ is characterized by

$$u_\sigma^*(x) = \begin{cases} u_\sigma(x) & , \; x \in X_\sigma \\ 0 & , \; \text{else} \end{cases} .$$

Define the ***restricted Voronoi box*** ω_σ^i around $x_\sigma^i \in X_\sigma$ by

$$\omega_\sigma^i = \{x \in \sigma | \text{dist}\,(x, x_\sigma^i) < \text{dist}\,(x, x') \quad \forall x' \in X_\sigma\}.$$

We call a simplex σ *non-degenerate* if all of the sets

$$\eta_\sigma^{ij} = \bar{\omega}_\sigma^i \cap \bar{\omega}_\sigma^j$$

are non-empty. In two space dimensions, this includes the case of rectangular simplices, where η_σ^{ij} can consist of one point but prevents σ from having obtuse angles.
In the sequel, we will assume, that Σ consists of non-degenerate triangles. In general, a box discretization can be well defined for more general meshes, but the compensation properties used in these cases do not hold in our case of coefficients varying from element to element. These possibilities to work in more general cases are discussed e.g. in [GG94].
Define the *node form factors*

$$m_\sigma^i = \text{meas}\, \omega_\sigma^i$$

for $i \neq j$ the *edge form factors*

$$e_\sigma^{ij} = \text{meas}\, \eta_\sigma^{ij}$$

and the *edge lengths*

$$h_\sigma^{ij} = |x_\sigma^i - x_\sigma^j|.$$

Now, for a given n and $p = p^n$, rewrite equation (2.2) as

$$A(p) + \bar{B}(p) = f \tag{3.1}$$

with

$$\begin{aligned}
\bar{B}(p) &= B(p)/(t^n - t^{n-1}) && (3.2)\\
A(p) &= -\nabla j(p) && (3.3)\\
j(p) &= a(x, p, |\nabla p - \rho(p)\gamma|)(\nabla p - \rho(p)\gamma) && (3.4)\\
f &= q + B(p^{n-1})/(t^n - t^{n-1}). && (3.5)
\end{aligned}$$

We look for a discretization which has the form

$$A_\Sigma(p) + B_\Sigma(p) = f \tag{3.6}$$

for $p \in V_\Sigma$ with

$$A_\Sigma(p) = \sum_{\sigma\in\Sigma} \pi_\sigma^* A_\sigma(\pi_\sigma p) \tag{3.7}$$

and

$$B_\Sigma(p) = \sum_{\sigma\in\Sigma} \pi_\sigma^* B_\sigma(\pi_\sigma p). \tag{3.8}$$

For $x \in X_\Sigma$ let $\Sigma^x = \{\sigma \in \Sigma | x \in X_\sigma\}$ be the set of simplices which have x as a vertex. Define the *Voronoi box* ω^x by

$$\bar{\omega}^x = \bigcup_{\sigma\in\Sigma^x} \bar{\omega}_\sigma^x. \tag{3.9}$$

To fulfil (3.1) for $p_\Sigma \in V_\Sigma$ approximately, we demand

$$\int_{\omega^x} (-\nabla j(p) + \bar{B}(p) - f)dx = 0 \quad \forall x \in X_\Sigma\,. \tag{3.10}$$

In fact this can be seen as a Galerkin-Petrov ansatz using piecewise linear basis functions and piecewise constant on ω^x test functions. Using Gauss' theorem we can rewrite (3.10) as

$$\int_{\partial\omega^x} (j(p)\cdot\nu)ds + \int_{\omega^x} (\bar{B}(p) - f)dx = 0 \quad \forall x \in X_\Sigma \tag{3.11}$$

where ν is the outer normal of ω^x. (3.9) yields for $x \in X_\Sigma$

$$\partial\omega_x = \overline{\left(\bigcup_{\sigma\in\Sigma^x} \partial\omega_\sigma^x \cap \sigma\right)} \cup \overline{\left(\bigcup_{\sigma\in\Sigma^x} \partial\omega_\sigma^x \cap \partial\Omega\right)}. \tag{3.12}$$

For homogeneous Neumann boundary conditions[1], using

$$\int_{\partial\omega_\sigma^x\cap\partial\Omega} (j(p)\cdot\nu)ds = 0$$

(3.11) now rewrites as

$$\sum_{\sigma\in\Sigma^x} \int_{\partial\omega_\sigma^x\cap\sigma} (j(p)\cdot\nu)ds + \int_{\omega_\sigma^x\cap\sigma} (\bar{B}(p) - f)dx = 0 \quad \forall x \in X_\Sigma, \tag{3.13}$$

and

$$\sum_{\sigma\in\Sigma} \pi_\sigma^* \left(\int_{\partial\omega_\sigma^i\cap\sigma} (j(p)\cdot\nu)ds\right)_{i=1\ldots d+1} + \sum_{\sigma\in\Sigma} \pi_\sigma^* \left(\int_{\omega_\sigma^i} (\bar{B}(p) - f)dx\right)_{i=1\ldots d+1} = 0 \tag{3.14}$$

as an equation in V_Σ.
For the integral in the second sum we choose the lowest order approximation

$$\int_{\omega_\sigma^i} (\bar{B}(p) - f)dx \approx m_\sigma^i \left(\bar{B}(p(x_\sigma^i)) - f(x_\sigma^i)\right).$$

Let j_σ^{ij} be an approximation of the flow through η_σ^{ij}. Then the partition

$$\partial\bar{\omega}_\sigma^i \cap \bar{\sigma} = \bigcup_{j\neq i} \eta_\sigma^{ij}$$

yields the approximation

$$\int_{\partial\omega_\sigma^x\cap\sigma} (j(p)\cdot\nu)ds \approx \sum_{j\neq i} j_\sigma^{ij} e_\sigma^{ij}.$$

Now, it remains to derive the flux approximations j_σ^{ij}. The fact that by definition, the faces η_σ^{ij} are orthogonal to the simplex edges $x_\sigma^i x_\sigma^j$ suggest the utilization of one-dimensional flux expressions along the simplex edges.
Let

$$\bar{\rho}_\sigma(p) = \frac{1}{d+1} \sum_{i=1}^{d+1} \rho(p(x_\sigma^i)).$$

[1] assumed here for simplicity

Define $\xi_\sigma(p) = |\nabla p - \bar{\rho}_\sigma(p)\gamma|$. This value is a naturally defined constant on σ. Let $a^i_\sigma(p) = a(x^i_\sigma, p(x^i_\sigma), \xi_\sigma(p))$.
In the incompressible case, we have to consider the usage of upwinding for the discretization of the convection term $\rho(p)\gamma$. For a given edge $x^i_\sigma x^j_\sigma$, let γ^{ij}_σ be the projection of the gravity vector onto that edge, and let

$$\rho^{ij,\mathrm{up}}_\sigma = \begin{cases} \rho(p(x^i_\sigma)), \gamma^{ij}_\sigma & < 0 \\ \rho(p(x^j_\sigma)), \gamma^{ij}_\sigma & > 0. \end{cases}$$

Now, we can define

$$j^{ij}_\sigma = \frac{a^i_\sigma(p) + a^j_\sigma(p)}{2}\left(\frac{p(x^i_\sigma) - p(x^j_\sigma)}{h^{ij}_\sigma} + \rho^{ij,\mathrm{up}}_\sigma \gamma^{ij}\right).$$

The local nonlinear operators

$$\begin{aligned} A_\sigma : V_\sigma &\to V_\sigma \quad \text{and} \\ B_\sigma : V_\sigma &\to V_\sigma \quad \text{and} \end{aligned}$$

then are defined by

$$A^i_\sigma(p) = \sum_{j \neq i} e^{ij}_\sigma j^{ij}_\sigma(p)$$

and

$$B^i_\sigma(p) = m^i_\sigma B(p(x^i_\sigma)),$$

respectively.
Together with (3.7) and (3.8) this completes the description of the space discretization scheme.
We have to note the following issues:

- Due to the dependence of a on $|\nabla p|$, in general no compensation of obtuse angles takes place. So the strong assumptions on the triangulation seem to be necessary. It might be possible to weaken them if this dependency is not present.

- For incompressible fluids, the upwinding does not take place. In [Fuh94a] this issue has been addressed by a change of variables.

- In contrary to [Fuh94b], this element based finite volume scheme is able to use information on element materials generated by mesh generators.

- All formulae given are valid for one-, two-, and three-dimensional simplices. This allows a dimension independent implementation in the sense that the application of the nonlinear operator on a vector, as well as the assembly of the Jacobian can be organized in loops over nodes and edges of a simplex and there is no necessity for switching to different code segments depending on the space dimension.

4 Solution of the discrete time step problem

To solve the discrete problem 3.6 we use Newton's method with line search and optionally affine invariant residual calculation [DH91]. In the Newton scheme, as a linear operator we use the linearization of the discrete nonlinear operator rather than the discretization of the linearization of the nonlinear operator. The advantage of this approach is the fact that we will have no interplay between the discretization error and the iteration error when controlling Newton's method, so that we can observe the quadratic convergence behaviour.

The linear problems are solved by a preconditioned CG resp. BICGstab method. As a preconditioner we can use an incomplete LU-factorization or algebraic multilevel methods [RS87, Fuh95].

5 Numerical examples

The porous medium equation. The porous medium equation, which models the gas flow through a porous medium, is an important benchmark for numerical methods for nonlinear parabolic equations.

We consider the initial value problem for

$$\frac{\partial p}{\partial t} - \nabla m p^{m-1} \nabla p = 0 \tag{5.1}$$

in $[-100, 100] \subset \mathbb{R}^1$ with Neumann boundary conditions. It obviously fits into the framework of equation (1.1).

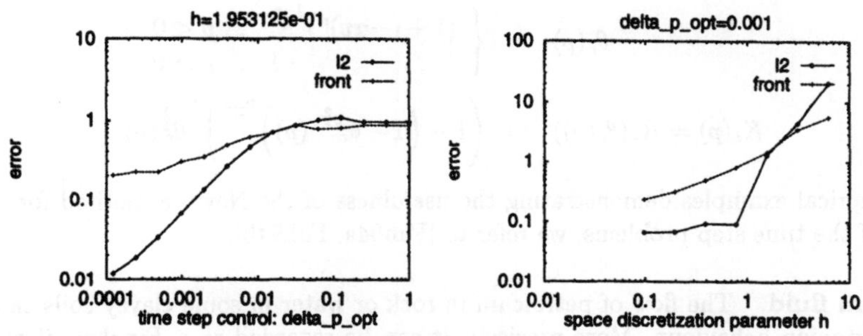

Figure 2: *Dependency of the solution error on the time step control parameter δ_p^{opt} (left) and the space discretization parameter h (right)*

We would like to compare the numerical solution to the *Barenblatt solution*, see e.g. [Gil92]

$$p_{\text{barenblatt}}(x,t) = (t+1)^{-\frac{1}{m+1}} \left(1 - \frac{m-1}{2m(m-1)}((t+1)^{-\frac{1}{m+1}}x)^2\right)_+^{\frac{1}{m-1}} \tag{5.2}$$

which has a support moving with finite speed.

So we use as the initial value for equation (5.1)

$$p(x,0) = p_{\text{barenblatt}}(x,0).$$

We solve until the support of the solution fills the whole space domain Ω.
In figure 2 we can see how the solution error (L^2 error in the space-time domain resp. error of the moving front) depends on the discretization parameters δ_p^{opt} and h.

Richard's equation. This elliptic- parabolic problem describes saturated/unsaturated flow in a porous medium. The constitutive relationships are described by assuming $\rho = 1$

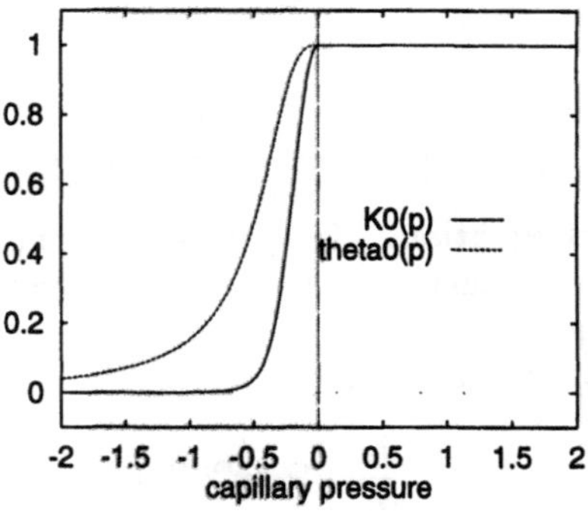

Figure 3: *van Genuchten's constitutive relationships*

and $\mu = 1$ and the van Genuchten ansatz [vG80]

$$\begin{aligned}
\theta_r(p) &= \begin{cases} (1+(-\alpha p)^n)^{\frac{1-n}{n}} & , \ p<0 \\ 1 & , \ p \geq 0 \end{cases} \\
K_r(p) = K_r(\theta_r(p)) &= \left(1-\left(1-\theta_r^{\frac{n}{n-1}}(p)\right)^{\frac{n-1}{n}}\right)^2 \theta_r^{\frac{1}{2}}(p).
\end{aligned}$$

For numerical examples demonstrating the usefulness of the Newton method for the solution of the time step problems, we refer to [Fuh94a, Fuh94b].

Bingham fluid. The flow of petroleum in rock or water in some clayey soils can show non-Newtonian behaviour. More precisely, it can be regarded as a *Bingham* flow which can be described by a degenerating viscosity term.
The evolution of such a fluid can stop at finite time. To demonstrate the possibilities of the numerical method, we use the following relationships which come from discussions with petroleum engineers. More involved expressions one finds in [WPW90].

$$\begin{aligned}
\rho(p) &= \rho_0 e^{\beta(p-p_0)} \\
F &= |\nabla p - \rho\gamma| \\
\frac{1}{\mu(F)} &= \begin{cases} 0 & , \ F < F_0 \\ F - F_0 & , \ F > F_0 . \end{cases}
\end{aligned}$$

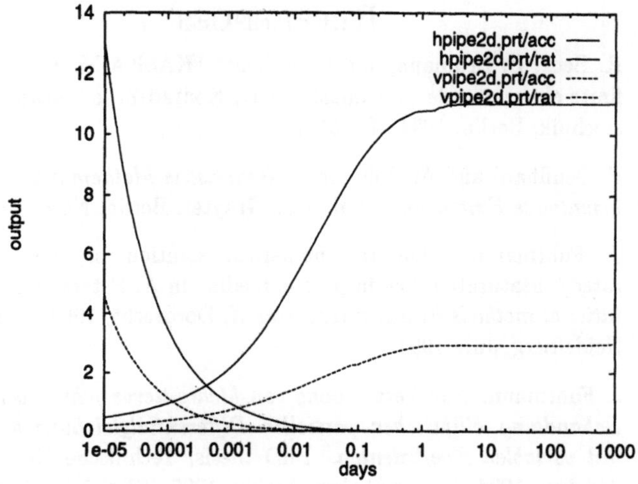

Figure 4: *Output rate and total yield of a horizontal and a vertical well (test data).*

Here, $\theta_r = K_r = 1$.
The parameters of the example below are still fictitious:

Ω	$[0,5] \times [0,1]$
γ	$(0,1)^T$
k	10^{-6}
β	10^{-2}
ρ_0	1
p_0	0

In the sub-domain $[0, 0.01] \times [0.5, 1]$ resp. $[0.49, 0.51] \times [0, 0.25]$ we define a vertical resp. horizontal "well" by setting k to 10^{-2}.
We proceed as follows: first apply a Dirichlet boundary condition with value 1 at the top of the domain and calculate a stationary solution letting $\mu = 1$. This stationary solution we take as an initial value. A constant initial value for the original problem makes no physical sense because of incompressibility and gravity. Then we remove the Dirichlet boundary condition at the top and apply a Dirichlet boundary condition with value 0 at the outlet of the "wells" ($[0, 0.1] \times 1$ or $0 \times [0.49, 0.51]$, respectively). We solve the initial boundary value problem for 1.1 and measure the amount of fluid leaving the domain through the well.
Fig. 4 shows that the output rate breaks down after a finite time though there is still fluid in the domain. This is a sign for the degenerate behaviour of the governing equation — the driving force F reaches the value F_0 in the whole domain and the fluid becomes immobile.

References

[BER94] R. Beck, B. Erdmann, and R. Roitzsch. KASKADE 3.0 - an object oriented finite element code. Technical report, Konrad-Zuse-Zentrum für Informationstechnik, Berlin, 1994. TR 95-4.

[DH91] P. Deuflhard and A. Hohmann. *Numerische Mathematik - eine algorithmisch orientierte Einführung*. Walter de Gruyter, Berlin, New York, 1991.

[Fuh94a] J. Fuhrmann. On the numerical solution of the equation of saturated/unsaturated flow in porous media. In A. Peters et al., editors, *Computational methods in water resources X*, Dordrecht, 1994. Kluwer. Proceedings Heidelberg, July 1994.

[Fuh94b] J. Fuhrmann. *Zur Verwendung von Mehrgitterverfahren bei der numerischen Behandlung elliptischer partieller Differentialgleichungen zweiter Ordnung mit variablen Koeffizienten*. PhD thesis, Technische Universität Chemnitz-Zwickau, 1994. Verlag Shaker, Aachen 1995, ISBN 3-8256-0522-0.

[Fuh95] J. Fuhrmann. A modular algebraic multilevel method. preprint no. 203, Weierstraß-Institut für Angewandte Analysis und Stochastik, Berlin, 1995.

[GG94] H. Gajewski and K. Gärtner. On the discretization of van Roosbroeck's equations with magnetic field. Technical report, ETH Zürich, Institut für integrierte Systeme, 1994. No. 94/14.

[Gil92] B.H. Gilding. Mathematical modelling of saturated and unsaturated groundwater flow. In S. Xiao, editor, *Flow and Transport in porous media*, pages 1–166, Singapore, 1992. World Scientific Publishing.

[JK91] W. Jäger and J. Kačur. Solution of porous medium type equations by linear approximation schemes. *Numerische Mathematik*, 60:407–427, 1991.

[Lan94] H. Langmach. Zur numerischen Simulation des reaktiven Schadstofftransports im porösen Medium mit Hilfe einer Box-Diskretisierung. In E. Holzbecher and G. Nützmann, editors, *Modellierung von Strömungs- und Ausbreitungsprozessen*, pages 47–62. Technische Universität Berlin, Institut für Wasserbau und Wasserwirtschaft, 1994. Mitteilung Nr. 126, ISSN 0409-1744.

[RS87] J.W. Ruge and K. Stüben. Algebraic multigrid. In S. McCormick, editor, *Multigrid methods*, volume 4 of *Frontiers in Applied Mathematics*, chapter 4, pages 73–130. SIAM, Philadelphia, 1987.

[Sch93] I. Schmelzer. 3D anisotropic grid generation with intersection-based geometry interface. Preprint, IMA, Minneapolis, 1993. no. 1180.

[vG80] M.Th. van Genuchten. A closed-form equation for predicting the hydraulic conductivity of unsaturated soils. *Soil Sci. Soc. Amer. J.*, 44:892–898, 1980.

[WPW90] Y.-S. Wu, K. Pruess, and P.A. Witherspoon. Flow and displacement of Bingham non-Newtonian fluids in porous media. In *60th California Regional Meeting, April 4-6, 1990*, pages 339–351. Society of Petroleum Engineers, 1990. SPE 20051.

FAST SOLVERS FOR NON-LINEAR FEM–BEM EQUATIONS

S.A. Funken & E.P Stephan

Institut für Angewandte Mathematik, University of Hannover
Welfengarten1, D–30167 Hannover, Germany

SUMMARY

This paper presents fast iterative solvers for coupled non-linear Finite Element and Boundary Element problems using a damped inexact Newton method á la Axellson and Kaporin. This method converges globally even if the second Gateaux-dervitative does not exist. The used solvers for the linear saddle point problems occuring in the modified Newton algorithm are optimal in the sense, that they are independent of the number of unknowns. These linear solvers are based either on preconditioned conjugate residual like methods, where no Schur Complement construction is required, or on an inner–outer iteration of Axelsson and Vassilevski. Both methods use multigrid of seperate positive semi–definite and negative definite parts of the coupled operator.
The efficiency of the solvers is shown by numerical experiments yielding fast convergence.

INTRODUCTION

We present fast iterative solvers for coupled non-linear Finite Element and Boundary Element problems using a damped inexact Newton method á la Axellson and Kaporin. This method converges globally even if the second Gateaux-dervitative does not exist [6]. The used solvers for the linear saddle point problems occuring in the modified Newton algorithm are optimal in the sense that they are independent of the number of unknowns. These linear solvers are based either on HMCR [4] (a preconditioned conjugate residual like method, where no Schur complement construction is required), or on an inner–outer iteration of Axelsson and Vassilevski [2]. Both methods use multigrid of seperate positive semi–definite and negative definite parts of the coupled operator.
The efficiency of the solvers is shown by numerical experiments yielding fast convergence.

THE INTERFACE PROBLEM

This section presents the interface problem and rewrites it equivalently with boundary integral operators as Problem (4) which will be solved by several efficient methods as described below.

For $\Omega_1 \subset \mathbb{R}^2$ being a bounded Lipschitz domain and $\Omega_2 := \mathbb{R}^2 \setminus \overline{\Omega}_1$ we consider the following problem involving the prescribed jumps across the interface $\Gamma := \partial\Omega_1$: *Given $f \in H^{-1}(\Omega)$, $u_0 \in H^{1/2}(\Gamma)$, $t_0 \in H^{-1/2}(\Gamma)$ and $p : \mathbb{R}^+ \to \mathbb{R}^+$ measurable and bounded. Find $u_1 \in H^1(\Omega_1)$, $u_2 \in H^1_{\mathrm{loc}}(\Omega_2)$, such that*

$$\begin{aligned} -\operatorname{div}\,(p(|\nabla u_1)| \cdot \nabla u_1) &= f \quad \text{in } \Omega_1 \\ -\Delta u_2 &= 0 \quad \text{in } \Omega_2 \end{aligned} \tag{1}$$

with

$$u_1 = u_2 + u_0, \quad p(|\nabla u_1|)\frac{\partial u_1}{\partial n} = \frac{\partial u_2}{\partial n} + t_0 \quad \text{on } \Gamma \tag{2}$$

and the regularity condition on u_2 at infinity with $a \in \mathbb{R}$

$$u_2(x) = a \log|x| + o(1) \quad \text{for } |x| \to \infty . \tag{3}$$

The interface problem (1)–(3) is equivalent to the following weak formulation which is the symmetric coupling method [3, 11].
Find $(u, \phi) \in H^1(\Omega_1) \times H^{-1/2}(\Gamma)$, such that

$$G\left(\begin{pmatrix} u \\ \phi \end{pmatrix}, \begin{pmatrix} v \\ \psi \end{pmatrix}\right) = H\begin{pmatrix} v \\ \psi \end{pmatrix} \quad \forall \begin{pmatrix} v \\ \psi \end{pmatrix} \in H^1(\Omega_1) \times H^{-1/2}(\Gamma) \tag{4}$$

with

$$\begin{aligned} G\left(\begin{pmatrix} u \\ \phi \end{pmatrix}, \begin{pmatrix} v \\ \psi \end{pmatrix}\right) &= 2 \cdot \textstyle\int_\Omega p(|\nabla u|)\nabla u . \nabla v \, dx + \langle W u + (K' - 1)\phi , v \rangle \\ &\quad + \langle \psi , (K-1)u - V\phi \rangle \end{aligned}$$

and

$$H\begin{pmatrix} v \\ \psi \end{pmatrix} = 2 \cdot \int_\Omega f v \, dx + \langle W u_0 + (K'+1)t_0 , v \rangle + \langle \psi , (K-1)u_0 - V t_0 \rangle .$$

Here, the brackets $\langle \cdot , \cdot \rangle$ denote the L^2-duality between the trace space $H^{1/2}(\Gamma)$ and its dual $H^{-1/2}(\Gamma)$, V represents the single-layer- and K the double-layer potential operator, K' its adjoint and W denotes the hypersingular integral operator, which is the normal derivative of the double-layer potential operator. For the definition of the Sobolev spaces $H^1(\Omega)$, $H^{1/2}(\Gamma)$ etc. see [10].

SOLVING METHODS

For simplicity we write the nonlinear problems (4) as

$$\mathcal{L}(x) = 0 , \quad \text{for } x \in H_h \times H_h^{-1/2} \tag{5}$$

where H_h, $H_h^{-1/2}$ are discretisations of $H^1(\Omega_1)$ and $H^{-1/2}(\Gamma)$, respectively. H_h denotes the space of continuous, piecewise linear functions on a regular triangulation of Ω_1, $H_h^{-1/2}$ denotes the space of piecewise constant functions on the discretisation of the interface Γ. Note that the choice of the right hand side in (5) is due to the definition of $\mathcal{L}$. The Jacobian of $\mathcal{L}$ is denoted by

$$L(x) := \mathcal{L}'(x) .$$

The existence of L^{-1} is ensured under appropriate conditions on p [11].

The Newton method transforms the nonlinear problem (5) into a sequence of linear ones. These linear problems are indefinite saddle point problems and have the following form:

$$\begin{pmatrix} A & B^T & 0 \\ B & (C+W) & (K^T - I) \\ 0 & (K-I) & -V \end{pmatrix} \begin{pmatrix} u_\Omega \\ u_\Gamma \\ u_\phi \end{pmatrix} = \begin{pmatrix} r_1 \\ r_2 \\ r_3 \end{pmatrix} \tag{6}$$

where u_Ω are the unknown nodal values of the finite element solution in Ω_1, u_Γ are the nodal values of u on the boundary Γ and u_ϕ represents an approximation for the co-normal derivative $p(|\nabla u_1|)\frac{\partial u_1}{\partial n}$ on Γ. The submatrix $\left(\begin{smallmatrix} A & B^T \\ B & C \end{smallmatrix}\right)$ is a FE–matrix and represents the discretisation of the interior Neumann problem in Ω_1 for the differential operator in (1). Here V, K, W are the discretisations of the integral operators denoted by the same letters in (4).

There are several fast solvers for such saddlepoint problems (e.g. HMCR [8], inner-outer iteration [7, 9], multigrid-method [5]) which have been used to solve efficiently the linear systems arising at each Newton step for the nonlinear system (5). In this way the costs in the standard Newton method are reduced. As a further disadvantage the standard Newton method is only locally convergent. These disadvantages do not occur in the damped inexact Newton method [1]. Its algorithm is as follows.
Since $\mathcal{L}$ is not necessarily differentiable, we chose another differentiable mapping $\mathcal{K}$ as an approximation to $\mathcal{L}$, which has to satisfy some relations. Let $K(v) := \mathcal{K}'(v)$ denote the Jacobian of $\mathcal{K}$.
Choose an initial guess x_0.

(i) For each linear system $K(x_k)s_k = -\mathcal{L}(x_k)$ compute s_k, s.t. for some ρ_k, $0 \le \rho_k < 1$ it holds
a) $\| \mathcal{L}(x_k)+K(x_k)s_k \| \le \rho_k \| \mathcal{L}(x_k) \|$. (After each HMCR-iteration check, if stopping criterion a) is satisfied.)

(ii) Normalize $s_k \to \hat{s}_k$

(iii) Search τ satisfying the inequality $\| \mathcal{L}(x_k + \tau\hat{s}_k) \| \le \sqrt{1-\frac{1}{2}c_k^2}\| \mathcal{L}(x_k) \|$. Let $x_{k+1} = x_k + \tau\hat{s}_k$. If "$\mathcal{L}(x_{k+1}) \neq 0$" go to (i).

If τ in (iii) increases at each step we decrease the parameter ρ_k in (i). Then, the linear system in the next Newton step will be solved with higher accuracy. The iterates x_k converge towards the exact solution x^* [1]. Hence, after j-steps the following iterates lie in the region of convergence of the standard Newton method, and the optimal damping parameter $\tau = 1$ is reached. This is shown in [6] for the above nonlinear FEM-BEM coupling problem (4) and corresponding problems in nonlinear elasticity.

NUMERICAL EXAMPLES

For the computations we consider Ω_1 to be the L-shaped domain with vertices $(0,0)$, $(s,0)$, (s,s), $(-s,s)$, $(-s,-s)$, $(0,-s)$ with $s = 0.25$, where the single layer potential operator V is positive definite. On a uniform mesh with triangular elements we use piecewise linear functions in Ω_1 and piecewise constant functions on Γ for discretisation. Into the right hand side of the nonlinear problem (4) we insert the data f, u_0, t_0, which are computed via (1), (2) from the harmonic functions

$$\begin{aligned} u_1(x,y) &= r^{\frac{2}{3}} \sin(\frac{2}{3}\alpha) \\ u_2(x,y) &= \frac{1}{2}\log\left((x+\frac{1}{8}) + (y-\frac{1}{8})\right) \end{aligned}$$

(in polar and Cartesian coordinates (r, α) and (x, y), respectively) and the function $p(t)$, which will be defined later. For preconditioning with the multigrid-method in the HMCR- or inner-outer-algorithm we take the smoother to be the Richardson method, where the damping parameter $\theta = 2$ and $\theta = 1$, are found to be optimal for the approximated inverse of $\left(\begin{smallmatrix} A & B^T \\ B & C \end{smallmatrix}\right)$ and V in our model problem. The single layer potential is further preconditioned by $P = I + Z$, where I is the identity operator and Z is the central difference approximation of the one-dimensional Laplacian [8].

In Fig. 1 a comparision of computing time for the Newton method and the damped inexact Newton method is given. The time to solve the problem (5) with $p(t) = 1 + \frac{5}{5t+1}$ for the inexact Newton method is $\approx 21\%$ of the time used by the standard Newton method.

The efficiency of the preconditioned HMCR-, inner-outer-algorithm and the multigrid method is illustrated in Fig. 2 which shows for all methods that the CPU-time increases linearly with the number of unknowns which is proportional to $\frac{1}{h}$. This underlines also the efficiency of the damped inexact Newton method together with these solvers, since the number of Newton iterations is independent of the number of unknowns – at least if $\tau = 1$.

Next we consider problem (5) with $p(t) = \frac{1}{100} \cdot \frac{t+100.05}{t+0.005}$. The standard Newton method with vanishing starting vector does not converge. By appropriate scaling of the new search direction $\hat{s}_k$ in the damped inexact Newton method one reaches the region of convergence of the standard Newton method. For the corresponding iterates we get damping parameter $\tau = 1$ in the damped inexact Newton method. Thus, we have the optimal damping parameter and quadratic convergence of the Newton method. A sequence of damping parameters for meshsize $h = \frac{1}{64}$ is shown in Fig. 3.

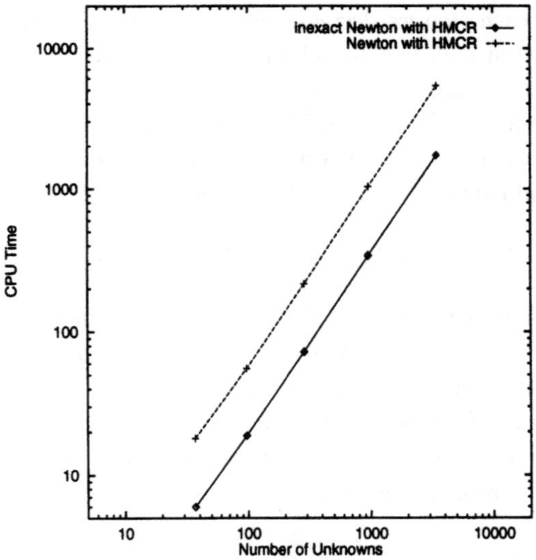

Figure 1: Computing time for damped inexact Newton and standard Newton

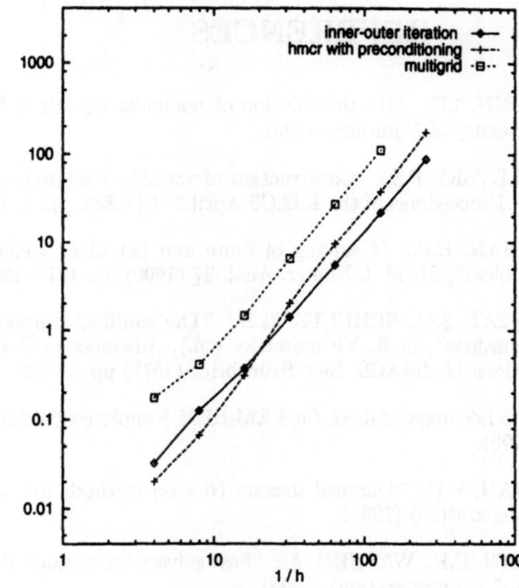

Figure 2: Computing time for a linear problem

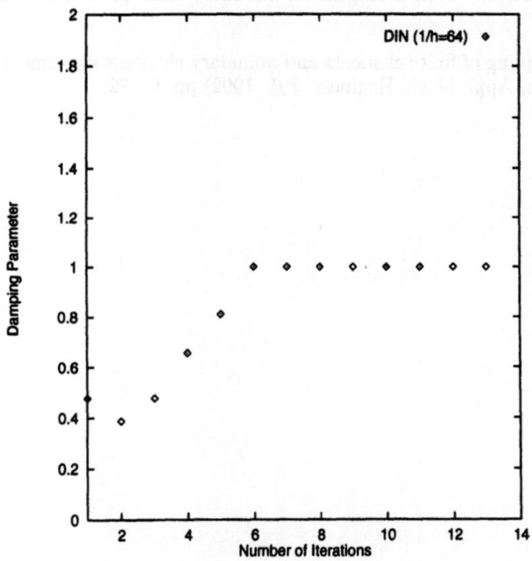

Figure 3: A sequence of damping parameters

REFERENCES

[1] AXELSSON, O., KAPORIN, I.E.: "On the solution of nonlinear equations for nondifferentiable mappings", Preprint University of Nijmegen (1995).

[2] AXELSSON, O., VASSILEVSKI, P.S.: "Construction of variable–step preconditioners for inner–outer iteration methods", Proceedings of the IMACS April 2–4 (1988), pp. 1–14.

[3] COSTABEL, M., STEPHAN, E.P.: "Coupling of finite and boundary element methods for an elastoplastic interface problem", SIAM J. Numer. Anal. 27 (1990) pp. 1212–1226.

[4] CHANDRA, R., EISENSTAT, S.C., SCHULTZ, M.H.: "The modified conjugate residual method for partial differential equations", in R. Vichnevetsky (ed.), Advances in Computer Methods for Partial Differential Equations *II*, IMACS, New Brunsbrick (1977) pp. 13–19.

[5] FUNKEN, S.A.: "Schnelle Lösungsverfahren für FEM-BEM Kopplungsgleichungen", Dissertation, Universität Hannover (1995).

[6] FUNKEN, S.A., STEPHAN, E.P.: "Damped inexact Newton methods for nonlinear FEM-BEM coupling problems" (in preparation) (1995).

[7] FUNKEN, S.A., STEPHAN, E.P., WATHEN, A.: "Fast solvers for coupled FEM-BEM equations *II*, inner / outer iterations" (in preparation) (1995).

[8] HAHNE, M., MAISCHAK, M., STEPHAN, E.P., WATHEN, A.: "Efficient preconditioners for coupled FEM-BEM equations", Numer. Meth. Part. Diff. Eqns. (submitted) (1994).

[9] HAHNE, M., STEPHAN, E.P., THIES, W.: "Fast Solvers for coupled FEM-BEM equations *I*", in W. Hackbusch and G. Wittum (eds.), Fast Solvers for Flow Problems, Notes on Numerical Fluid Mechanics 49 (1995) pp. 121–130.

[10] LIONS, J.L., MAGENES, E.: "Non-homogeneous boundary value problems and applications *I* ", Springer, Berlin (1972).

[11] STEPHAN, E.P.: "Coupling of finite elements and boundary elements for some nonlinear interface problems", Comp. Meth. Appl. Mech. Engineer. 101 (1992) pp. 61–72.

A MULTIGRID-HOMOGENIZATION METHOD

M. GRIEBEL AND S. KNAPEK
INSTITUT FÜR ANGEWANDTE MATHEMATIK
ABTEILUNG FÜR WISSENSCHAFTLICHES RECHNEN UND NUMERISCHE SIMULATION
UNIVERSITÄT BONN
D-53115 BONN

SUMMARY

For problems with strongly varying or discontinuous diffusion coefficients, we present a method to approximately determine the effective diffusion coefficient on the coarse scale level. It is based on techniques used also in multigrid, i.e. matrix-dependent prolongations and the construction of coarse grid operators by means of the Galerkin approximation. In numerical experiments, we compare our multigrid-homogenization method with homogenization, renormalization and averaging approaches.

INTRODUCTION

Exact solutions for problems, which model locally strong varying phenomena on a micro-scale level, require that all length scales appearing in the problem are completely resolved. In numerical simulations however, due to reasons of storage requirements and numerical complexity, this demand can often not be met, i.e. the grid for numerical simulation can not be chosen fine enough. However, in many practical applications, the fine-scale details of the solution are not of interest, but merely a coarse-scale solution is sought. Therefore, it is necessary, to work with averaged equations, which describe directly the large-scale behavior of the problem under consideration. Thus, the first step in the direction of a numerical treatment of the coarse scale problem is the determination of equations which model the influence of the unresolved fine scales. A typical example for such kind of problem is the diffusion equation with strongly varying diffusion coefficient as it arises from Darcy's law in reservoir simulation. There, the diffusion coefficient models the local permeability of the medium.

In the context of reservoir simulation, the development of so called upscaling techniques has been motivated by the development of geostatistical reservoir description algorithms. These algorithms result in fine scale descriptions of reservoir properties, such as permeability, on grids of cells far too fine to be used as a computational grid in simulations. Typically the order of magnitude between the geostatistical and the computational grid size is about a factor of 100 and more. The use of upscaling techniques is to assign values for permeability given on the geological grid to the much coarser simulation grid in such a way, that the gross features of the flow are preserved. Here the term effective diffusion or effective permeability is of importance. The effective permeability of a heterogenous medium is the permeability of an equivalent homogeneous medium. Equivalent means in this context that the total flow through the homogeneous medium is (for the same boundary conditions) the same as through the heterogenous medium.

In the earlier times of numerical simulation of reservoir performance this upscaling process was done by simply averaging the (measured or randomly generated) fine-scale diffusion coefficients using the arithmetic, geometric or harmonic mean. However, this approach turned out to be invalid for systems with strong permeability variations. Therefore, more advanced techniques are used nowadays, of which we chose renormalization and homogenization for comparison purposes. For a review of upscaling techniques see [Chr] and the references therein.

The discretization of an elliptic differential equation can be interpreted as a certain kind of averaging or filtering, where smaller scales than the respective grid size are eliminated. In multigrid methods, discretizations on successive coarser grids are needed. To obtain them, besides direct discretization on these grids, the Galerkin coarse grid discretization method is used frequently. It involves the discretization on the finest grid only and produces the operators and equations on the coarser levels by only using an interpolation operator between two successive grids. This corresponds basically to just an averaging of the fine grid equations. Now, in the multigrid field, there exist methods to determine the interpolation involved in the Galerkin coarsening in a *matrix-dependent* way. This results in stable and "physically meaningful" coarse grid operators which are necessary for good convergence rates of the corresponding multigrid method also in the case of interface problems and singular perturbed operators.

Thus, on the one hand, methods known from modelling (homogenization, renormalization) could be used for the determination of coarse grid operators in multigrid methods, by directly discretizing the homogenized or renormalized continuous equations, and using them as coarse grid operators [ER93, EL93, Kna95, Neu96]. On the other hand, matrix-dependent Galerkin approximations used in some multigrid methods to obtain robust solvers, lead to coarse grid operators, that directly describe the behavior of the coarse scale solution. By interpreting them as discrete coarse scale operators they lead to approximations of the effective diffusion equation. This is illustrated in Figure 1.

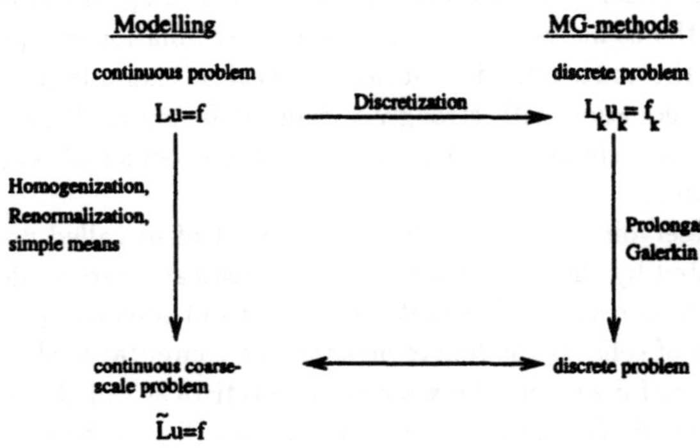

Fig. 1 Relation between mathematical modelling methods and multigrid coarsening

In this paper, we present the idea to use such multigrid Galerkin coarsening methods to derive a technique for modelling the effective, i.e. coarse-scale diffusion coefficient.

First, we consider a simple diffusion problem and describe the approaches and problems related to the mathematical modelling of coarse scale equations. Then, we consider the multigrid method and describe how matrix-dependent prolongations and Schur complement approximations lead to energy dependent averaging procedures and to averaged equations. Finally, we give some results from numerical experiments for the diffusion equation with different types of diffusion fields, compare our new method with averaging, renormalization and homogenization and discuss its properties, advantages and limitations.

THE PROBLEM OF APPROXIMATING THE EFFECTIVE DIFFUSION

We consider the model problem

$$\begin{array}{ll} -\nabla(D\nabla u) = f & \text{in } \Omega = [0,1]^2 \\ u = 0 & \text{on } \delta\Omega \end{array} \quad \text{where } D = \begin{pmatrix} d_{11}(x,y) & d_{12}(x,y) \\ d_{21}(x,y) & d_{22}(x,y) \end{pmatrix}. \tag{1}$$

In porous media flow, under the assumption $\vec{g} = 0$ (zero gravity), the conservation of mass, the slight incompressibility assumption $u = u_0 e^{cp}$ and the restriction to the stationary case, this equation results from Darcy's law $\vec{q} = -\frac{D}{\mu}(\nabla p - u\vec{g})$.[1]

There, the diffusion coefficient matrix D describes the permeability of the medium. In general, D is not continuous, highly varying and anisotropic. For isotropic media, D is a diagonal matrix with diag$(D) =: d \cdot I$, where d is a scalar function[2].

For locally strong varying D, a problem arises when it comes to numerical simulation: Due to storage requirements and numerical complexity, the grid Ω_k used in the discretization process can in general not be made fine enough to fully resolve the diffusion coefficient D. However, since fine scale details of the solution are usually not of interest, only a coarse scale solution is sought anyway. Therefore, the task of mathematical modelling is now to find a coarse scale problem describing directly the coarse scale solution of the problem. Thus, we are left with the problem to determine a so-called *effective* diffusion coefficient $\tilde{D}$, where

- $\tilde{D}$ is slowly varying or even constant,
- $\nabla\tilde{D}\nabla$ is a direct description of the problem on the coarse scale,
- $\tilde{D}$ takes the influence of the unresolved fine scales as good as possible into account.

Replacing D by $\tilde{D}$ in (1) then gives a continuous coarse scale problem, whose solution directly describes u on a coarse scale.

One way to gain an approximation of the effective diffusion coefficient is by simple averaging. There, for example the arithmetic mean

$$d_{i,j}^{arith} = \frac{1}{|U|}\int_U d_{i,j}(x,y)d\Omega, \quad U \subset \Omega, \tag{2}$$

or the geometric or harmonic mean can be applied (component-wise). However, these simple methods are not very accurate – especially in the case of strongly varying diffusion coefficients – and therefore are of little or even no use. This can easily be seen from the example of a diagonal matrix D for which any averaging procedure must result in again an effective diagonal matrix $\tilde{D}$. However, for a diffusion coefficient involving layers in diagonal direction (which still can be described on the fine scale by a diagonal D) this is not more adequate, as $\tilde{D}$ must contain non-diagonal entries which describe diffusion in the diagonal direction.

Another, more sophisticated approach, is the so-called renormalization method, described in [Kin89]. The idea is, to locally compute an approximation of the effective diffusion coefficient by treating it as a resistance network (consider a diffusion coefficient of the form $D = \alpha \cdot I$ with I the identity matrix, then the inverse of the diffusion is taken

[1] Here, p denotes the pressure, u the density, μ the dynamic viscosity of the fluid, $\vec{g}$ describes the gravity and $\vec{q}$ the volume of the fluid transported per time unit and area unit, f denotes a sink or source term.

[2] If D is a bounded, symmetric and elliptic function, that is $\exists\alpha \in \mathrm{R}^+ : d_{ij}(x)\xi_i\xi_j \geq \alpha\xi_i\xi_j \forall x \in \Omega, \xi = (\xi_i) \in \mathrm{R}^n$ and $d_{ij} = d_{ji}, (1 \leq i,j \leq n)$ and if f in $L^2(\Omega)$, then (1) is an elliptic boundary value problem.

as resistance, that is $R \equiv 1/\alpha$). However, a severe drawback of this approach is, that only diagonal matrices for D can be used. Again the resulting $\tilde{D}$ is diagonal and, therefore, this method also fails for layered structures in the diagonal directions.

Finally we consider the method of homogenization. For a general description of this approach see for example [Bab76, San80, BP89, JKO94] or [BLP87]. In the simplest case, one assumes a periodic structure of the diffusion coefficient (with the parameter ϵ as a characteristic length). If u_ϵ is the solution of

$$-\nabla(D^\epsilon(x)\nabla u_\epsilon) = f(x) \text{ in } \Omega, \quad (3)$$

$$u_\epsilon = 0 \text{ on } \partial\Omega \quad (4)$$

with D^ϵ periodic with characteristic length ϵ, then the homogenization approach is to find the equations describing u_ϵ as $\epsilon \searrow 0$, and to find u_0, the limit of u_ϵ. In contrast to simple averaging and to renormalization, the homogenized diffusion coefficient $\tilde{D}$ is in general no longer a diagonal matrix.

MATRIX-DEPENDENT MULTIGRID COARSENING AS AN APPROXIMATE HOMOGENIZATION METHOD

Now we turn to our approach for the computation of an effective diffusion coefficient. It is based on methods used already for some time in robust multigrid algorithms which use matrix-dependent prolongations. The main point is, that the coarse grid operators built with matrix-dependent prolongations by means of the Galerkin approximation can be viewed as some sort of discrete homogenized operators, see also Figure 1. From the coarse grid limit operator, approximations of the effective diffusion coefficient can be read of, which are (in the periodic case) in 1D the exact homogenized diffusion coefficients, and in 2D approximations of the homogenized coefficient (depending on the strategy how the matrix-dependent prolongation is chosen).

In the following we shortly review multigrid and then discuss matrix-dependent prolongations. We give a general principle on which all schemes we know of can be based. This shows their relationship to incomplete factorization and approximate block Gaussian elimination.

Then we shortly describe the procedure to compute the multigrid homogenized diffusion coefficient and discuss advantages and limitations of the method.

Multigrid discretization

Multigrid is an efficient method for solving partial differential equations. There, a sequence of grids

$$\Omega_0 \subset \Omega_1 \subset \Omega_2 \subset \ldots \subset \Omega_{k-1} \subset \Omega_k$$

is set up. Furthermore, mappings between the associated grid function spaces are needed. Here, R_l^{l-1} denotes the restriction from level l to $l-1$ and P_{l-1}^l denotes the prolongation from level $l-1$ to l which is a (possibly weighted) interpolation. Then, given a discretization L_k on the finest level k, coarse grid operators $L_l, l = k-1, ..0$, must be determined. Beside direct discretization on level l, often the Galerkin approximation is used, i.e. L_{l-1} is computed from L_l by the application of restriction and prolongation:

$$L_{l-1} = R_l^{l-1} L_l P_{l-1}^l, \text{ where mostly } R_l^{l-1} := (P_{l-1}^l)^T. \quad (5)$$

Thus, the coarse grid operators are (recursively) determined by the discretization on the finest grid L_k and the prolongations P^l_{l-1}.

It is well known that, for singular perturbed problems and operators with locally highly varying or discontinuous coefficient functions, the performance and robustness of a multigrid solver depends on the right choice of the coarse grid operators, and thus, in the context of (5), on the fine grid discretization and the choice of the prolongations only. For example, for jumps in the coefficient matrix D, standard prolongation and restriction lead to a deterioration of the convergence rate. A remedy to this problem is the choice of matrix-dependent prolongations, which will be described later. They lead via (5) to coarse grid operators, which describe the large scale behavior of the problem more properly. Note, that in this way the Galerkin approximation can then be interpreted as a discrete method for calculating averaged equations, which also model the influence of the scales smaller than the grid size of the actual level.

Matrix-dependent prolongations

The often used standard prolongations

$$P^l_{l-1}{}^* = \begin{bmatrix} 1/2 & 1 & 1/2 \end{bmatrix}, \text{ and } P^k_{k-1}{}^* = \begin{bmatrix} 1/4 & 1/2 & 1/4 \\ 1/2 & 1 & 1/2 \\ 1/4 & 1/2 & 1/4 \end{bmatrix}$$

(here in stencil notation for the 1D- and 2D-case) involving (bi-)linear interpolation, correspond to standard nodal basis functions on every level. Now, a short calculation shows that, with this choice in the Galerkin approach (5), just an *arithmetic averaging* of D is performed. As mentioned already, arithmetic averaging is however not suited for the computation of the effective diffusion coefficient in the case of strongly varying or discontinuous D. Therefore, we will not use standard prolongation in the multigrid coarsening process. Instead, we rely on the information stored in the matrix L_l to construct a *locally weighted* prolongation P^l_{l-1} that resembles an operator-induced interpolation which leads via (5) to an implicitly weighted averaging of D.

Matrix-dependent prolongations P^l_{l-1} can be derived using a general principle: They can be interpreted as transformations to perform (approximately) a block Gaussian elimination of the operator L_l, which will be explained in the following. Let there be given a discrete problem $L_l u_l = f_l$ on a grid Ω_l. Partitioning of L_l in coarse grid and fine-without-coarse grid parts gives

$$\begin{pmatrix} L_{11} & L_{12} \\ L_{21} & L_{22} \end{pmatrix} \begin{pmatrix} u_1 \\ u_2 \end{pmatrix} = \begin{pmatrix} f_1 \\ f_2 \end{pmatrix} \tag{6}$$

where u_1 denotes the unknowns in $\Omega_l \setminus \Omega_{l-1}$ and u_2 the unknowns in the next coarser grid Ω_{l-1}. Now, the optimal matrix-dependent prolongation arises from the exact block Gaussian elimination of the outer diagonal blocks L_{12} and L_{21}, i.e. from transforming L_l to

$$\begin{pmatrix} I & -L_{11}^{-1}L_{12} \\ 0 & I \end{pmatrix}^T \begin{pmatrix} L_{11} & L_{12} \\ L_{21} & L_{22} \end{pmatrix} \begin{pmatrix} I & -L_{11}^{-1}L_{12} \\ 0 & I \end{pmatrix} = \begin{pmatrix} L_{11} & 0 \\ 0 & L_{22} - L_{21}L_{11}^{-1}L_{12} \end{pmatrix}. \tag{7}$$

The associated matrix-dependent prolongation operator between grid $l-1$ and l is then defined by

$$P^l_{l-1} = \begin{pmatrix} -L_{11}^{-1} L_{12} \\ I \end{pmatrix} \tag{8}$$

and the coarse grid operator obtained via the Galerkin approach (5) is then

$$L_{l-1} = (P^l_{l-1})^T L_l P^l_{l-1} := L_{22} - L_{21} L_{11}^{-1} L_{12} \tag{9}$$

which is just the well known Schur complement of L_l with respect to Ω_{l-1}.

Of course, this approach can usually not be used for the construction of a coarse grid operator, since, in general, L_{11} is not a diagonal matrix, and therefore, L_{l-1} would be no local operator anymore. The result would be an inefficient multigrid algorithm. However, in the 1D case, this is different: Let L_l be given by the 3-point stencil

$$L^*_k(x_i) = [\ L_{-1}(x_i) \quad L_0(x_i) \quad L_1(x_i)\ ]. \tag{10}$$

The partitioning of the associated matrix according to (6) now results in just a diagonal matrix L_{11}. Then L_{11}^{-1} is diagonal and, consequently, the prolongation P^l_{l-1} and the coarse grid operator L_{l-1} are local operators. In stencil notation, we obtain from (8)

$$P^l_{l-1}{}^*(x_{i+1}) = \left[\ -\frac{L_1(x_i)}{L_0(x_i)} \quad 1 \quad -\frac{L_{-1}(x_{i+2})}{L_0(x_{i+2})}\ \right] \tag{11}$$

and by the Galerkin approach (9)

$$L_{l-1}{}^*(x_{i+1}) = \left[\ -\frac{L_{-1}(x_i)L_{-1}(x_{i+1})}{L_0(x_i)} \quad \frac{L_{-1}(x_i)L_{-1}(x_{i+1})}{L_0(x_i)} + \frac{L_1(x_{i+1})L_1(x_{i+2})}{L_0(x_{i+2})} \quad -\frac{L_1(x_{i+1})L_1(x_{i+2})}{L_0(x_{i+2})}\ \right]$$

for $x_{i+1} \in \Omega_{l-1}$.

For the special case of the diffusion equation $(d(x)u(x)_x)_x = f(x)$, where $d(x)$ is assumed to be piecewise constant on Ω_k having jumps only at the grid points, we obtain, by discretizing with finite differences or finite elements on Ω_k, the stencil

$$L_k{}^*(x_{IF+1}) = \frac{1}{h_k}\left[\ -d^k_{IF+1} \quad d^k_{IF+1} + d^k_{IF+2} \quad -d^k_{IF+2}\ \right] \tag{12}$$

with $d^k_{IF+1} = d(x_{IF+1} - h_k/2)$, compare also Figure 2.

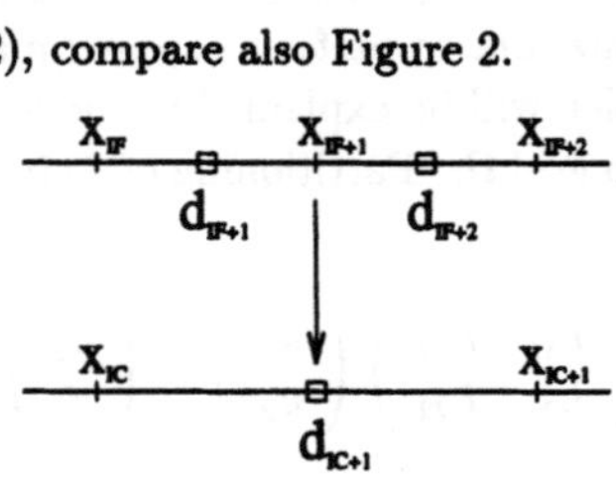

Fig. 2 1D case: fine and coarse grid

Now, using the corresponding prolongation (11)

$$P^k_{k-1}{}^*(x_{IC}) = \left[\ \frac{d^k_{IF}}{d^k_{IF-1}+d^k_{IF}} \quad 1 \quad \frac{d^k_{IF+1}}{d^k_{IF+1}+d^k_{IF+2}}\ \right] \text{ for } x_{IC} \in \Omega_{k-1}, \tag{13}$$

we obtain

$$L_{k-1}{}^*(x_{IC}) = \frac{2}{h_{k-1}}\left[\ -\frac{d^k_{IF-1}d^k_{IF}}{d^k_{IF-1}+d^k_{IF}} \quad \frac{d^k_{IF-1}d^k_{IF}}{d^k_{IF-1}+d^k_{IF}} + \frac{d^k_{IF+1}d^k_{IF+2}}{d^k_{IF+1}+d^k_{IF+2}} \quad -\frac{d^k_{IF+1}d^k_{IF+2}}{d^k_{IF+1}+d^k_{IF+2}}\ \right] \tag{14}$$

for $x_{IC} \in \Omega_{k-1}$. Then, comparing (14) with (12) gives the coarse grid diffusion coefficient on the interval $[x_{IC}, x_{IC+1}]$:

$$d_{IC+1}^{k-1} = 2\frac{d_{IF+1}^{k} d_{IF+2}^{k}}{d_{IF+1}^{k} + d_{IF+2}^{k}} \tag{15}$$

which is just the harmonic mean of the two corresponding fine grid values. In other words: The direct discretization on the coarse grid Ω_{k-1} using the *local harmonic average* of the fine grid diffusion coefficient would give the same coarse grid operator. Now, if we apply this approach recursively for the sequence of successively coarser grids, in the end, we obtain on Ω_0 the global harmonic mean as effective diffusion coefficient. Here, we can use periodic conditions on the boundary as well. Then, we obtain the *same* result as with homogenization.

However, in the two-dimensional case, the exact matrix-dependent prolongation cannot be used anymore, since now, the unknowns belonging to the points from $\Omega_k \setminus \Omega_{k-1}$ are coupled to the unknowns belonging to coarse grid points and neighbored fine grid points by the entries in $\begin{pmatrix} L_{11} & L_{12} \end{pmatrix}$. Then, L_{11} is no longer a diagonal matrix as it was in the one-dimensional case. Consequently, the matrix-dependent prolongation and the coarse grid operators are no local operators anymore and they would involve the computation of L_{11}^{-1} which is i.g. a full matrix.

Therefore, instead of the optimal matrix-dependent prolongation, approximations to it are used, which result in local operators. They can be gained using the following general approach: Substitute the set of equations associated to the grid points $\Omega_k \setminus \Omega_{k-1}$ by an altered set of equations, i.e. substitute $\begin{pmatrix} L_{11} & L_{12} \end{pmatrix}$ by some $\begin{pmatrix} \tilde{L}_{11} & \tilde{L}_{12} \end{pmatrix}$, which meets the following demands:

- $\tilde{L}_{11}$ is 'easy' to invert,
- $\tilde{L}_{11}^{-1}\tilde{L}_{12}$ is a 'good' approximation of $L_{11}^{-1}L_{12}$, and
- $\begin{pmatrix} -\tilde{L}_{11}^{-1}\tilde{L}_{12} \\ I \end{pmatrix}$ is a local operator.

Then, the associated prolongation operator $\tilde{P}_{k-1}^{k}$ is defined by

$$\tilde{P}_{k-1}^{k} = \begin{pmatrix} -\tilde{L}_{11}^{-1}\tilde{L}_{12} \\ I \end{pmatrix}. \tag{16}$$

Thus, the approximation of the optimal matrix-dependent prolongation is the exact optimal matrix-dependent prolongation of an altered discretization, namely

$$\tilde{L}_k = \begin{pmatrix} \tilde{L}_{11} & \tilde{L}_{12} \\ L_{21} & L_{22} \end{pmatrix}. \tag{17}$$

Now, we consider simple examples how to construct an approximate matrix-dependent prolongation. To this end, we proceed as follows: If $l > 0$, then every grid Ω_l can be divided into four disjunct grids:

$$\begin{aligned}
\Omega_l^{00} &:= \{(x,y) \in \Omega_l : x = ih_l, y = jh_l, i,j \in \mathbb{Z}, i \text{ even}, j \text{ even}\} = \Omega_{l-1} \\
\Omega_l^{10} &:= \{(x,y) \in \Omega_l : x = ih_l, y = jh_l, i,j \in \mathbb{Z}, i \text{ odd}, j \text{ even}\} \\
\Omega_l^{01} &:= \{(x,y) \in \Omega_l : x = ih_l, y = jh_l, i,j \in \mathbb{Z}, i \text{ even}, j \text{ odd}\} \\
\Omega_l^{11} &:= \{(x,y) \in \Omega_l : x = ih_l, y = jh_l, i,j \in \mathbb{Z}, i \text{ odd}, j \text{ odd}\}.
\end{aligned} \tag{18}$$

Then we can partition the matrix L_l as

$$L_l = \begin{pmatrix} L_{11} & L_{12} \\ L_{21} & L_{22} \end{pmatrix} \begin{pmatrix} l_{11} & l_{12} & l_{13} \\ l_{21} & l_{22} & l_{23} \\ l_{31} & l_{32} & l_{33} \end{pmatrix} = \left(\begin{pmatrix} \begin{pmatrix} l_{11} & l_{12} \\ l_{21} & l_{22} \end{pmatrix} & \begin{pmatrix} l_{13} \\ l_{23} \end{pmatrix} \\ \begin{pmatrix} l_{31} & l_{32} \end{pmatrix} & l_{33} \end{pmatrix} \right). \tag{19}$$

Here, $\begin{pmatrix} l_{11} & l_{12} & l_{13} \end{pmatrix}$ corresponds to the equations belonging to the unknowns associated to the points in Ω^{11}, $\begin{pmatrix} l_{21} & l_{22} & l_{23} \end{pmatrix}$ corresponds to the equations belonging to the unknowns associated to the points in $\Omega^{10} \cup \Omega^{01}$ and $\begin{pmatrix} l_{31} & l_{32} & l_{33} \end{pmatrix}$ corresponds to the equations belonging to the unknowns associated to the points in $\Omega^{22} = \Omega_{l-1}$. Now, if L_l can be described by a 9-point stencil then l_{11} and l_{33} are diagonal matrices.

Surely, the easiest way for a substitution is to replace L_{11} by its diagonal $diag\{L_{11}\}$.

Another possibility, which is obviously more accurate, is to choose

$$\tilde{L}_{11} = \begin{pmatrix} l_{11} & l_{12} \\ 0 & \tilde{l}_{22} \end{pmatrix} \tag{20}$$

(remember that l_{11} is diagonal!) with $\tilde{l}_{22}$ a diagonal matrix or an upper triangular matrix. Then, the inverse is given by

$$\tilde{L}_{11}^{-1} = \begin{pmatrix} l_{11}^{-1} & -l_{11}^{-1} l_{12} \tilde{l}_{22}^{-1} \\ 0 & \tilde{l}_{22}^{-1} \end{pmatrix} \tag{21}$$

and the prolongation can be defined by

$$\tilde{P}_{l-1}^{l} = \begin{pmatrix} -\tilde{L}_{11}^{-1} \begin{pmatrix} l_{13} \\ \tilde{l}_{23} \end{pmatrix} \\ I \end{pmatrix} = \begin{pmatrix} -l_{11}^{-1}(l_{13} - l_{12}\tilde{l}_{22}^{-1}\tilde{l}_{23}) \\ -\tilde{l}_{22}^{-1}\tilde{l}_{23} \\ I \end{pmatrix}.$$

Note that this is the optimal prolongation of the altered discretization

$$\tilde{L}_l = \begin{pmatrix} l_{11} & l_{12} & l_{13} \\ 0 & \tilde{l}_{22} & \tilde{l}_{23} \\ l_{31} & l_{32} & l_{33} \end{pmatrix}. \tag{22}$$

One way to obtain a diagonal $\tilde{l}_{22}$ is to perform the following substitutions, which involve 'directional' lumping:

$$\begin{bmatrix} L_{-1,1} & L_{0,1} & L_{1,1} \\ L_{-1,0} & L_{0,0} & L_{1,0} \\ L_{-1,-1} & L_{0,-1} & L_{1,-1} \end{bmatrix} \longrightarrow \left[\begin{array}{c|c|c} 0 & 0 & 0 \\ \hline L_{-1,0} & L_{0,0} & L_{1,0} \\ +L_{-1,-1} & +L_{0,-1} & +L_{1,-1} \\ +L_{-1,1} & +L_{0,1} & +L_{1,1} \\ \hline 0 & 0 & 0 \end{array}\right] \text{ in } \Omega^{10}, \tag{23}$$

$$\begin{bmatrix} L_{-1,1} & L_{0,1} & L_{1,1} \\ L_{-1,0} & L_{0,0} & L_{1,0} \\ L_{-1,-1} & L_{0,-1} & L_{1,-1} \end{bmatrix} \longrightarrow \begin{bmatrix} 0 & L_{0,1} + L_{-1,1} + L_{1,1} & 0 \\ 0 & L_{0,0} + L_{-1,0} + L_{1,0} & 0 \\ 0 & L_{0,-1} + L_{-1,-1} + L_{1,-1} & 0 \end{bmatrix} \text{ in } \Omega^{01}. \tag{24}$$

These substitutions lead to the prolongation described in [Den82]. Note that the lumping introduced in (23,24) in Ω^{01} and Ω^{10} is aligned with the coordinate directions.

Further prolongations can easily be constructed in a similar way. Such matrix-dependent prolongations are described, for example, in [ABDP81, Den83, Den82, FG93, Fuh94, Re93a, Re93b] and [AV89, AV90, AV91, Zee90]. We state that all the matrix-dependent schemes we know of can be based on the principle mentioned above and can be denoted also in form of a substitution (17). This shows their relationship to incomplete factorization and approximate block Gaussian elimination. Since, for our simple model problem (1), most of these methods coincide with the scheme due to Dendy [Den82], we will restrict ourselves to it in the numerical experiments.

The resulting coarse grid operator L_{l-1} built from a 9-point stencil matrix L_l and the associated matrix-dependent prolongation P_{l-1}^{l} by means of the Galerkin approximation (9) are again given by 9-point stencils and our matrix-dependent coarsening procedure can be repeated recursively on coarser grids.

In the two-dimensional case, for periodic boundary conditions, we no longer have the exact equivalence of homogenization and recursive matrix-dependent Galerkin coarsening as we had in the one-dimensional case. Now we introduced an error compared with homogenization due to the substitution (17) which is necessary to keep the coarse grid operators in the recursive process local.

As long as the diffusion coefficient is of tensor product type, or is near to it, the lumping procedure (23,24) works well, as can be seen from the experiments below. However, in the worst case, multigrid homogenization with (23,24) reduces to computing the arithmetic mean. See example 4 below.

The multigrid homogenized diffusion coefficient

In this section we describe, how we compute an approximation of the effective diffusion coefficient (in the following called multigrid-homogenized (MG-homogenized) diffusion coefficient). We first consider the periodic and then the non periodic case.

Periodic case: Let D be a periodic diffusion coefficient on an infinitely large domain, and imagine the diffusion coefficient to be sufficiently resolved by the numerical grid. We use bilinear finite elements to gain an discrete operator, which can be described by a nine point stencil in every grid point. Then it is possible to apply the matrix dependent Galerkin coarsening procedure recursively until we have (due to the periodicity of D) in every coarse grid point the same discrete operator, that is the same 9-point stencil, which is invariant under further Galerkin coarsening steps. This coarse scale limit operator is our so-called MG-homogenized operator. It is obvious, that due to the periodicity of D it suffices to compute the coarse grid operator for a finite domain, that is for only a few periods of the diffusion coefficient.

From this discrete large scale operator, we compute the approximation of the effective diffusion coefficient (D^{MG-hom}) by interpreting the stencil of the MG-homogenized operator as a linear combination of the finite element stencils associated to I, ∂_x, ∂_y, ∂_{xx}, ∂_{yy}, ∂_{xy}, ∂_{xxy}, ∂_{xyy} and ∂_{xxyy} which form a basis of the nine point stencil. Then we read of the entries of D^{MG-hom} from the coefficients of ∂_{xx}, ∂_{yy} and ∂_{xy}. It should be mentioned, that it is in general not possible to describe the coarse scale limit operator as a linear combination of the finite element stencils of $\partial_{xx}, \partial_{yy}$ and ∂_{xy} only, as we will see in the following section.

Altogether, we obtain the following scheme:

1. Discretize the diffusion equation on a domain Ω of several periods of D
2. Apply a matrix dependent Galerkin coarsening step to the fine grid operator
3. Repeat step 2 until in every grid point the same invariant stencil (invariant under Galerkin coarsening) is obtained
4. Interpret this stencil as linear combination of the finite element stencils associated to I, ∂_x, ∂_y, ∂_{xx}, ∂_{yy}, ∂_{xy}, ∂_{xxy}, ∂_{xyy} and ∂_{xxyy} and read of the entries of D^{MG-hom} which are defined as the coefficients of ∂_{xx}, ∂_{yy} and ∂_{xy} in the linear combination

Non periodic case: In the case of a randomly distributed diffusion coefficient, we apply Galerkin coarsening recursively on the sequence of grids $\Omega_k, \Omega_{k-1}, \ldots \Omega_0$. We then determine the corresponding approximation of the effective diffusion coefficient from the stencil belonging to the middle grid point on the second coarsest level Ω_1. In this way, the method is independent of the boundary conditions. Then we again interpret this stencil as a linear combination of the finite element stencils associated to the basis of the 9-point stencil given above and read of the entries of D^{MG-hom} as in 4 above.

Note that if the diffusion coefficient D can not be completely resolved on the finest grid, then the result of MG-homogenization is dependent on the size of the finest grid, as the discretization on this grid is in itself some kind of averaging.

NUMERICAL EXPERIMENTS

In the following, we compare the different methods to obtain an effective diffusion coefficient for our model problem (1). Here, if not indicated otherwise, we consider the periodic setting for the computation of the approximation D^{MG-hom} of the effective diffusion coefficient. We use bilinear finite elements for discretization and repeat the Galerkin coarsening process using Dendys matrix-dependent prolongations.

Example 1: Domain with inclusion

Now, we define the diffusion coefficient of our first model problem by

$$D(x,y) = \begin{cases} \alpha \cdot I & \text{in subdomain } Y_I \\ I & \text{in subdomain } Y_{II}. \end{cases}$$

This is illustrated in Figure 3 (left). By mean of periodicity, structures like in Figure 3 (right) are associated, see also [ABK91].

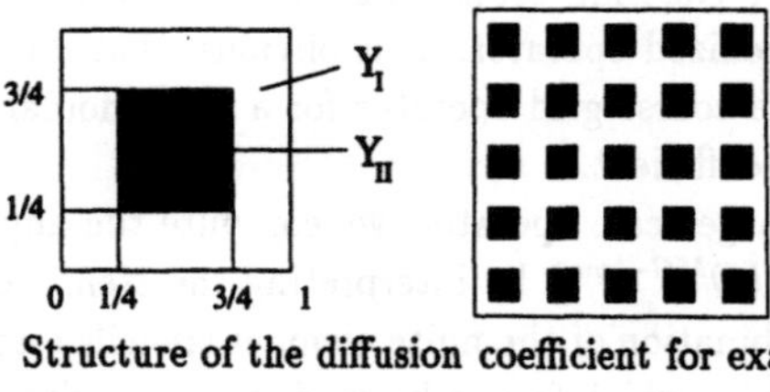

Fig. 3 Structure of the diffusion coefficient for example 1

Because of the symmetry of the problem and the diagonal structure of D, we obtain from all methods an approximation of the effective diffusion operator $\nabla \tilde{D} \nabla$ where

$$\tilde{D} = d \cdot I, \quad d \in \mathbb{R}.$$

The value for d depends on the value α and the respective method under consideration. For two different α, the results are given in Table 1. We clearly see that the approach based on recursive matrix-dependent Galerkin coarsening gives values which are closest to the exact values obtainable for this periodic problem by homogenization. All other approaches give worse values.

Table 1 Problem 1: Comparison of the approximations of the effective diffusion coefficient

α	d^{hom}	d^{MG-hom}	d^{ren}	d^{arith}	d^{geo}	d^{harm}
10	6.52	6.70088	6.16712	7.75	5.6234	3.0769
100	59.2	61.73646	54.24763	75.25	31.6228	3.8835

Note that the computed values are not depending on the grid size of the finest level used, if the jumps of the diffusion coefficient coincide with the grid lines. However, for more general problems involving a diffusion coefficient which is not constant on each cell of the finest grid, the resulting approximations of the effective diffusion coefficient depend on the fine grid size.

The good results of MG-homogenization as well as those of renormalization are due to the tensor product form of the diffusion coefficient. Therefore, MG-homogenization, as well as the corresponding multigrid solvers work well in this case.

Example 2: Layered medium

Now, we consider a diffusion coefficient which is of tensor product type and depends on the x-coordinate only, i.e.

$$D = \begin{pmatrix} d(x) & 0 \\ 0 & d(x) \end{pmatrix} \text{ where } d(x) = \begin{cases} d_1 & \text{in subdomain } Y_I \\ d_2 & \text{in subdomain } Y_{II}, \end{cases} \quad \text{with } d_1, d_2 \in \mathbf{R}^+.$$

This resembles a layered structure of the medium which is aligned to the y-direction, see Figure 4.

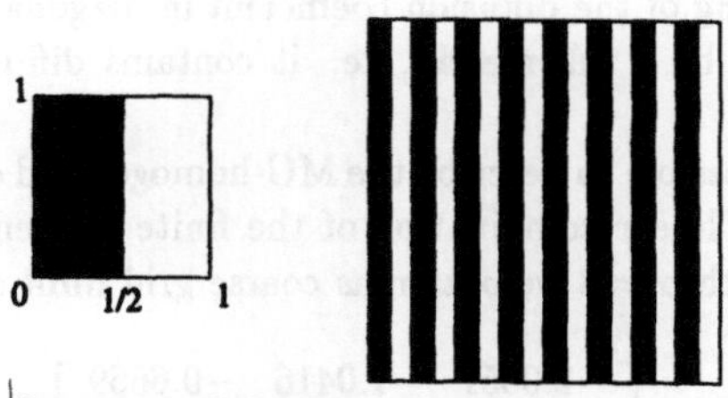

Fig. 4 Structure of the diffusion coefficient for example 2

Then, homogenization results in

$$d_{11}^{hom} = \frac{2d_1d_2}{d_1+d_2}, \quad d_{22}^{hom} = \frac{1}{2}(d_1+d_2), \quad \text{and} \quad d_{12} = d_{21} = 0, \tag{25}$$

i.e. in x-direction, we obtain the harmonic average of d_1 and d_2, and, in y-direction, we get the arithmetic average of d_1 and d_2.

For the MG-homogenization, we have to consider the discretization and the corresponding matrix-dependent prolongation. Here, the stencil of the finite element discretization operator L_k in any point is

$$L_k^* = \begin{bmatrix} L_{-1} & (L_{-1}+L_1)/2 & L_1 \\ L_{-1} & -4(L_{-1}+L_1) & L_1 \\ L_{-1} & (L_{-1}+L_1)/2 & L_1 \end{bmatrix}$$

Plugging this into the formulae of Dendy's approach, we directly see the following: The corresponding matrix-dependent prolongation reduces in y-direction to just the standard interpolation and thus results in arithmetic averaging, whereas, in x-direction, we obtain the harmonic average of the diffusion coefficient like in the 1D example (15). Thus, for this problem, homogenization and MG-homogenization give the *same* values. Furthermore, the result (25) is also obtained with renormalization.

Example 3: Diagonally layered structure

In our third considered example, we now set the diffusion coefficient to

$$D(x,y) = \begin{cases} I & \text{if } (x,y) \in Y_I \\ \alpha \cdot I & \text{if } (x,y) \in Y_{II} \end{cases}$$

where the subdomains Y_I and Y_{II} are indicated in Figure 5 (left), see also [SV93]. Then, by means of periodicity, layered structures with diagonal direction like the one given in 5 (right) are built.

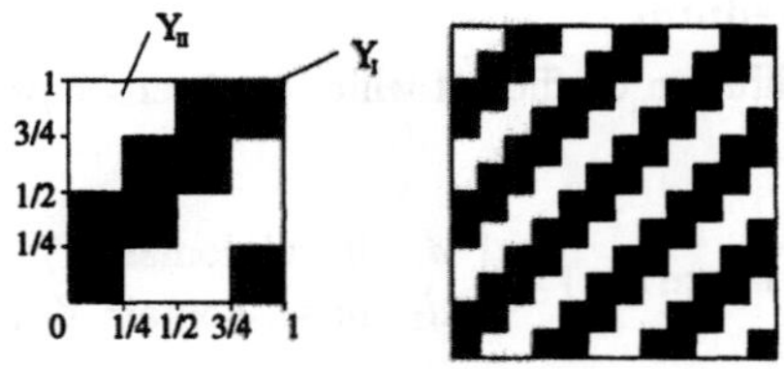

Fig. 5 Subdomains for diffusion coefficient of problem 3

Because of the layering of the diffusion coefficient in diagonal direction, the effective diffusion coefficient must be a full matrix, i.e. it contains diffusion in the off-diagonal entries of $\tilde{D}$ as well.

Here it is not more possible to describe the MG-homogenized operator we obtain with Galerkin coarsening as a linear combination of the finite element stencils associated to $\partial_{xx}, \partial_{yy}$ and ∂_{xy} only. With $\alpha = 8$ we obtain as coarse grid limit operator

$$A = \begin{bmatrix} -2.0551 & -1.0416 & -0.6669 \\ -1.0416 & 9.6105 & -1.0416 \\ -0.6669 & -1.0416 & -2.0551 \end{bmatrix},$$

which is a linear combination of the finite element stencils corresponding to ∂_{xx}, ∂_{yy}, ∂_{xy} and ∂_{xxyy}. Reinterpretating this as a continuous equation, we have the continuous MG-homogenized equation

$$-\nabla D^{MG-hom}\nabla + 0.1065 \cdot h_k^2 \frac{\partial^4}{\partial x^2 \partial y^2},$$

with

$$D^{MG-hom} = \begin{pmatrix} 3.76 & 1.39 \\ 1.39 & 3.76 \end{pmatrix},$$

where h_k is the mesh width. We see that through the Galerkin coarsening an additional term of higher order has been introduced.

The approximations that we obtained with the different methods were

$$D^{hom} = \begin{pmatrix} 2.98 & 0.94 \\ 0.94 & 2.98 \end{pmatrix}, \quad D^{ren} = \begin{pmatrix} 2.44 & 0 \\ 0 & 2.44 \end{pmatrix},$$

$$D^{arith} = \begin{pmatrix} 4.5 & 0 \\ 0 & 4.5 \end{pmatrix}, \quad D^{geo} = \begin{pmatrix} 2.83 & 0 \\ 0 & 2.83 \end{pmatrix}.$$

The approaches based on renormalization and the arithmetic mean (as well as the other simple averaging schemes) are not able to produce diffusion entries in the off-diagonal and, thus, give a completely wrong picture of the coarse scale equation.

An error in the MG-homogenized diffusion coefficient compared to the homogenized diffusion coefficient can be observed. This is due to the lumping (23,24). However, MG-homogenization gives off-diagonal diffusion terms, like homogenization, and therefore preserves the gross features of the flow quite reasonable.

Example 4: Checkerboard diffusion

In this example the diffusion coefficient is chosen in a red-black or checkerboard type, as indicated in figure 6. The diffusion coefficient is set according to

$$D = \begin{pmatrix} d(x) & 0 \\ 0 & d(x) \end{pmatrix} \text{ where } d(x) = \begin{cases} d_1 & \text{in subdomain } Y_I \\ d_2 & \text{in subdomain } Y_{II}, \end{cases} \quad \text{with } d_1, d_2 \in \mathbb{R}^+.$$

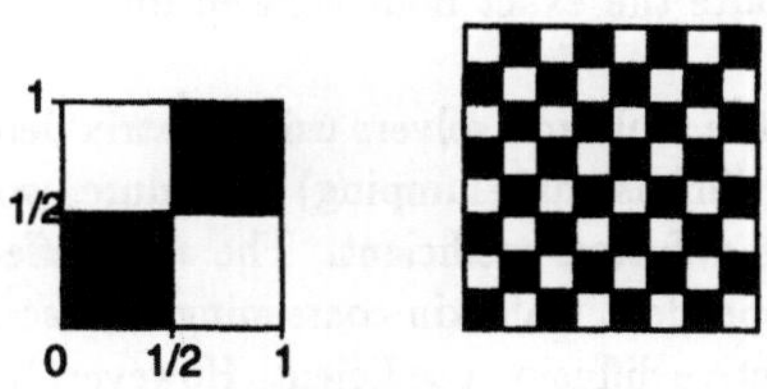

Fig. 6 Structure of the diffusion coefficient for example 4

Due to the lumping performed in (23,24) the MG-homogenization reduces to the arithmetic mean $(d_1 + d_2)/2 \cdot I$, and the renormalized diffusion coefficient is given by $8(d_1 + d_2)d_1d_2/(4d_1d_2 + 3(d_1 + d_2)^2) \cdot I$. The homogenized diffusion coefficient is however $\sqrt{d_1 \cdot d_2} \cdot I$. So, in this specific example, MG-homogenization and renormalization fail completely.

Example 5: Randomly distributed diffusion

The last example involves an isotropic diffusion coefficient $D(x, y) = d(x, y) \cdot I$ where the values of $d(x, y)$ are randomly chosen over Ω and assumed to be piecewise constant on fine grid cells.

We used a distribution with very large jumps, namely d is distributed according to

$$d = z^{-ln(a)}, \ z \in \mathcal{X}, \mathcal{X} \text{ equally distr. in } (0,1).$$

For this special distribution, the exact effective diffusion coefficient is the geometric mean of the diffusion coefficient which is given by a. The results obtained with the different methods are given in Table 2.

It can be clearly seen, that MG-homogenization is superior to the other methods.

Table 2 Problem 5: Results for random diffusion coefficients

a	2	5	10	20
arith. mean	3.26	9.1e3	7.5e8	5.7e15
harm. mean	1.68	2.60	3.27	3.80
renormalization	1.88	3.73	5.82	8.76
MG-homogenization	2.12	5.59	11.49	23.63
geom. mean	2	5	10	20

CONCLUDING REMARKS

We considered Galerkin coarse grid operators as discrete large scale operators. This view enabled us to compute approximations to the effective diffusion coefficient, using techniques well known from the multigrid context. The method can be applied to both, the periodic and the non-periodic case. The straightforward interpretation of matrix dependent Galerkin coarsening as discrete homogenization – as performed in this paper – yields in many cases results far better than other upscaling techniques. However it is i.g. not possible, to compute the exact homogenized diffusion coefficient using 9-point stencils.

It is well known, that the multigrid solvers using matrix dependent Galerkin coarsening with some kind of one dimensional (lumping) procedure, work well only for problems with a tensor product like diffusion coefficient. The same effect can be observed when interpreting the matrix dependent Galerkin coarsening as discrete method of computing approximations to the effective diffusion coefficient. However, because of the need to keep the coarse grid operators local, some kind of approximation procedure has to be used. Methods better adopted to the situation of non tensor product like diffusion coefficients and new upscaling techniques using the framework and ideas presented here are still to be constructed.

It is interesting, that techniques which are successfully used in multigrid algorithms are also applicable in other areas, where multiscale questions arise. The further development of the ideas and the framework presented here, especially regarding the obvious deficiencies of the method, is the purpose of further research.

ACKNOWLEDGEMENTS

We thank F. Bornemann from the ZIB-Berlin for his hints and comments.

REFERENCES

[ABDP81] Alcouffe R.E., Brandt A., Dendy J.E., Painter J.W.: *The multi-grid method for the diffusion equation with strongly discontinuous coefficients*, SIAM J. Sci. Stat. Comput. Vol. 2, 1981, 430-454.

[ABK91] Amaziane B., Bourgeat A., Koebbe A.J.: *Numerical simulation and homogenization of two-phase flow in heterogeneous porous media*, Transport in porous media 6, 519-547, 1991.

[AV89] Axelsson O., Vassilevski P.S.: *Algebraic multilevel preconditioning methods. I*, Numer. Math. 56, 157-177, 1989.

[AV90] Axelsson O., Vassilevski P.S.: *Algebraic multilevel preconditioning methods, II*, Siam J. Numer. Anal., Vol. 27, No. 6, 1569-1590, Dec. 1990.

[AV91] Axelsson O., Vassilevski P.S.: *Construction of variable preconditioners for inner-outer iteration methods*, Report 9107, Department of Math., Katholische Universität Nijmegen, Niederlande, 1991.

[Bab76] Babuska I.: *Homogenization and its application. Mathematical and Computational Problems*, Num. Sol. of Part. Diff. Eqn.-III, B. Hubbard, ed., Academic Press, N.Y., 89-116, 1976.

[BP89] Bakhvalov N., Panasenko G.: *Homogenisation: Averaging processes in periodic media*, Mathmatics and its Applications, Kluwer Academic Publishers, 1989.

[BLP87] Bensoussan A., Lions J.L., Papanicolaou G.: *Asymptotic analysis for periodic structure*, North-Holland, Amsterdam, 1987.

[Chr] Christie M.A.: *Upscaling*, review article, T. Petroleum Technology, to appear.

[Den83] Dendy J.E.: *Black box multigrid for nonsymmetric problems*, Appl. Math. Comput., Vol. 13, 261-283, 1983.

[Den82] Dendy J.E.: *Black box multigrid*, J. Comp. Phys., Vol. 48, 366-386, 1982.

[ER93] Edwards M.G., Rogers C.F., *Multigrid and renormalization for reservoir simulation*, Multigrid Methods, Vol. IV , Springer Lectures in Mathematics, SpringerVerlag, 1993.

[EL93] Engquist B., Luo E.: *Multigrid methods for differential equations with highly oscillatory coefficients*, Copper Mountain Conference 1993.

[FG93] Fuhrmann J., Gärtner K.: *On matrix data structures and the stability of multigrid algorithms*, Multigrid Methods, Vol. IV, Springer Lectures in Math., Springer-Verlag, 1993.

[Fuh94] Fuhrmann, J.: *Zur Verwendung von Mehrgitterverfahren bei der numerischen Behandlung elliptischer Differentialgleichungen mit variablen Koeffizienten*, Aachen: Shaker, 1995 (Berichte aus der Mathematik), Chemnitz-Zwickau, Techn. Univ., Diss., 1994.

[JKO94] Jikov V.V., Kozlov S.M., Oleinik O.A.: *Homogenization of Differential Operators and Integral Functionals*, Springer, 1994.

[Kes79] Kesavan S.: *Homogenization of Elliptic Eigenvalue Problems: Part 2*, Appl. Math. Optim. 5, 153-167, 1979.

[Kin89] King P.R.: *The use of renormalization for calculating effective permeability*, Transport in porous media 4, 37-58, 1989.

[Kna95] Knapek S.: *Multiskalenverfahren bei der Modellierung, Diskretisierung und Lösung von Diffusionsproblemen*, Diplomarbeit, Institut für Informatik, TU München, 1995.

[Neu96] Neuss N.: *Homogenisierung und Mehrgitter*, Bericht N96/7, Institut für Computeranwendungen der Universität Stuttgart, Mai 1996.

[San80] Sanchez-Palencia E.: *Non-homogenious media and vibration theory*, Lecture Notes in Physics 127, Springer-Verlag, 1980.

[SV93] Santosa F., Vogelius M.: *First order correctors to the homogenized eigenvalues of a periodic composite medium*, Siam J. Appl. Math., Vol. 53, No. 6, 1636-1668, 1993.

[Re93a] Reusken A.: *Multigrid with matrix-dependent transfer operators for convection-diffusion problems*, Multigrid Methods, Vol. IV, Springer Lectures in Math., Springer-Verlag, 1993.

[Re93b] Reusken A.: *Multigrid with matrix-dependent transfer operators for a singular perturbation problem*, Computing 50, 199-211, 1993.

[Zee90] De Zeeuw P.M.: *Matrix-dependent prolongations and restrictions in a blackbox multigrid solver*, J. of Comp. and Appl. Math. 33, 1-27, 1990.

A Parallel 2D Operator Splitting Method for the Navier–Stokes and Transport Equations

R. Hinkelmann, W. Zielke
Institut für Strömungsmechanik und Elektronisches Rechnen im Bauwesen
Appelstr. 9A, 30167 Hannover, Germany

SUMMARY

A 2D algorithm for free surface flow and salinity transport in coastal areas on a 'multiple instruction multiple data' (MIMD) parallel computer with 128 processors is presented. Operator splitting in the basic equations leads to a separated treatment of the hyperbolic and parabolic terms, which are then combined by coupling the Method of Characteristics with the Finite Element technique. Different iterative solvers based on conjugate gradient methods and different preconditioners are used for solving the two sets of linear equations. The data of the unstructured grids are distributed among the processors with Recursive Spectral Bisection in combination with the Kernighan–Lin heuristic. The performance of the implemented algorithm is demonstrated using a field case study of the German estuary 'Darß–Zingster Boddenkette'.

1. INTRODUCTION

In recent years the use of massively parallel computer systems for large scale scientific and engineering problems has won more and more on importance. Among different hardware concepts, the 'multiple instruction multiple data' (MIMD) architecture has proved to be the most promising for parallel computers, especially when problems such as solving large sets of equations are dealt with. A MIMD parallel computer consists of a number of autonomous processors, each with its own memory connected with a fast communication network. A program is executed on each processor but with different data. In addition to a serial program, a parallel one must be enlarged by supplementary data handling administration and by communication routines which control the flow of data between the different processors. New basic concepts must be developed in order to minimize the communication and to optimize the load balance among the processors. In the field of supercomputing, a parallel computer is characterized by a much better cost performance ratio when compared with a vector computer. There seems to be much room for increasing the performance on parallel machines, if more and more processors are added, while increasing the performance is physically limited for vector computers.

In chapter 2, the developed mesh partitioning techniques are introduced. The basic equations and the parallel numerical algorithm are discussed in chapter 3. Chapter 4 describes the parallelized iterative conjugate gradient solvers and preconditioners. The parallel code is applied to a field case study of the German estuary 'Darß–Zingster Boddenkette' (Baltic Sea) in chapter 5. Chapter 6 contains an outlook on further research work.

2. Data Partitioning Techniques

Many large scale computational problems are based on unstructured computational discretisations. Important examples are unstructured grid calculations based on finite volume or finite element methods in computational fluid dynamics or strutural analysis. One of the key difficulties when implementing such large scale unstructured problems on a distributed memory machine is the question of how to efficiently partition the underlying computational domain. The inter–processor communication has to be minimized and the work load balanced in such a way that every processor is equally loaded. This problem is much more complex for unstructured grids when compared with structured ones. Numerous methods based on graph partitioning techniques or heuristics are mentioned in the literature. A very good overview is given by Farhat et al. [7] and Hendrickson, Leland [4].

One of the most promising techniques is Simon's Recursive Spectral Bisection (RSB, [10]), which minimizes the number of common edges on the intersection lines when a graph is subdivided into two subgraphs with the same number of nodes. RSB is based on finding the Fiedler vector, which is the eigenvector corresponding to the second smallest eigenvalue of the graph connectivity matrix (Laplace matrix). The eigenvector is computed by inverse vector iteration for the Rayleigh quotient. The nodes are sorted according to the eigenvector size and half is assigned to each subdomain. This scheme is then repeated recursively. If the data partitioning is determined in a preprocessing stage and does not change during the simulation time – which is the case in our chosen problem –, RSB generally leads to the best load balance and a minimal communication time. If e.g. adaptive algorithm are dealt with, where the work load changes with the simulation time and a dynamic data redistribution is necessary, one has to take into account that the computation of the Fiedler vector consumes significant CPU time. In such cases other methods can be more appropriate, see e.g. Bastian [2].
Using a special storage technique for the system matrix ('element by element' technique, see chapter 4), the solver run time is dominated by loops running over the element number. Naturally, the same effect occurs while evaluating the system matrix. To make sure that the number of elements is the same on each processor, the graph is transformed into a dual graph. A dual graph is generated, when every element is replaced by its center point and these nodes are connected according to the element neighborhood conditions of the graph. RSB is then applied to the dual problem. The 'element by element' technique causes a good load balance, because the most CPU time consuming parts in the time loop, namely building up the system matrix and solving the sets of equations, are performed by loops running over the element number.
The mesh partitioning for the dual graph often results in evaluating the intersection lines in the form of a 'saw tooth' pattern. Therefore, the Kerninghan–Lin heuristic is implemented for a further improvement of this solution. Keeping the difference in element numbers of two subgraphs under a given tolerance, the Kernighan–Lin heuristic switches elements from one subgraph to the other when the number of common nodes can be decreased.

The developed algorithm for the data distribution generates an input file for each subdomain, so that the complete input and output handling runs in parallel. A mesh partitioning for one of the grids used in chapter 5 with about 8900 nodes and 16000 elements in 128 subdomains is demonstrated in figure 1. There the intersection lines and a small part of their surroundings is left white.

Figure 1: Mesh partitioning in 128 subdomain

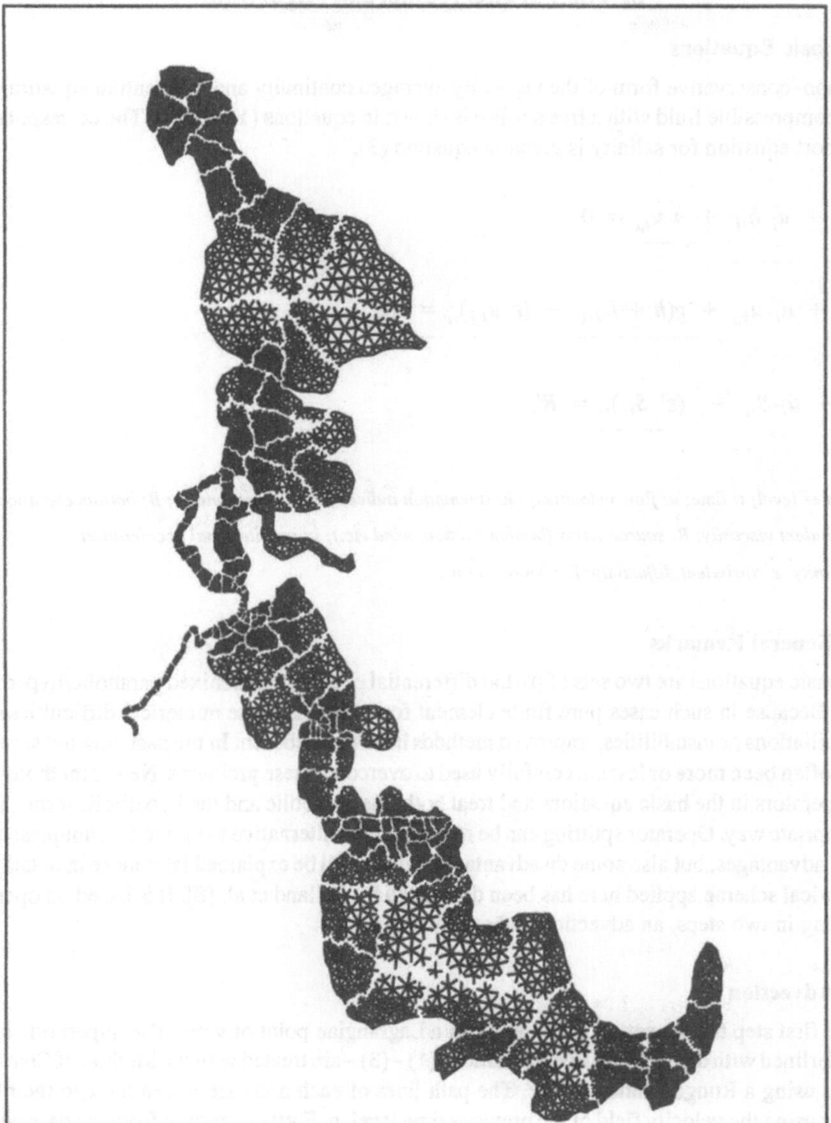

Figure 1: Mesh partitioning in 128 subdomain

3. The Parallel Numerical Algorithm

3.1. Basic Equations

The non–conservative form of the vertically averaged continuity and momentum equations for an incompressible fluid with a free surface is shown in equations (1) and (2). The corresponding transport equation for salinity is given in equation (3).

$$\underline{h_{,t}} + u_i\, h_{,i} + \underline{h\, u_{i,i}} = 0 \qquad (1)$$

$$\underline{u_{k,t}} + u_i\, u_{k,i} + \underline{g(h+B)_{,k}} - \underline{(\varepsilon\, u_{k,i})_{,i}} = \underline{R_k} \qquad (2)$$

$$\underline{S_{,t}} + u_i\, S_{,i} - \underline{(\varepsilon'\, S_{,i})_{,i}} = \underline{R'} \qquad (3)$$

h: water level; t: time; u: flow velocities; i,k: summation indices for x,y-components; B: bottom elevation;
ε: turbulent viscosity; R: source terms (bottom friction, wind etc.); g: gravitational acceleration
S: salinity; ε': turbulent diffusivity; R': source term .

3.2. General Remarks

The basic equations are two sets of partial differential equations of a mixed parabolic/hyperbolic type. Because in such cases pure finite element formulations cause numerical difficulties such as oscillations or instabilities, improved methods have to be chosen. In the past, upwind schemes have often been more or less successfully used to overcome these problems. Newer methods split the operators in the basic equations and treat both, the parabolic and the hyperbolic terms, in an appropriate way. Operator splitting can be regarded as an alternative to upwind techniques; it has some advantages, but also some disadvantages, which will be explained later more in detail. The numerical scheme applied here has been developed by Galland et al. [8]. It is based on operator splitting in two steps, an advection and a diffusion step.

3.3. Advection

In the first step the advection is treated from a Lagrangian point of view. The hyperbolic terms – underlined with dashed lines in the equations (1) – (3) – are treated with the Method of Characteristics using a Runge–Kutta scheme. The path lines of each node are traced back to their base points using the velocity field of the previous time level, n. Further, starting from the base points, the values are interpolated to the nodes to obtain the solution for the next time level, n+1. This method is explicit and can therefore be completely calculated in parallel. Concerning the parallization, two aspects have to be taken into account. First, the nodes on the intersection lines (see fig. 1) are traced back on every bordering processor, but depending on the flow conditions the base points are only found on one processor. Hence communication between the bordering processors is necessary after each advection step to update the results on the intersection lines. Second, when tracing back an inner node, it is possible that such a node leaves its subdomain and then isn't found on any processor. Since this only happens rarely, a time step restart is performed in this case with half of the time step. Naturally the time step is doubled afterwards. The results after the advection are $\overline{h}$, $\overline{u}_i$ and $\overline{S}$.

3.4. Diffusion, Propagation and Source Terms

Taking the solution of the advection step as the initial condition, the diffusion, propagation and source terms are computed with an Eulerian approach in the diffusion step. The parabolic terms – underlined with full lines in the equations (1) – (3) – are treated with a Finite Element Method based on a variational formulation. The space is modeled with linear triangles and the time discretization is implicit. The following scheme is obtained on the element level by making some simplifications such as a linearization by neglecting second order terms in time:

Continuity Equation

$$\frac{dh_M^{n+1}}{\Delta t} \int \Psi_M \Psi_L d\Omega - \theta h_M^n \int (\Psi_M \Psi_J),_i \Psi_L \, d\Omega \; u_{iL}^{n+1} \quad =$$

$$\frac{d\bar{h}_M}{\Delta t} \int \Psi_M \Psi_L \, d\Omega - \theta h_M^n \int \Psi_M \Psi_J \Psi_L \, N \, d\Gamma \; u_{iL}^{n+1} \; +$$

$$(\theta - 1) \, h_M^n \left[\int \Psi_M \Psi_J \Psi_L \, N \, d\Gamma - \int (\Psi_M \Psi_J),_i \Psi_L \, d\Omega \right] u_{iL}^n \, . \qquad (4)$$

Momentum Equation k

$$dh_M^{n+1} \, \theta g \int \Psi_{M,k} \Psi_L d\Omega + \left[\frac{1}{\Delta t} \int \Psi_M \Psi_L \, d\Omega + \int \Psi_{M,i} \, D \Psi_{L,r} \, N \, d\Omega \right] u_{kL}^{n+1} \quad =$$

$$(R_{kM} + \frac{\bar{u}_{kM}}{\Delta t}) \int \Psi_M \Psi_L \, d\Omega - g(h_M^n + B_M) \int \Psi_{M,k} \Psi_L d\Omega \, . \qquad (5)$$

Salinity Transport Equation

$$S_M^{n+1} \left[\frac{1}{\Delta t} \int \Psi_M \Psi_L \, d\Omega + \theta \int \Psi_{M,i} \, T \Psi_{L,r} \, N \, d\Omega \right] \quad =$$

$$(St_M + \frac{\bar{S}_M}{\Delta t}) \int \Psi_M \Psi_L \, d\Omega + (1 - \theta) \, S_M^n \int \Psi_{M,i} \, T \Psi_{L,r} \, N \, d\Omega \, . \qquad (6)$$

$dh^{n+1} = h^{n+1} - h^n$; $d\bar{h} = \bar{h} - h^n$; Δt: time step; Ψ: shape function; Ω: domain; θ: implification factor; Γ: boundary; N: normal vector; r: summation index for x,y components; M,L,J: summation index 1,2,3 .

Taking the boundary conditions into account one gets two decoupled sets of linear equations. The first, for the flow equations (7), is non-symmetric, decoupled for u_x and u_y and has three degrees of freedom per node. The second, for the transport equation (8), is symmetric, has only one degree of freedom per node and requires the results of the first set of equation in its solution.

$$
\begin{vmatrix} A_1 & -B_1 & -B_2 \\ C_1^T & A_2 & 0 \\ C_2^T & 0 & A_3 \end{vmatrix}
\begin{vmatrix} dh^{n+1} \\ u_x^{n+1} \\ u_y^{n+1} \end{vmatrix}
=
\begin{vmatrix} R_1 \\ R_2 \\ R_3 \end{vmatrix}
\qquad (7)
$$

$$
\begin{vmatrix} A_4 \end{vmatrix} \begin{vmatrix} S^{n+1} \end{vmatrix} = \begin{vmatrix} R_4 \end{vmatrix} \qquad (8)
$$

3.5. Assessment of the Numerical Algorithm

The advection step must be calculated only once for the flow and transport equations, when both are computed with the same time step. Although the Method of Characteristics is explicit there is, from stability considerations, theoretically no limitation on the time step. Therefore it is possible to choose time steps several times greater than those specified by the Courant number (see chapter 5).

In the advection step some damping is caused by the interpolation of the traced back nodes from their base points to the grid nodes. Further, the Method of Characteristics does generally not guarantee mass conservation, so that a mass correction scheme has to be used. Among other things the linearization in the diffusion step only leads to first order accuracy in time.

4. Parallel Iterative Preconditioned Solvers

4.1. Storage Technique, Parallel Matrix-Vector and Scalar Product

The system matrix is stored using the 'element by element' technique (see Lepeintre [6]). First the element matrices are evaluated. Then only the diagonal terms are assembled on the processor level and the off diagonal terms remain - not assembled - in an element list. No further storage for the system matrix is necessary, but to perform efficient matrix-vector products some additional storage is required. In the matrix-vector product only operations with non-zero terms occur. The solver CPU time is clearly dominated by matrix-vector products, which are mostly computed by loops running over the element numbers. Naturally, the system matrix construction is also done in loops running over the element number. Therefore data partitioning for the dual graph leads to a good synchronization and load balance in the time loop (see chapter 2).

Each processor calculates with its data its part of the system matrix and the right-hand terms completely in parallel and without communication. But note that on the intersection lines, every processor has only a part of the complete information. For the parallel matrix-vector product just the right-hand terms have to communicate, so that they obtain the complete values. In the scalar product the nodes on the intersection lines must be multiplied with a factor depending on the number of processors to which they belong. Again every processor computes its part and eventually a global communication leads to global summation.

The additional storage requirements caused by supplementary parallel data administration is only weakly dependent on the number of processor and for 128 processors on the order of 20 %.

Calculations were performed on both, nCUBE2 respectively nCUBE2S machines. Like on many other common parallel computers sending data happens asynchronly, i.e. non blocking, while receiving data is synchron. First, each processor sends data to a neighbor, whose processor number is one higher, i.e. the increment (which is explained later) is one. Then it receives data from its neighbor, whose processor number is one lower, makes data operations (add up or compare with zero) and sends back the results. Next, it receives the data back from the processor whose processor number is one higher. This is then repeated in a kind of level structure with all other processors, where communication is necessary by increasing the increment in the processor number difference.

4.2. Iterative Solvers

The matrix resulting from the flow equations (7) is non-symmetric. Therefore the 'BiConjugate Gradient Stabilized Method' (BiCGSTAB) is chosen. The symmetric transport equation is solved using the 'Conjugate Gradient Method' (CG). Both techniques are not explained here in detail. An excellent overview for such iterative solvers is given in Barret et al. [1]. In the iteration loop of BiCGSTAB two matrix-vector and five scalar products are computed with the communication described above. CG only needs one matrix-vector and two scalar products in the iteration loop.

4.3. Preconditioners

A preconditioner improves the condition number of a linear system, i.e. it decreases the iteration number which is necessary to obtain the solution. If the system Ax=B is transfered into $M^{-1}Ax=M^{-1}B$ with the preconditioning matrix M, then both systems have the same solution, but one generally obtains this solution faster with the transfered system. The problem consists of finding M^{-1} which is an approximation of the inverse matrix of A. In this paper three different kinds of preconditioners which need no or only little communication are investigated. This type is very efficient for a parallel code.

☞ diagonal preconditioner (dia): The inverse matrix of A is approximated by the inverted values of the diagonal terms of A.

☞ 'element by element' preconditioner (ebe): The inverse matrix of A is approximated by a matrix which contains the inverted local element matrices. In the flow equations this method can only be applied to the main diagonal blocks A_1, A_2 and A_3. A detailed description is found in Hughes [5].

☞ additive Schwarz method preconditioner (aSm): Using a domain decomposition approach, the global inverse matrix of A is approximated by a local inverse matrix of A in each subdomain. In the transport equation, this matrix is calculated completely in parallel with CG (without communication) in a secondary iteration. A Schur complement technique is used for the 3x3 block structure of the flow equations, because it enables the decoupled computation of the three degrees of freedom per node. Thus, CG can be applied three times – just as in the transport equation.

$$
\begin{vmatrix} A_1 & -B_1 & -B_2 \\ C_1^T & A_2 & 0 \\ C_2^T & 0 & A_3 \end{vmatrix} \begin{vmatrix} dh^{n+1} \\ v_x^{n+1} \\ v_y^{n+1} \end{vmatrix} = \begin{vmatrix} R_1 \\ R_2 \\ R_3 \end{vmatrix} \Rightarrow \begin{vmatrix} S & 0 & 0 \\ C_1^T & A_2 & 0 \\ C_2^T & 0 & A_3 \end{vmatrix} \begin{vmatrix} dh^{n+1} \\ v_x^{n+1} \\ v_y^{n+1} \end{vmatrix} = \begin{vmatrix} \mathbf{R_s} \\ R_2 \\ R_3 \end{vmatrix}
$$

$$
S = A_1 + B_2 A_3^{-1} C_2^T + B_1 A_2^{-1} C_1^T
$$

$$
\mathbf{R_s} = R_1 + B_2 A_3^{-1} R_3 + B_1 A_2^{-1} R_2 \,. \tag{9}
$$

The computation of the whole Schur complement in the preconditioning stage is not efficient because it takes too much CPU time. Therefore, some approximations were made. Further explanations are given in Hackbusch [3].

Averaging the values on the intersection lines after preconditioning has improved the convergence behaviour and reduced the CPU time for all three techniques.

5. Examples

The performance of the implemented algorithm is demonstrated using a field case study of the German estuary 'Darß–Zingster Boddenkette'. The object was to calibrate a numerical model in order to provide input for evaluating environmental and water quality changes associated with anthropogenic influences such as dike construction and channels directly connected to the Baltic Sea. For the parallel computing, two grids were chosen. The one in figure 1 contains about 8900 nodes and 16000 elements, while that in figure 2 has 2200 nodes and 4000 elements. The measured surface elevations and salinities at position B19 on the open boundary to the Baltic Sea, the river discharge as well as measured wind velocities were specified as the boundary conditions. A comparison between measurements and numerical simulations computed with the coarse grid for three positions can be seen in figure 3. Detailed information is found in Stückrad et al. [11].

For the parallel computations 1000 time steps of 15 seconds with the coarse grid and 100 time steps of 6 seconds with the fine grid were investigated. The Courant number in both cases was ≈ 6. Further analysis have shown that longer simulation periods have only a very small influence on the parallel run time behaviour. In figure 4 the parallel efficiences, i.e. the ratio of the CPU time for one processor to the CPU time for p processors multiplied with p, are shown for the diagonal preconditioning. The computations with the coarse grid were only undertaken up to 32 processors, because there are already less than 100 nodes on each processor and the parallel overhead over proportionally increases. The calculations with the fine mesh could not run on one processor, because the available 8 MByte RAM per processor were not sufficient. Naturally, the parallel effiency decreases with increasing processor numbers, if the problem size, i.e. the grid, is kept constant. The parallel overhead decreases with increasing problem size. Achieving about 50% parallel effiency on 128 processors for an unstructured grid code is quiet reasonable for such a problem size.

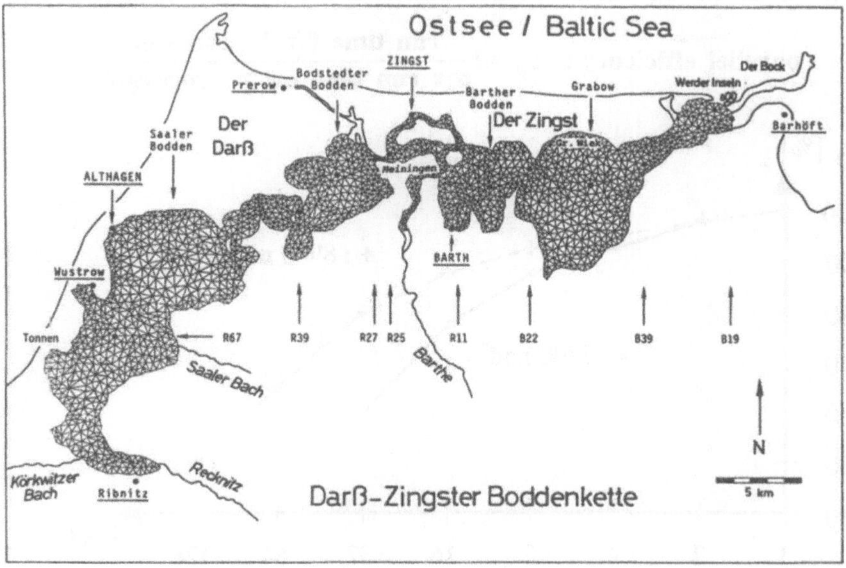

Figure 2: Computational grid and measurement locations

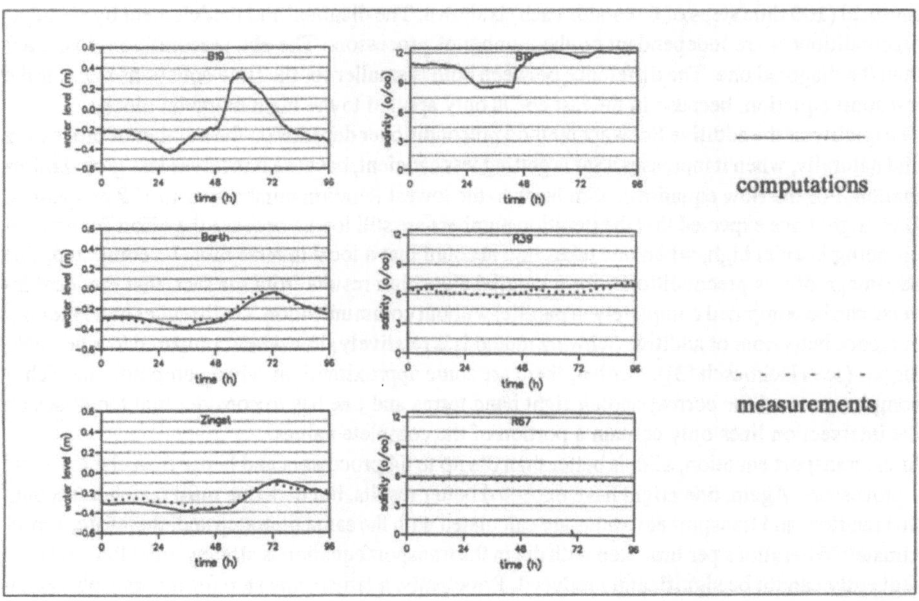

Figure 3: Time series of water level and salinity

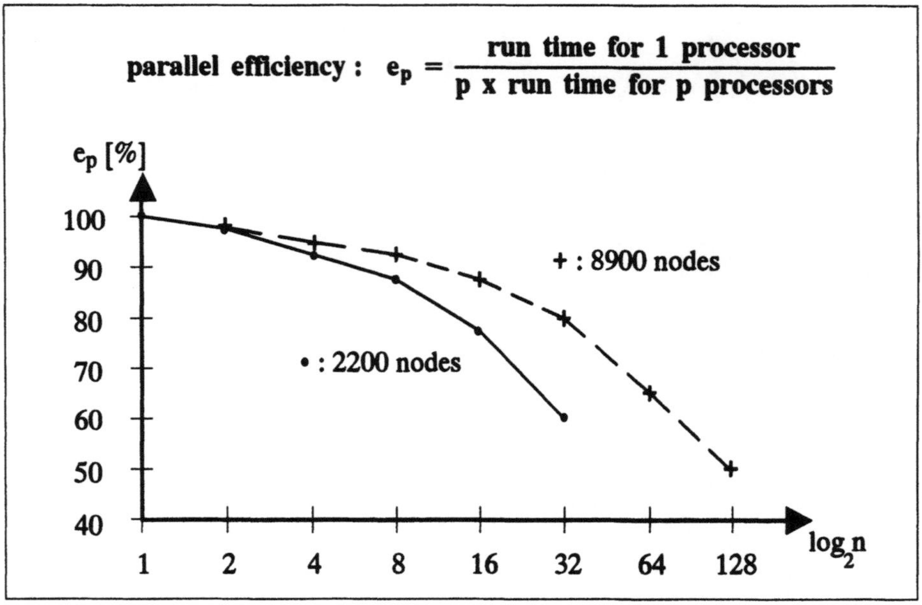

Figure 4: Parallel efficiences for two different grids (diagonal preconditioning)

In figure 5 the influence of different preconditioners on the iteration numbers calculated for the fine grid (100 time steps of 6 seconds each) is shown. The diagonal and the 'element by element' preconditioners are independent on the number of processors. The ebe preconditioner is better than the diagonal one. The difference between both is smaller for the flow equations than for the transport equation, because in the last ebe is only applied to the main diagonal blocks.
The quality of the additive Schwarz method preconditioner depends on the number of processors and naturally, when it increases aSm is getting less efficient, because it receives less global information. For the flow equations, aSm leads to the lowest iteration numbers up to 128 processors. One might have expected that the iteration numbers are still lower, because the effort for preconditioning is quiet high, when one takes into account that a local inverse must be computed. The advantage of this preconditioner for a parallel algorithm results from the fact, that the local inverse can be computed completely in parallel without communication. On the other hand the convergence behaviour of additive Schwarz methods is relatively slow when compared to other techniques (see Hackbusch [3]). Further, there are some approximations when computing the Schur complement and the corresponding right hand terms and one has to consider that the nodes on the intersection lines only contain a portion of the complete values.
In the transport equation, aSm is better than dia up to 64 processors and better than ebe for 2 and 4 processors. Again, one might have expected better results, but here one must take into account that the flow and transport equations are calculated with the same time step and, therefore, approximately 6 iterations per time step with dia in the transport equation is already quiet low and consequently cannot be significantly reduced. Principally, a larger time step for the transport equation is possible, which would probably lead to better results for aSm.

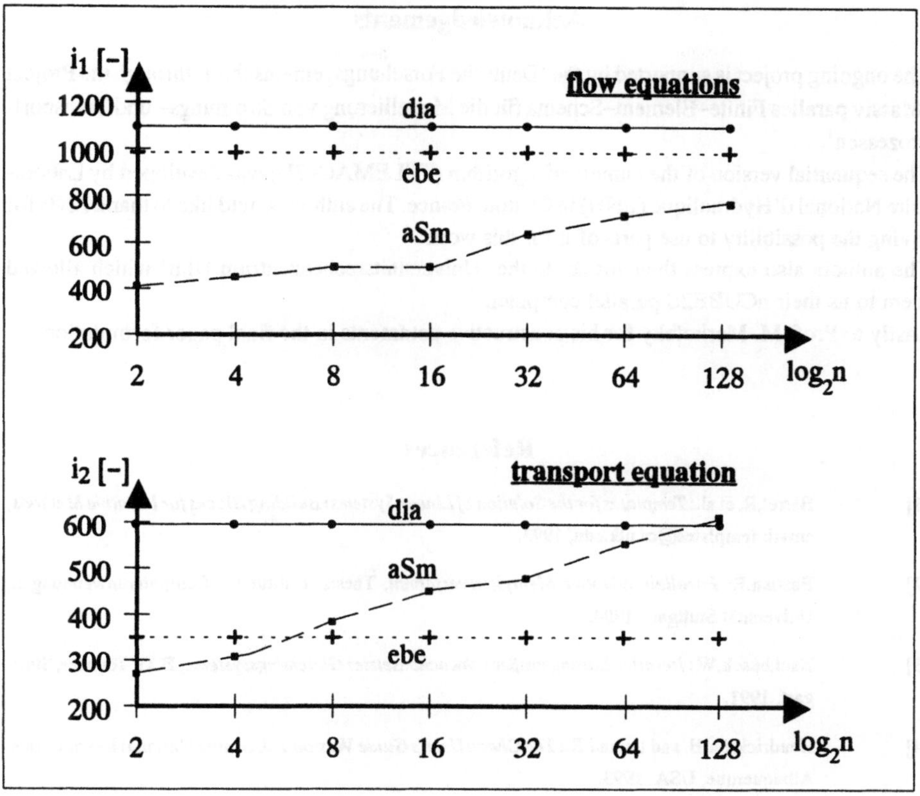

Figure 5: Iteration numbers for different preconditioners (mesh with 8900 nodes)

It must be mentioned that the CPU time could not be reduced with aSm because the domain decomposition approach in the preconditioner also has some disadvantages. That means, if there is only 'action' in one or a few subdomains, the corresponding processors need much time to compute their local inverses, while all the other processors are waiting. This leads to an adverse load balance when compared with the other preconditioners.

6. Further Work

The code will be ported to the parallel supercomputer Cray T3D using a PVM (Parallel Virtual Machine) interface, which has already been successfully tested in a small workstation cluster. Many problems in coastal areas cannot be realistically treated with 2D models because of the assumption of constant vertical profiles. Therefore, we need a 3D model which, however, normally has hugh requirements concerning CPU time and storage. The development of a parallel 3D code is presently underway.

Acknowledgements

The ongoing project is supported by the 'Deutsche Forschungsgemeinschaft' through the Project 'Massiv paralles Finite-Element-Schema für die Modellierung von Strömungs- und Transportprozessen'.
The sequential version of the numerical algorithm (TELEMAC-2D) was developed by Laboratoire National d'Hydraulique (LNH) in Chatou, France. The authors would like to thank LNH for giving the possibility to use parts of it for this work.
The authors also express their thanks to the 'Universitätsrechenzentrum Ulm' which allowed them to us their nCUBE2S parallel computer.
Lastly to Prof. M. Markofsky for his constructive comments in the final paper formulation.

References

[1] Barret,R. et al.: *Templates for the Solution of Linear Systems: Building Blocks for Iterative Methods*, email: templates@cs.utk.edu, 1993.

[2] Bastian,P.: *Parallele adaptive Mehrgitterverfahren*, Thesis, Institut für Computeranwendungen, Universität Stuttgart, 1994.

[3] Hackbusch,W.: *Iterative Lösung großer schwachbesetzter Gleichungssysteme*, B. G. Teubner, Stuttgart, 1991.

[4] Hendrickson,B. and Leland,R.: *The Chaco User's Guide Version 1.0*, Sandia National Laboratories, Albuquerque, USA, 1993.

[5] Hughes,J.R., Ferencz,M. and Hallquist,O.: *Large Scale Vectorized Implicit Calculations in Solid Mechanics on a Cray X-MP/48 Utilizing EBE Preconditioned Conjugate Gradients*, Computer Methods in Applied Mechanics and Engineering 61, North-Holland, 1987.

[6] Lepeintre,F.: *Bibliotheque BIEF - Note de Principe et Descriptif Informatique*, EDF report HE-43/92-16, Chatou, France, 1992.

[7] Farhat,C., Lanteri,S. and Simon,H.D.: *TOP/DOMDEC - A Software Tool for Mesh Partitioning and Parallel Processing*, Department of Aerospace Sciences and Center for Space Structures and Controls, University of Colorado and NASA Ames Research Center, Moffett Field, Colorado, USA, 1994.

[8] Galland,J.C., Goutal,N. and Hervouet,J.M.: *Telemac: A New Numerical Model for Solving Shallow Water Equations*, Advanced Water Resources, Vol. 14, No. 3, 1991.

[9] Kernighan,B.W. and Lin,S.: *An Efficient Heuristic Procedure for Partitioning Graphs*, The Bell System, Technical Journal 49, 1970.

[10] Simon,H.D.: *Partitioning of Unstructured Problems for Parallel Processing*, NASA Ames Research Center, Moffett Field, Colorado, USA, 1994.

[11] Stückrad,H., Hinkelmann,R. and Zielke,W.: *Hydrodynamisch-numerische Modellierung der Darß-Zingster Boddenkette*, Deutsche Hydrographische Zeitschrift, Heft 2, 1995.

DUAL-POROSITY MODELLING OF CONTAMINANT TRANSPORT IN FRACTURED POROUS FORMATIONS: THE EFFECT OF SPATIAL VARIATIONS OF MATRIX BLOCK PROPERTIES

D. Jansen, J. Birkhölzer, J. Köngeter

Institute of Hydraulic Engineering and Water Resources Management
Aachen University of Technology (RWTH)
Mies-van-der-Rohe-Str. 1, 52056 Aachen, Germany
Fon: 0241 805266, Fax: 0241 8888348, e-mail: dietmar@iwwpc.iwwl.rwth-aachen.de

SUMMARY

Contaminant transport in fractured porous formations is often simulated with dual-porosity models. In such models the heterogeneous formation is separated into two coupled continua, one representing the fractures, one representing the rock matrix. The main objective of the present study is to investigate the effect of spatial variations of matrix block properties (especially such as size and shape) and to determine effective continuum parameters for the matrix continuum. Numerical simulations are performed with a newly developed code applying a Lagrangian-Eulerian algorithm for the transport simulation in stochastically generated discrete fracture systems combined with a sophisticated Finite Element formulation to account for molecular diffusion in matrix blocks of arbitrary size, shape and material properties. To check the accuracy of the continuum approach, breakthrough curves are compared of two different sets of simulation runs, one considering the exact matrix properties, the second using different averaged effective continuum parameters.

INTRODUCTION

Fractured porous formations are typified by a high permeable fracture system and a matrix system with very low permeability but high storativity. Due to the different response times in the fractures and in the matrix regional transport takes place in the fractures and is of advective-dispersive character. Transport in the matrix blocks is of diffusive type; it depends on the concentration in the adjacent fractures and is therefore of local nature.

Contaminant transport in such systems is often simulated with dual-porosity models. Central to dual-porosity models is the assumption that the heterogeneous rock formation can be separated into two homogeneous media (continua), one continuum representing the interconnected fracture system, a second continuum representing the porous matrix blocks [1]. In dual-porosity models the conductivity of the matrix blocks is neglected and only matrix-diffusion perpendicular to the matrix-fracture interface is considered.

Essential to the principle on homogenization for heterogeneous media is the definition of equivalent model parameters for both media, capable of describing the correct physical behavior of the system. Numerous studies have been performed e.g. by *Long et al.* [6], [7] in the past to check if a continuum representation of fracture networks is valid. If so equivalent continuum parameters can be derived and the model error associated to this averaging process may be estimated. However, only little work has been done to address this task with regard to the matrix blocks of a given subdomain, which may considerably vary in size, shape or material properties. In most cases of dual-porosity modelling, the matrix continuum parameters are only roughly estimated and error analysis is not performed. Due to the distribution of the fractures the geometric parameters (block size and shape) can be extremely heterogeneous in the matrix system. These heterogeneities are important for the transport behavior of the matrix system.

In many works dual-porosity models with rather simple approximation of the matrix block geometry are suggested. *Tang et al.* [12] developed an analytical solution for matrix diffusion from a single fracture into a matrix block of infinite volume. *Bibby* [2] and *Sudicky and Frind* [11] extended this model to a system of parallel fractures with matrix blocks of finite volume. For the three dimensional approach Warren and Root [13] investigated cubes of unique size arranged in a regularly orthogonal fracture system. For the same set of fractures *Huyakorn et al.* [4] suggested sphere shaped matrix blocks, considering that both cubes and spheres have the same surface to volume ratio. For those simple block geometries it is straightforward to determine analytical expressions for the cross-section area available for matrix diffusion in a single block at a certain distance from the matrix-fracture interface (fig. 2). *Pruess and Karasaki* [10] introduced the concept of a 'Proximity-Function' as the central geometrical quantity of the matrix system in a heterogeneous fractured porous volume. In this approach the total available cross-section area for matrix diffusion in a certain volume of fractures and rock is determined with a statistical procedure. The concept of a 'Proximity-Function' can be applied for block geometries of arbitrary shape. *Neretnieks and Rasmuson* [8] developed the concept of a 'Pseudobody' to account for blocks of various size and shape in a certain volume. The basic idea of the 'Pseudobody' approach is to divide all single blocks in the considered domain into a number of shells and to assume that all shells at the same distance from the matrix-fracture interface behave similar. The shells at the same distance from the matrix-fracture interface can therefore be accumulated in one fictive matrix block, the so-called 'Pseudobody'. This method is capable of accounting for blocks of different size and simple shape with a certain frequency in the considered domain.

APPROACH

Models for fractured porous formations can be distinguished into (1) discrete, (2) continuum and (3) hybrid approaches (fig. 1). Discrete models offer a discrete representation of the fractures and of the matrix blocks. This approach implies extremely high discretization effort and computer capacity and is therefore limited to small investigation areas and detailed problems. For sufficiently densely fractured aquifers a continuum approach is suitable to model the transport processes of interest. The averaged representation of the fracture and the matrix system in one (single continuum) or more (multiple continuum, e.g. dual-porosity) continua allows for modelling large scale problems. Essential to the continuum approach is the reliable determination of equivalent continuum parameters for each continuum. Another model type can be valuable in modelling sparsely fractured formations where a continuum representation for the fracture system is not possible. For such problems a discrete model for the fracture system coupled with a continuum model for the matrix system can be developed. This models can be called 'hybrid models'. Because of the partially continuum representation of the matrix-fracture system hybrid models require less computer capacity than purely discrete models and the consideration of larger investigation areas is possible.

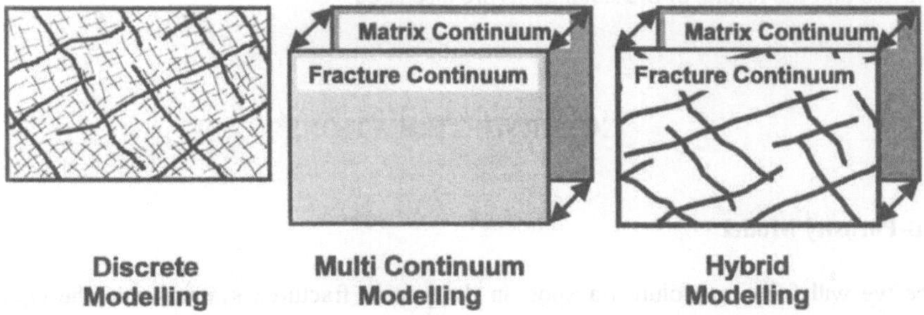

Figure 1: Models for fractured porous formations

For continuum representation of heterogeneous rock formations (e.g. dual-porosity models) averaging procedures for the fracture system as well as for the matrix system are necessary. In many studies the equivalent continuum parameters of fracture systems are derived by determining the heterogeneous behavior of the system using discrete models. In most of the case studies concerning dual-porosity models the matrix system has been strongly idealized to a set of matrix blocks of uniform size and shape and detailed averaging procedures have not been applied.

In our study the same methodology used for homogenization of the fracture system is adopted for the matrix system to determine effective continuum parameters. For the matrix blocks hydrogeologic . parameters (porosity, diffusion coefficient) and geometric parameters (interface-function, maximum penetration depth) have to be averaged. First we study how the overall transport in stochastically generated *discrete* matrix-fracture systems is affected by spatial variations of the matrix properties. A number of simulation runs is performed with varying block sizes and shapes. These simulations with both discrete modelling of fracture and matrix system are now repeated with *averaged* parameters (effective continuum parameters) for the matrix system, using different averaging techniques. The fracture system is still modeled discretely. Breakthrough curves of both simulation sets are compared and analyzed to check the accuracy of the effective continuum representation of the matrix system and to estimate the error of the average solutions, respectively.

Due to the combination of discrete (fractures) and continuum (matrix blocks) modelling a hybrid model was applied. The hybrid model is capable of addressing directly the simulation errors due to averaging procedures for the matrix system, since the discrete representation of the fracture system is naturally free from any homogenization errors.

In the following paragraph the general governing equations for the dual-porosity model are presented in order to develop the governing equations for the hybrid model used in the present study. Then the new model applying a Lagrangian-Eulerian algorithm for the regional transport in stochastically generated discrete fracture systems combined with a sophisticated Finite Element formulation [3] to account for molecular diffusion in matrix blocks of arbitrary size, shape and material properties is described. Then the configuration of the numerical study are presented and the results of the simulations are discussed.

GOVERNING EQUATIONS

Dual-Porosity Model

Since we will focus on solute transport in the porous fractured system, only the equations concerning the transport problem are presented. The flow problem is similar but less complicated and we assume the flow problem to be solved already. The general transport equations for dual-porosity models considering advection, dispersion, molecular diffusion and sinks or source terms can be written as follows.

For the fracture system (superscript F), for a unit volume of the whole system:

$$R^F n^F \frac{\partial C^F}{\partial t} + q_i^F \frac{\partial C^F}{x_i} - \frac{\partial}{\partial x_i} D_{ij}^F \frac{\partial C^F}{\partial x_j} + Q^F + W^D = 0 \,. \tag{1}$$

For the matrix system (superscript M), for a unit volume of rock matrix:

$$\left(1 - n^F\right) \cdot \left[R^M n^M \frac{\partial C^M}{\partial t} - \frac{\partial}{\partial x_i} D_{mol}^M \frac{\partial C^M}{\partial x_j} + Q^M \right] - W^D = 0 \tag{2}$$

where W^D is the diffusive mass exchange between the fracture and the matrix per unit volume of the rock matrix:

$$W^D = -\left(1 - n^F\right) \Omega\, n^M D_{mol}^M \left. \frac{\partial C^M}{\partial s} \right|_{s=0} \,. \tag{3}$$

C $[ML^{-3}]$ is the solute concentration, q_i $[LT^{-1}]$ is the DARCY-velocity, D_{ij} $[L^2T^{-1}]$ is the hydrodynamic dispersion tensor, D_{mol} $[L^2T^{-1}]$ is the coefficient of molecular diffusion, n [-] is the porosity, Q $[ML^{-3}T^{-1}]$ is a sink or source term, R [-] is the retardation coefficient, Ω $[L^{-1}]$ is the surface to volume ratio of the matrix blocks and s [L] is the penetration depth into the matrix block perpendicular to the matrix-fracture interface (s=0 at the matrix-fracture interface and $s=s_{max}$ at the block center). Due to the dual-porosity approach (1)-(3) are describing the relevant transport processes in coupled continua and therefore all values are averaged values determined for a representative elementary volume.

Hybrid Model

For discrete representation of the fractures and the matrix blocks (1)-(3) can be simplified for a two-dimensional area of interest. The governing equation for the fractures (one-dimensional pipes) is given in the EULERIAN formulation by:

$$R^F n^F \frac{\partial C^F}{\partial t} + q^F_{x'} \frac{\partial C^F}{\partial x'} - \overline{D}^F \frac{\partial^2 C^F}{\partial x'^2} + \overline{Q}^F + \overline{W}^D = 0 \tag{4}$$

and in the LAGRANGIAN formulation by:

$$R^F \frac{DC^F}{Dt} = \overline{D}^F \frac{\partial^2 C^F}{\partial x'^2} - \overline{Q}^F - \overline{W}^D \tag{5}$$

and

$$\frac{DC^F}{Dt} = \frac{\partial C^F}{\partial t} + \frac{\partial C^F}{\partial x'} \cdot \frac{Dx'}{Dt} \tag{6}$$

where C^F is the concentration of a moving control volume along a characteristic described by

$$q^F_{x'} \frac{1}{R^F} = \frac{Dx'}{Dt} \tag{7}$$

where q_s [LT-1] is the velocity in the direction x' of the considered fracture and x' [L] is the local coordinate for the fracture. (2) is replaced by a one-dimensional diffusion equation for the matrix blocks

$$R^M n^M \frac{\partial C^M}{\partial t} - \frac{1}{A(s)} \cdot n^M D^M_{mol} \frac{\partial}{\partial s}\left(A(s) \frac{\partial C^M}{\partial s} \right) = 0 \tag{8}$$

where A(s) $[L^2]$ is the interface area available for diffusive mass flux in the matrix block at a certain distance s from the matrix-fracture interface (fig. 2). After solving (8) the solute exchange is obtained by applying FICK's law at the matrix-fracture interface

$$\overline{W}^D = 2 \cdot n^M D^M_{mol} \left. \frac{\partial C^M}{\partial s} \right|_{s=0} . \tag{9}$$

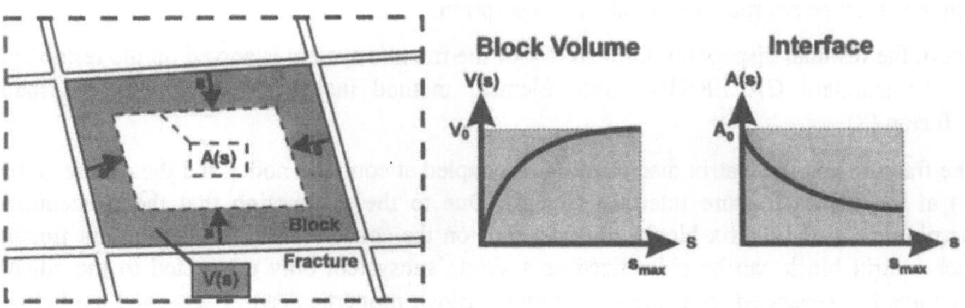

Figure 2: One-dimensional matrix diffusion and interface-function A(s)

THE LAGRANGIAN-EULERIAN FINITE ELEMENT SCHEME

In the present study a hybrid model has been developed based on the work of *Karasaki* [5] and *Birkhölzer* [3]. The developed model is capable of addressing directly the simulation errors due to averaging procedures for the matrix system, since the discrete representation of the fracture system is naturally free from any homogenization errors. To avoid numerical dispersion a Lagrangian-Eulerian scheme is applied to solve the transport problem for the fracture system, described by (5) and (7). The diffusive mass flux problem in the matrix blocks described by (8) is solved by a GALERKIN Finite-Element approach. The model for the matrix system allows for representation of each single block and also for adopting averaging procedures in the sense of continuum modelling.

The applied Lagrangian-Eulerian Finite Element scheme can be separated into three steps. The steps concerning the Lagrangian-Eulerian method for the fracture system follow the scheme presented by *Neuman* [9] and was extended to the matrix diffusion part. First, the concentration is decomposed into a pure advection and a pure dispersion part:

$$C^F = C^F_{adv} + C^F_{disp} \,. \tag{10}$$

Second, the advective transport in the fractures is performed by particle tracking methods (Single Step Reverse Particle Tracking - SRPT, Continuous Forward Particle Tracking - CFPT) for moving nodes instead of particles (fig. 3). In the vicinity of steep concentration gradients the Continuous Forward Particle Tracking technique is applied (11). To avoid numerical dispersion new Eulerian grid points are created in the vicinity of sharp concentration fronts and the basic Eulerian grid is locally refined. In certain distance from sharp fronts the Single Reverse Particle Tracking technique is applied. In this technique the nodes n of the Eulerian grid are treated as fictive nodes reaching there position x_n at time t_{k+1} in the considered time step. Their starting point x_p at time t_k is calculated by mapping back their characteristic (12).

$$x_p(t_{k+1}) = x_p(t_k) + \int_{t_k}^{t_{k+1}} \frac{v'}{R^F} Dt \tag{11}$$

$$x_p(t_k) = x_n - \int_{t_k}^{t_{k+1}} \frac{v'}{R^F} Dt \tag{12}$$

where v' may be retarded due to ad- and desorption.

Third, the residual dispersion problem (5) for the fracture system is solved on the refined grid by the standard GALERKIN Finite Element method including one-dimensional matrix diffusion (8).

The fracture and the matrix discretization is coupled at common nodes and the exchange term (9) at the matrix-fracture interface (fig. 4). Due to the assumption that the concentration distribution in the matrix blocks only depends on the concentration of the adjacent fractures each matrix block can be discretized as a single subsystem only connected to the adjacent fractures but separated from the other blocks. Block properties can be assigned explicitly to each block (discrete approach) or averaged properties can be distributed (average approach).

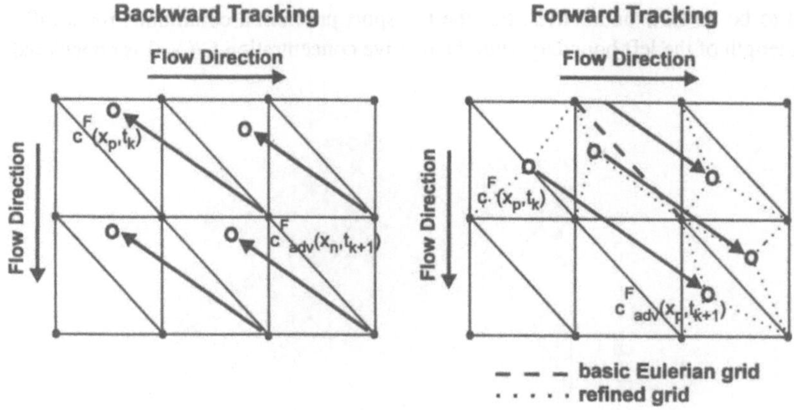

Figure 3: General idea of particle tracking techniques illustrated for 2D

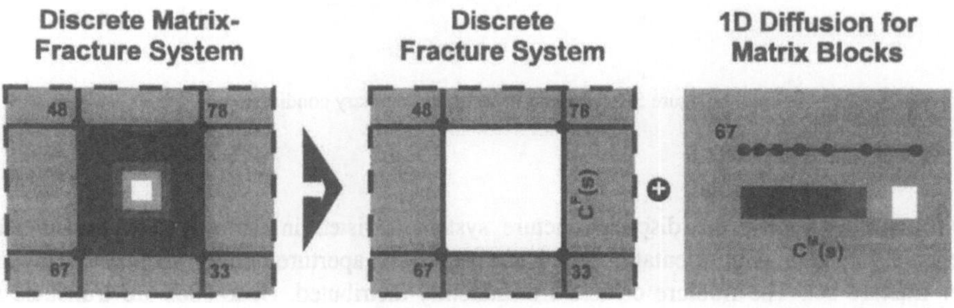

Figure 4: Coupling of matrix fracture system

NUMERICAL SIMULATIONS

In this study we consider a square shaped investigation area with a side length of 10 m (fig. 5). For this investigation area we assume constant hydrogeologic parameters in the matrix system. Therefore we can focus on the geometric heterogeneities of the matrix blocks due to the random distribution of the fractures.

Numerical tracer experiments are performed to investigate different averaging approaches. First the steady state flow field in the discrete fracture system is calculated by a GALERKIN Finite Element Method neglecting fluid flux in the matrix blocks. A constant regional gradient of I=0.01 is assumed from the left to the right (fig. 5). The upper and lower boundaries are assumed to be no flow boundaries. For the transport problem a continuous mass inflow over the total length of the left boundary with the relative concentration $C/C_0=1$ is considered.

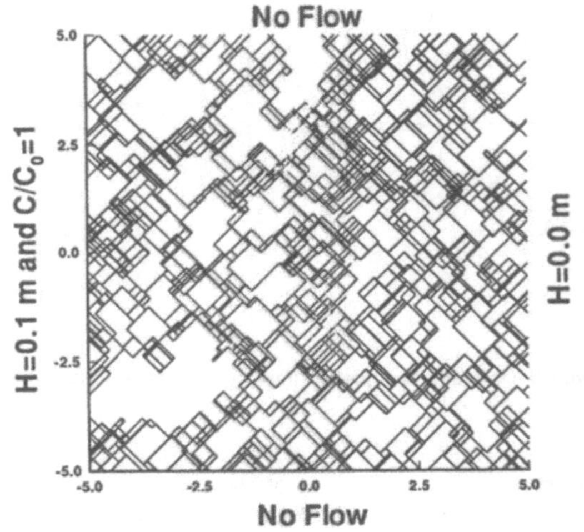

Figure 5: Investigation area and boundary conditions

The configuration of the discrete fracture system is listed in table 1. The fractures are generated by giving the orientation, the basic length, the aperture and the number of fractures per fracture set. The fracture centers are randomly distributed. Dead-ends are truncated for simplicity.

Table 1: Configuration for fracture-system

Fracture System: 2 sets of 600 fractures each; centers are randomly distributed; no dead-ends			
Orientation	Θ	[Deg]	45 / 135
Length	L	[m]	1.00
Aperture	a	[m]	$1.00*10^{-04}$
Porosity	n^F	[-]	1.00
Long. Dispersion Coefficient	α	[m]	0.05
Molecular Diffusion	D^F_{mol}	[m²/s]	$2.0*10^{-07}$
Retardation Coefficient	R^F	[-]	1.00

The corresponding configuration of the matrix system is summarized in table 2. Each block is identified by a closed polygon of connected fractures representing the matrix-fracture interface, the maximum penetration depth and a polynom for the interface-function A(s).

Table 2: Configuration for matrix-system

Matrix System: 2338 Blocks			
Total Volume	V_0	[m³/m]	100
Total Matrix-Fracture Interface	A_0	[m²/m]	1559.53
k_f-Value	k_f^M	[m/s]	0.00
Specific Storage Coefficient	S_S^M	[1/m]	0.00
Porosity	n^M	[-]	0.05
Molecular Diffusion	D^M_{mol}	[m²/s]	$4.00*10^{-08}$
Retardation Coefficient	R^M	[-]	1.00

A numerical reference case was simulated with discrete representation of both fractures and matrix blocks. The simulation results for this detailed simulations are assumed to represent the real transport behavior of the heterogeneous system and were interpreted using breakthrough curves at the right boundary. Then, four averaging cases (AV1-AV4) are considered typified by different average parameters for block shape and size (table 3).

Table 3: Matrix geometries for simulation runs

Case	Description	Block Shape	Max. Penetration Depth $\bar{s}_{max}$ [m]	Matrix-Fracture Interface $\bar{A}(s=0)$ [m²/m]
DISC	Reference Case (Discrete Modelling of each single block)			
AV1	square blocks due to A_{tot}/V_{tot} ratio	square	$1.28*10^{-01}$	$0.102*10^{+01}$
AV2	arithmetic mean for s_{max}	rectangular	$5.13*10^{-02}$	$0.154*10^{+01}$
AV3	geometric mean for s_{max}	rectangular	$2.59*10^{-02}$	$0.528*10^{+00}$
AV4	Proximity-Approach	not explicitly defined	$1.04*10^{+00}$	$0.156*10^{+04}$

A widely used approach in dual-porosity studies is to assume uniform block shape and size for the matrix continuum. In average case AV1 to AV3 the matrix system is represented by uniform blocks to check the validity of this approach. For AV1 a square shaped average block is calculated based on the total surface to volume ratio A_0/V_0. The size of the average square block is explicitly defined by the total matrix-fracture interface A_0 and the total block volume V_0 in the area of interest. For AV2 and AV3 the distribution of block size was statistically analyzed and the arithmetic mean and the geometric mean of the maximum penetration depth, respectively, were chosen as possible representative size for rectangular shaped blocks (fig. 7). For AV4 the concept of a 'Proximity-Function' [10] was applied to calculate an interface-function for the whole investigation area including all blocks and considering the heterogeneous character of the matrix system.

RESULTS

The results of the numerical tracer tests are compared based on the breakthrough curves (BTC) (fig. 6). Generally, due to the mass losses from the fractures into the matrix blocks a considerable increase of the residence time in the porous system can be observed compared to the BTC derived when neglecting the matrix diffusion.

Considering the BTC of the discrete reference case for early time steps, the delay in the arrival time of the tracer is the effect of the strong mass flux from the fractures into the matrix block due to the steep gradient at the matrix-fracture interface (9). After the small blocks are completely contaminated the gradient at their matrix-fracture interface vanishes and no further matrix diffusion takes places in the small blocks. Therefore, the contaminant entering the right boundary increases. For later time steps also the blocks of medium size are contaminated but there are still mass losses into the large blocks. In our study those large blocks are not totally contaminated at the end of the simulation and a long tailing of the BTC can be observed. After 1.2E+08 seconds the BTC still has not reached the level of the input concentration.

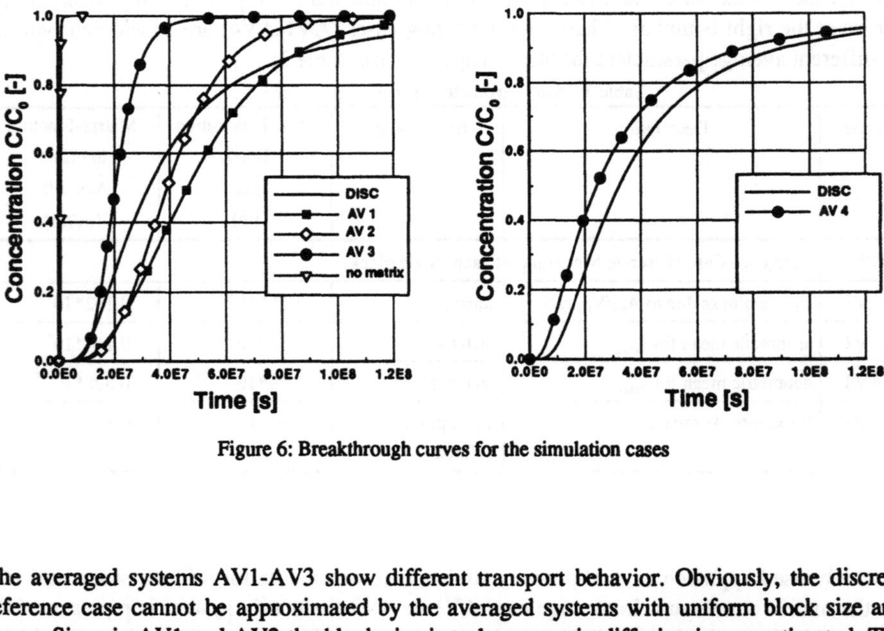

Figure 6: Breakthrough curves for the simulation cases

The averaged systems AV1-AV3 show different transport behavior. Obviously, the discrete reference case cannot be approximated by the averaged systems with uniform block size and shape. Since in AV1 and AV2 the block size is to large matrix diffusion is overestimated. The effect of the small blocks is not considered. For AV3 the block size is to small to account for the effect of matrix blocks of medium size. The BTC is to fast. Due to the lack of the large blocks the averaged systems in AV1 to AV3 are not capable of describing the strong tailing effect.

AV4 shows the best agreement with the reference case. In AV4 the 'Proximity-Function' approach was applied to account for the geometric matrix heterogeneities in a continuum approach. The result is one proximity-function Prox(s) considering the distribution of block geometries (fig. 8). Therefore, the effects of the small and of the large blocks can be addressed.

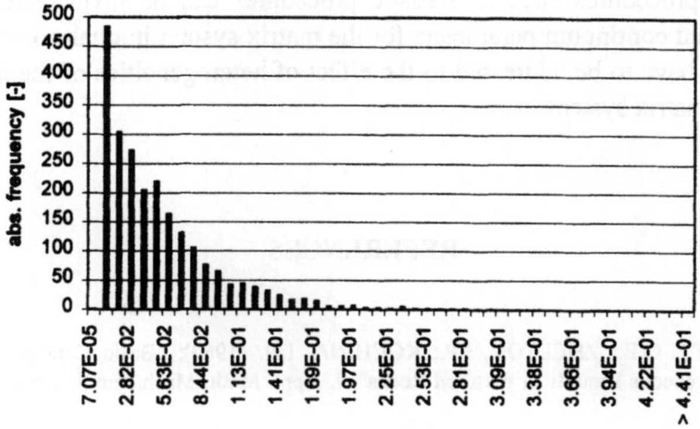

Figure 7: Distribution of block size

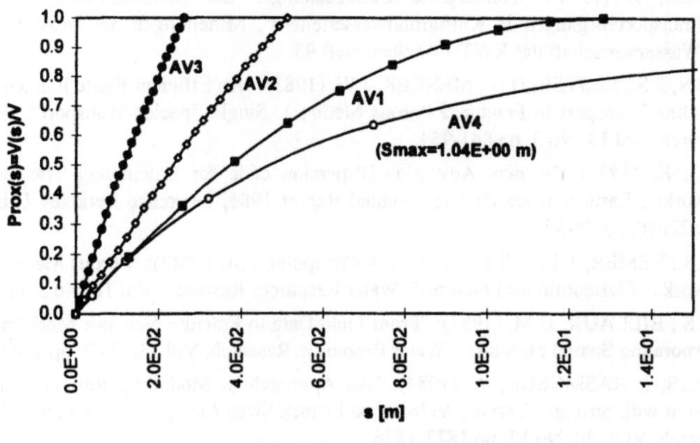

Figure 8: Proximity-functions Prox(s) for simulation cases AV1-AV4

CONCLUSIONS

Numerical tracer experiments were performed using a Lagrangian-Eulerian Finite Element code to investigate the effects of matrix heterogeneities in fractured porous formations. Hydrogeological parameters were assumed to be constant. Only the heterogeneities of the geometric properties (block size and shape) were addressed. A model with discrete representation of both fracture and matrix system served as reference case to compare different averaging approaches for the matrix system. Three averaging approaches were studied neglecting matrix heterogeneities and one averaging approach considering the matrix heterogeneities by using the concept of the 'Proximity-Function'. The first three models are not capable of describing the transport behavior of the discrete system. Only the averaging case considering the matrix heterogeneities shows good qualitative agreement with the reference case.

Other averaging procedures such as statistic procedures can be investigated to determine adequate equivalent continuum parameters for the matrix system in dual-porosity approaches. The same studies have to be addressed to the effect of heterogeneities of the hydrogeological properties of the matrix system.

REFERENCES

[1] BARENBLATT, G.E.; ZHELTOV, I.P.; KOCHINA, I.N. (1960): "Basic Concept on the Theory of Homogeneous Liquids in Fissured Rocks", J. Appl. Math. Mech., Engl. Transl., Vol 20, pp 852-864.

[2] BIBBY, R. (1981): "Mass Transport of Solutes in Dual-Porosity Media", Water Resources Research, Vol 17, No 4, pp 1075-1081.

[3] BIRKHÖLZER, J. (1994): "Numerische Untersuchungen zur Mehrkontinuumsmodellierung von Stofftransportvorgängen in Kluftgrundwasserleitern", Mitteilungen des Instituts für Wasserbau und Wasserwirtschaft der RWTH Aachen, Heft 93.

[4] HUYAKORN, P.S.; LESTER, B.H.; MERCER, J.W. (1983): "An Efficient Finite Element Technique for Modeling Transport in Fractured Porous Media: 1. Single Species Transport", Water Resources Research, Vol 19, No 3, pp 841-954.

[5] KARASAKI, K. (1987): "A new Advection-Dispersion code for Calculating Transport in Fracture Networks", Earth Science Division Annual Report 1986, Lawrence Berkeley Laboratory Report LBL-22090, pp 55-57.

[6] LONG, J.C.S.; REMER, J.S.; WILSON, C.R.; Witherspoon, P.A. (1982): "Porous Media Equivalents for Networks of Discontinous Fractures", Water Resources Research, Vol 18, No 3, pp 645-658.

[7] LONG, J.C.S.; BILLAUX, D.M. (1987): "From Field Data to Fracture Network Modeling: An Example Incorporating Spatial Structure", Water Resources Research, Vol 23, No 7, pp 1201-1216.

[8] NERETNIEKS, I; RASMUSON, A. (1984): "An Approach to Modelling Radionuclide Migration in Medium with Strongly Varying Velocity and Block Sizes Along the Flow Path", Water Resources Research, VOL 20, No 12, pp 1823-1836.

[9] NEUMAN, S.P. (1984): "Adaptive Eulerian-Lagrangian Finite Element Method for Advection-Dispersion", Int. Journal for Numerical Methods in Engineering, Vol 20, pp 321-337.

[10] PRUESS, K.; KARASAKI, K. (1982): "Proximity Functions for Modeling Fluid and Heat Flow in Reservoirs with Stochastic Fracture Distribution", Proceedings Eighth Workshop Geothermal Reservoir Engineering, Stanford University, pp 219-224.

[11] SUDICKY, E.A.; FRIND, E.O. (1982): "Contaminant Transport in Fractured Porous Media: Analytical Solution for a System of Parallel Fractures", Water Resources Research, Vol 18, No 6, pp 1634.1642.

[12] TANG, D.; FRIND, E.O.; SUDICKY, E.A. (1981): "Contaminant Transport in Fractured Porous Media: Analytical Solution for a Single Fracture", Water Resources Research, Vol 17, No 3, pp 555-564

[13] WARREN, J. E.; ROOT, P.J. (1963): "The Behavior of Naturally Fractured Reservoirs", Soc. Petrol. Eng. J., Vol 3, pp 245-255.

An Aligned 3D-Finite-Volume Method for Convection-Diffusion Problems

Klaus Johannsen

Institut für Computeranwendungen 3, Universität Stuttgart, Pfaffenwaldring 27, D-70569 Stuttgart

SUMMARY

A new construction for the finite volume method is presented. It is applicable to 2D and 3D unstructured meshes and uses finite volumes aligned to a given velocity. In the case of a stationary scalar convection-diffusion equation it provides a better approximation than standard finite volumes. In the special case of grid-alignment the discretization of the convective part results in a two-point-star if full upwinding is used. Numerical results are presented demonstrating the improved approximation properties.

Key Words: convection-diffusion-equation, finite volumes, grid-alignement

1. Introduction

The aim of this paper is to present a new discretization method for convection-diffusion equations. The method can handle unstructured 2D meshes (triangles) as well as unstructured 3D meshes (tetrahedra). It is based on finite volumes (FV) introduced by Baliga/Patankar in 1980 [1]. If standard FVs (see [1, 5, 6, 7]) are used, the method suffers from strong numerical diffusion even in the case when convection is aligned with the grid. To overcome this problem, a construction of finite volumes aligned with the convecting velocity is presented which reduces the numerical diffusion significantly. In the case of grid-alignment, there is no numerical diffusion. With these aligned finite volumes (AFV) the continuous equation is discretized using a basis of nodal linear finite elements. If a full upwind scheme is used in combination with the AFVs to overcome the problem of spatial oscillations, the formulation leads to a monotone scheme which was already proposed in 1986 by Schneider/Raw [6] for the 2D case. The use of a conventional skewed upwinding (i.e., taking into account the behaviour of the solution within one triangle/tetrahedron) leads to a non-monotone scheme similar to the skewed upwind schemes considered in [5, 6] (and even though not dealing with upwinding of integration-points but with an evaluation of special finite elements: [1]). In this case the AFVs lead to similar solutions even though the sparse matrix structure is reduced. Numerical results on the base of the AFVs for the unstructured 3D case are shown and compared with standard finite volumes (using full upwinding in both cases).

Throughout this paper we consider the scalar convection-diffusion equation

$$-\varepsilon\Delta u + v \cdot \nabla u = f \quad \text{in } \Omega \tag{1}$$

with Dirichlet boundary conditions and Ω a subdomain of $\Re^2$ or $\Re^3$. First we discretize (1) by means of the standard FV method. Next we show the construction of the *AFVs* for the 2D case (triangles). Using full upwinding the scheme reduces to a method proposed by Schneider/Raw. In section 2 we show the construction for the 3D case (tetrahedra). After a discussion of some aspects of the AFVs (section 3) we present some 3D examples on the convection-diffusion equation with discontinuous Dirichlet boundary conditions in section 4. Finally we make some concluding remarks in section 5.

2. The Finite Volume Method

We shall first present the standard finite volume discretization of (1) following the construction in [4] for the 2D case. For simplicity, we assume the domain Ω to be a polygon. For every triangulation *T* we define a dual mesh consisting of control volumes b_i for every node i of the mesh. The control volumes are chosen as polygons with a boundary consisting of two straight lines in each related triangle. The lines are defined by the midpoints of the edges and the barycentre of the triangles. Fig. 1 shows a typical triangulation with its dual mesh.

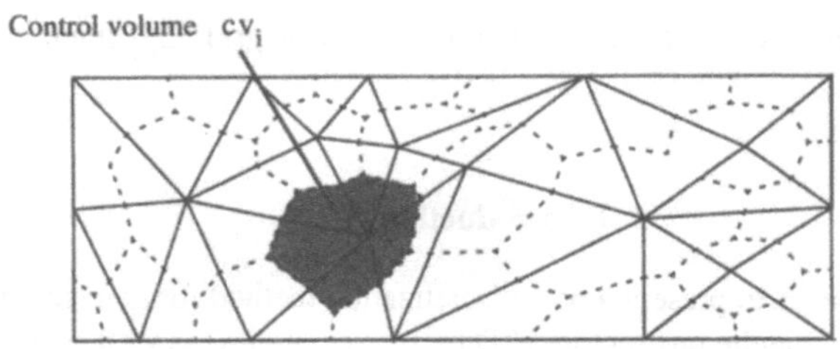

Figure 1: Triangulation with its dual mesh

Following [4] we use piecewise linear ansatz function represented in terms of the corresponding nodal basis $\{\varphi_i\}$. Integration of the convection-diffusion equation (1) over a control volume b_i and Gauss' Theorem leads to

$$\sum_j u_j \left[\int_{\partial b_i} (-\varepsilon \nabla \varphi_j) \cdot d\sigma + \int_{\partial b_i} (v \cdot \varphi_j) \cdot d\sigma \right] = \int_{b_i} f \, dx \,. \tag{2}$$

This is the discretized form of (1) with the disrete solution

$$u = \sum_j u_j \varphi_j \,. \tag{3}$$

Assuming ε to be constant, the flux integral for the diffusive part in (2) (the first term in square brackets on the left-hand side of (2)) is carried out by using the midpoint rule for every straight line of the boundary of the control volume:

$$\int_{\partial b_i} \nabla\varphi_j \cdot d\sigma = \sum_k l_k n_k \cdot \nabla\varphi_j(mp_k) \tag{4}$$

where the sum on the right-hand side runs over all straight lines forming the boundary of b_i, l_k denotes the length of line k, n_k the outer normal vector of the control volume in the midpoint ξ_k of the line. This discretization of the Laplace operator leads to a second order approximation (see [4]).
Assuming the velocity v to be constant, we approximate the flux integral for the convective part (second term on the left-hand side of (2)) by

$$\int_{\partial b_i} (v\varphi_j) \cdot d\sigma \approx v \cdot \sum_k l_k n_k \cdot \varphi_j(\xi_k(mp_k)) . \tag{5}$$

Again l_k, n_k are the length and the outer normal of the line k. The choice of the „upwind point" $\xi_k(mp_k)$ determines the discretization scheme:

a) $\xi(mp_k) = mp_k$ leads to a centered scheme (6)

b) $\xi(mp_k) = su_k$ leads to a skewed upwind scheme (7)

c) $\xi(mp_k) = fu_k$ leads to a full upwind scheme . (8)

The different possibilities (6) - (8) are illustrated by an example in Fig. 2 showing a part of a control volume (shaded area). For the general case we refer to [1, 6].

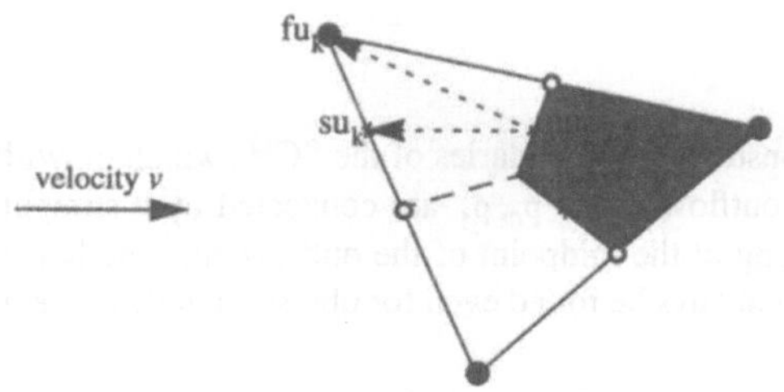

Figure 2: Upwind integration points

Since we are only considering problems with Dirichlet boundary conditions we shall not discuss the construction of the discretization for nodes lying on the boundary of the domain Ω. It is straightforward.

2.1 The Aligned Finite-Volumes in 2D

In this subsection we shall construct the aligned finite volumes (AFVs) for the 2D case. It is clear from the previous discussion that the construction of the dual mesh for a triangulation T can be done by partitioning each triangle of T into three closed sets called subcontrol volumes (SCV). These three subcontrol volumes cover the triangle completely and intersect at most on

their boundaries (see Fig.2). The idea is to construct SCVs whose boundaries within the triangle are maximally aligned with the velocity v. The idea of this method is due to P. Bastian [2].
We have to consider three cases. For each case the midpoints of the edges of the triangle lie on the boundary of at least two of the three SCVs.

a The triangle has two inflow- and one outflow-edge, w.r.t. the velocity v.
b The triangle has one inflow- and two outflow-edges, w.r.t. the velocity v.
c The triangle has one inflow-, one outflow-edge and one edge is parallel to the velocity v.

Case *a*
We construct the boundaries of the SCVs which lie within the triangle. First the midpoints of the two inflow-edges p_0, p_2 are connected by a straight line. Then we connect this line by a line starting at the midpoint of the outflow-edge p_1 being parallel to the velocity v. This second line can always be found even for obtuse triangles. The desired partition is shown in Fig.3.

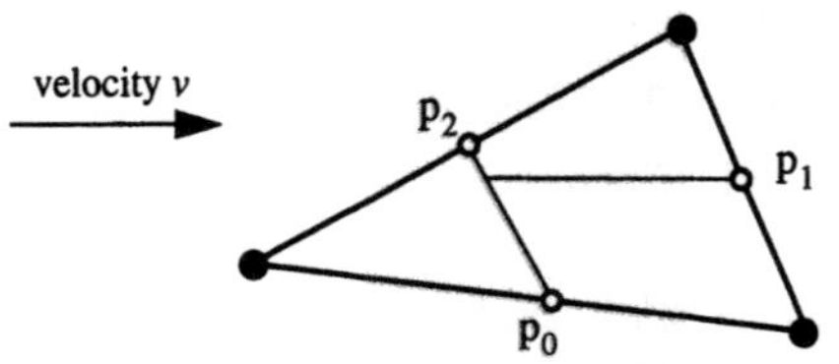

Figure 3: SCVs for a triangle with two inflow- and one outflow-sides

Case *b*
Like in the previous case we construct the boundaries of the SCVs which lie within the triangle. First the midpoints of the two outflow-edges p_0, p_1 are connected by a straight line. Then we connect this line by a line starting at the midpoint of the outflow-edge p_2 being parallel to the velocity v. This second line can always be found even for obtuse triangles. The desired partition is shown in Fig.4.

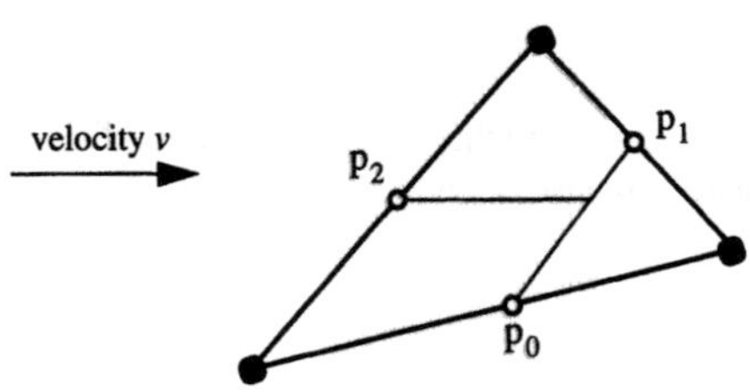

Figure 4: SCVs for a triangle with one inflow- and two outflow-sides

Case *c*

Again we construct only the boundaries of the SCVs which lie within the triangle. First the midpoints of the inflow-edge p_2 and the outflow-edge p_1 are connected by a straight line. This line is parallel to the velocity v. Then we connect the midpoint of that line and the midpoint of the aligned edge p_0 by a straight line. The desired partition is shown in Fig.5.

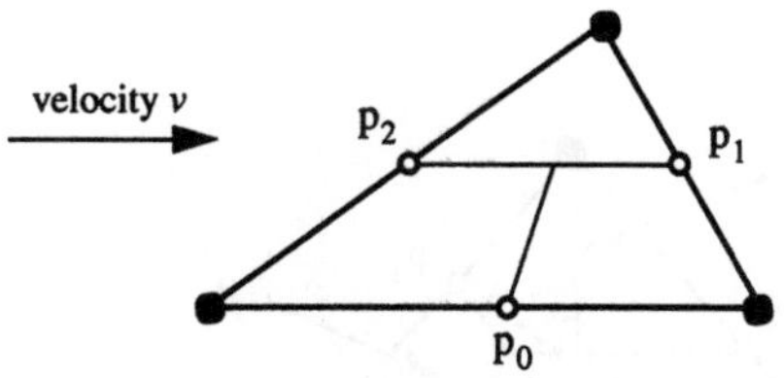

Figure 5: SCVs for an aligned triangle

Remarks:

+ The discretization of the Laplace operator both with standard FV and AFV lead to the same discrete operator. This is due to the fact that the linear ansatz functions are harmonic within the triangles and that in both cases the boundary of the control volumes have the midpoints of the edges of the triangles in common (see [4]).
+ The upwind schemes (5) - (8) are well defined for the AFVs.
+ The AFV discretization of the convective part of (1) leads to a two-point-star for the full upwind scheme ((5), (8)). Hence, there is no numerical diffusion.

Comparison with the Method proposed by Schneider/Raw

In 1986 Raw and Schneider proposed a scheme to discretize convection on the basis of barycentric finite volumes (called *Skewed, Positive Influence Coefficient Upwinding Procedure* [6]), and demonstrated their scheme on quadrilateral meshes. Fig. 6 illustrates their construction. They used the standard barycentric finite volumes with a special upwinding procedure. The solid lines in Fig. 6 show the boundaries of the SCVs lying in a triangle (see also Fig. 2). The flux from SCV d_1 to SCV d_2 is defined to pass through the line $\overline{bp_1}$. The idea is that only upwind-information from $\overline{p_2n_1}$ determines the flux.

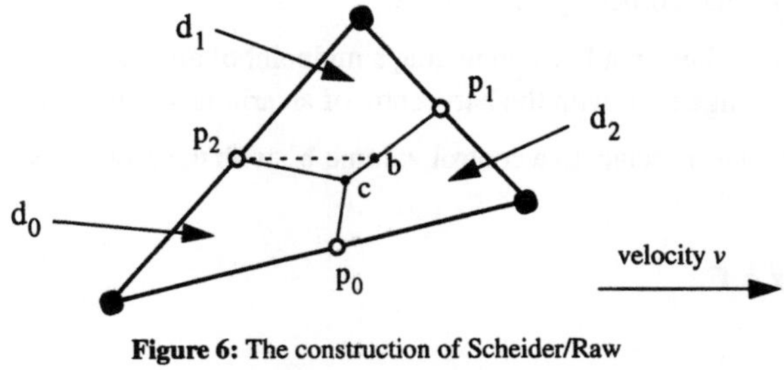

Figure 6: The construction of Scheider/Raw

The flux from d_0 to d_2 is defined not only by the flux passing through $\overline{p_0c}$ but additionally by the flux over $\overline{p2c}$ (≡ flux over $\overline{cb}$). Finally they define the flux from d_0 to d_1 to be zero (the flux which enters over $\overline{p2c}$ leaves over $\overline{cb}$). To get a monotone scheme they used full upwinding (5), (8)). Thus their scheme is equivalent to the aligned finite volumes (redrawn in Fig. 7) using the standard way to determine the convective fluxes). In Fig. 7 the centernode c has been moved from the barycentre to a downwind point of p_2, here coinciding with the centernode of the AFV (see Fig. 4).

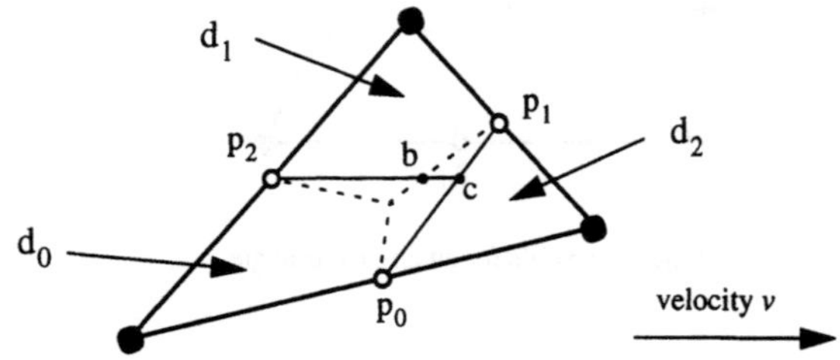

Figure 7: Equivalence of the AFV with the scheme of Scheider/Raw

For this special case the scheme proposed by Schneider/Raw is equivalent to the AFV method with full upwinding. The other cases can be proved the same way.

2.2 The Aligned Finite-Volumes in 3D

We shall now construct aligned finite volumes for tetrahedra in 3D. The construction is similar to the one in 2D. The idea is to define subcontrol volumes whose boundaries within a tetrahedron are maximally aligned with the velocity.

For simplicity we assume the domain Ω to be a polyhedron. Hence it can be covered by a triangulation

$$T = \{t_1, t_2, \ldots, t_n\}. \tag{9}$$

Let

$$C_t := \{x \in \overline{\Omega} \mid x \text{ is a corner of } t\} \quad \text{for} \quad t \in T \tag{10}$$

$$C := \bigcup_{t \in T} C_t \tag{11}$$

$$T_x := \{t \in T \mid t \text{ has corner } x\} \quad \text{for} \quad x \in C \tag{12}$$

$$a_t := \{x \in \partial t \mid x \text{ lies on a line connecting a midpoint of an arbitrary edge of } t \text{ with the barycentre of an arbitrary side of } t\}. \tag{13}$$

With every $x \in C$ we associate to a control volume b_x with the properties

$$b_x \subset \Omega \tag{14}$$

$$b_x = \overline{b_x} \quad \forall x \in C \tag{15}$$

$$\partial b_x \text{ is piecewise smooth } \quad \forall x \in C \tag{16}$$

$$x \in b_x \quad \forall x \in C \tag{17}$$

$$b_x \text{ and } b_y,\ \forall x, y \in C, x \neq y \text{ intersect at most on their boundary} \tag{18}$$

$$b_x \subset \bigcup_{t \in T_x} t \quad \forall x \in C \tag{19}$$

$$\bigcup_{x \in C} b_x = \Omega \tag{20}$$

$$\partial b_x \cap \partial t \subset a_t \quad t \in T_x \ \wedge \ x \in C_t \,. \tag{21}$$

We define the dual mesh *B* of *T* by

$$B := \{b_x \mid x \in C\} \tag{22}$$

the subcontrol volume by

$$d_{t,x} := (t \cap b_x \,,\ t \in T, x \in C_t, b_x \in B\ ) \tag{23}$$

and the separation surface of two corners of a tetrahedron by

$$s_{t,x,y} := d_{t,x} \cap d_{t,y} \quad \text{for} \quad t \in T, \quad x, y \in C_t, x \neq y \,. \tag{24}$$

For each tetrahedron *t* there are six separation surfaces. Each of them corresponds to that edge of *t* having the two corners from definition (24) as endpoints.
Note that the intersection

$$\partial t \cap b_x \quad \text{for} \quad t \in T, x \in C$$

is welldefined by the definitions (9) - (24) above. The separation surfaces within each tetrahedron need to be fixed in order to uniquely specify a finite volume method. The standard finite volumes in 3D (see [3]) satisfy the conditions (9) - (24). Fig. 8 shows one of the six separation surfaces defining the standard finite volumes.

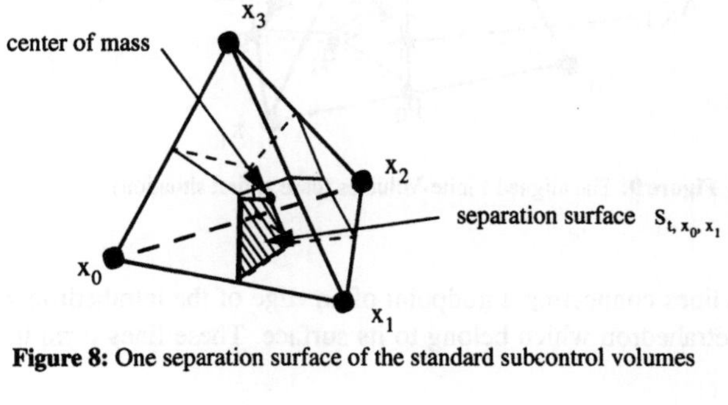

Figure 8: One separation surface of the standard subcontrol volumes

The separation surfaces defining the standard finite volumes are flat quadrilaterals. The corner points of each of them are: the center of mass, the midpoint of the corresponding edge and the barycentres of the two sides containing the edge.
The discretization (5) - (8) of the convective part of (1) leads to a full discretization-star (usually a 14 point-star for unstructured tetrahedra-meshes). This is due to the fact that it is only for special cases (the velocity is parallel to one of the separation surfaces) that the convective flux through a separation surface vanishes.
The aligned finite volumes are defined by a different choice of the separation surfaces. Their construction is similar to the one for the 2D case. We have to consider three different situations for which we show the construction in detail.

<u>*a*</u> The tetrahedron has two inflow- and two outflow-sides, w.r.t. the velocity.
<u>*b*</u> The tetrahedron has one inflow- and three outflow-sides, w.r.t. the velocity.
<u>*b'*</u> The tetrahedron has three inflow- and one outflow-sides, w.r.t. the velocity.
<u>*c*</u> One or more sides of the tetrahedron are parallel to the velocity.

Case *a*
To illustrate the construction, the projection of a tetrahedron on the plane perpendicular to the velocity v is shown in Fig. 9. The velocity is supposed to point out of the drawing plane. Therefore the surfaces (x_0, x_1, x_3) and (x_1, x_2, x_3) are outflow-surfaces, the surfaces (x_0, x_1, x_2) and (x_0, x_2, x_3) are inflow-surfaces. The points p_0, p_1, p_2 and p_3 are the corners of a flat quadrilateral:

$$S := \{x \in t | x \text{ is point of the quadrilateral with the corners } p_0, p_1, p_2, p_3\} . \qquad (25)$$

The four surfaces indicated by arrows ($A_{0,1}, A_{1,2}, A_{2,3}$ and $A_{0,3}$) are closed subsets of S. Note that no one of the points $q_0, q_1, q_2, q_3, p_4, p_5$ belongs to S.

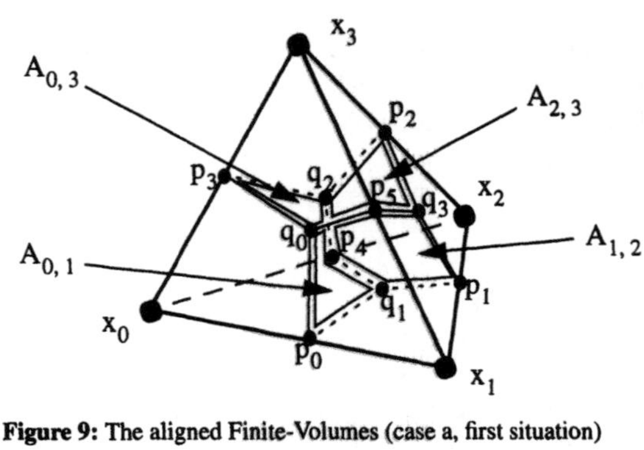

Figure 9: The aligned Finite-Volumes (case a, first situation)

There are twelve lines connecting a midpoint of an edge of the tetrahedron with the barycentre of a side of the tetrahedron which belong to its surface. These lines form the set a_t defined in (13).

Now we define a plane to each of these lines

$$B_{i,j} := \{x \in t \mid \exists y \in \overline{p_i q_j} \ \wedge \ \exists z \in S \ \wedge \ \exists \alpha \in [0,1]$$

with

$$x = \alpha y + (1-\alpha) z \ \wedge \ y\text{-}z \parallel v \},$$

$$(i,j) \in \{ (0,0), (0,1), (1,1), (1,3), (2,2), (2,3), (3,0), (3,2), (4,1), (4,2), (5,0), (5,3) \}$$

and

v velocity.

Now we define the separation surfaces to be

$$s_{t,x_0,x_1} := A_{0,1} \cup B_{0,0} \cup B_{0,1}$$

$$s_{t,x_0,x_2} := B_{4,1} \cup B_{4,2}$$

$$s_{t,x_0,x_3} := A_{0,3} \cup B_{3,0} \cup B_{3,2}$$

$$s_{t,x_1,x_2} := A_{1,2} \cup B_{1,1} \cup B_{1,3}$$

$$s_{t,x_1,x_3} = B_{5,0} \cup B_{5,3}$$

$$s_{t,x_2,x_3} := A_{2,3} \cup B_{2,2} \cup B_{2,3}\,.$$

Note that no convective flux passes through the surfaces $B_{i,j}$. Hence only $A_{0,1}$, $A_{1,2}$, $A_{2,3}$ and $A_{0,3}$ are relevant if convection is discretized. In that case only four of the six possible couplings between the corners are used.

The construction shown in Fig. 9 is one of two possibilities which can occur in the case of two inflow- and two outflow-sides of the tetrahedron. The second one is illustrated by Fig. 10. The velocity is chosen again to be perpendicular to the drawing plane. In addition to the situation in Fig. 9 a new coupling between x_0 and x_2 arises from $A_{0,2}$. $A_{0,2}$ is the subset of Q common to b_{x_0} and b_{x_2}. In principle the construction of the separation surfaces remains the same.

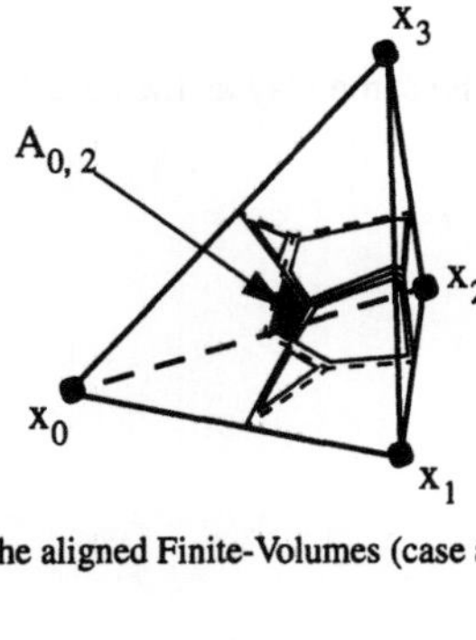

Figure 10: The aligned Finite-Volumes (case a, second situation)

Case *b*

In this case we consider the situation of one inflow- and three outflow-sides of the tetrahedron. Fig. 11 illustrates the situation. It shows again the projection of the tetrahedron to the plane perpendicular to the velocity, which points out of the drawing plane. As in case *a*, we start with the definition of the area S. This time it is not possible to define S to be planar.

Let

$$x_{cm} = 0.25\,(x_0 + x_1 + x_2 + x_3)\,,$$

$$S_{i,j} := \{x \in t \mid x \text{ belongs to the triangle with the corners } x_i, x_j, x_{cm}\}$$

and

$$S := S_{0,1} \cup S_{1,2} \cup S_{2,0}\,.$$

The six surfaces indicated by arrows $(A_{0,1}, A_{0,2}, A_{0,3}, A_{1,2}, A_{1,3}, A_{2,3})$ are closed subsets of S.

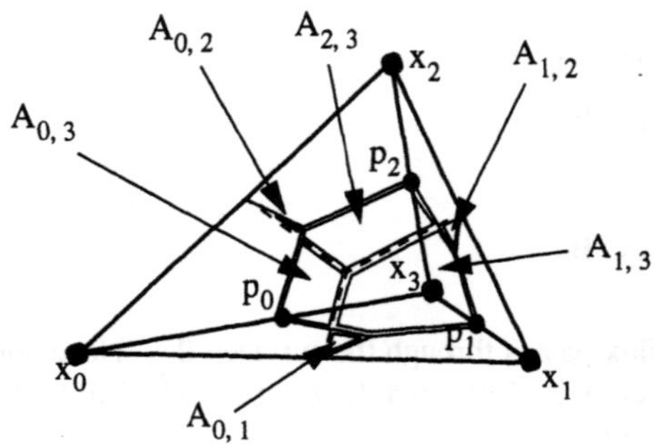

Figure 11: The aligned Finite-Volumes (case b)

The rest of the construction is similar to case *a*.

In this case all six possible connection are used in the discretization of the convection, but three of them are subdominant.

Case b' and c

The case *b'* can be handled exactly the same way as the case *b*. The case c can be regarded as a limiting case of *b*.

3. Aspects of the Aligned Finite Volumes

Using the *AFVs* for the diffusion equation linear ansatz functions are harmonic on each tetrahedron and therefore a discretization of the Laplace operator with finite volumes which satisfy (9) - (24) leads to a matrix which is the same as that obtained for the finite element method with piecewise linear functions. Investigating the approximation of the right-hand side, one can prove first order approximation of the solution in $H_0^1(\Omega)$ (see [4] for the 2D case). In general, we cannot fulfill the equilibrium condition in 2D

$$\mu(t \cap b_x) = \frac{1}{3}\mu(t) \qquad \text{for} \quad x \in C_t, \ \ t \in T$$

and therefore we do not get second order approximation of the solution in $L_2(\Omega)$.

In case that one edge of the tetrahedron is aligned to the convecting velocity only one coupling is generated by that tetrahedron discretizing the convection. That means having a grid which is completely aligned to the velocity, the discretization of the convection results in a two-point star if full upwinding is used. Hence we do not get any no numerical diffusion. Even without grid-alignment numerical diffusion is reduced compared to the standard finite volumes.

4. Numerical results

In this section we compare the standard and the aligned finite volume discretization. Since we are interested in the convection dominated case, we investigate the limit $\varepsilon \to 0$ of (1) on the unit cube

$$\Omega = (0, 1) \times (0, 1) \times (0, 1) \tag{26}$$

using Dirichlet boundary conditions and full upwinding for both schemes. Therefore, the Dirichlet boundary values at the outflow effect the solution only in a boundary layer of thickness h. Hence, the limit $\varepsilon \to 0$ does not cause any numerical difficulties. The calculations were carried out on a y-aligned grid. We consider situations with constant velocity pointing in an arbitrary direction. The mean over a representative set of directions gives then an idea of the approximation properties in the general case. This is examined in the examples 3 and 4.
Since we always deal with discontinuous boundary conditions, the interpolation order of the considered problem is 0.5 in L_2. Therefore no order of approximation larger than this value can be expected.

Example 1

The discretization of equation (1) with $\varepsilon = 0$ on domain (26) was carried out with the velocity and the boundary conditions

$$v = (0, 1, 0)^T$$

$$u_0(x) = \begin{pmatrix} 1 & \text{for } \|x - (1/2, 0, 1/2)\| < 0.15 \\ 0 & \text{otherwise} \end{pmatrix} \quad \text{for} \quad x \in \partial\Omega .$$

The standard finite volumes show a large amount of numerical diffusion whereas the *AFVs* lead to a solution without any numerical diffusion (Fig. 12). This was expected since the grid is aligned with the velocity.

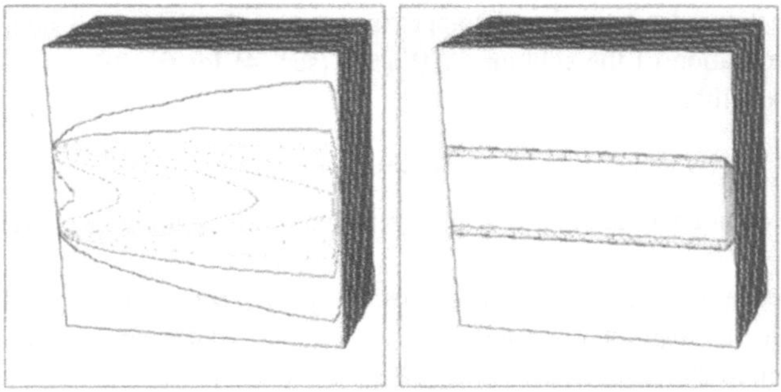

Figure 12: Solution obtained by standard (left) and aligned (right) finite volumes

Example 2

The discretization of equation (1) with $\varepsilon = 0$ on domain (26) was carried out with the velocity and the boundary conditions

$$\vec{v} = (0, 1, 1/4)^T$$

$$u_0(x) = \begin{pmatrix} 1 & \text{for } \|x - (1/2, 0, 3/8)\| < 0.15 \\ 0 & \text{otherwise} \end{pmatrix} \quad \text{for} \quad x \in \partial\Omega .$$

Now the aligned finite volumes also suffer from numerical diffusion, but in compared with standard finite volumes the numerical diffusion is reduced (Fig. 13).

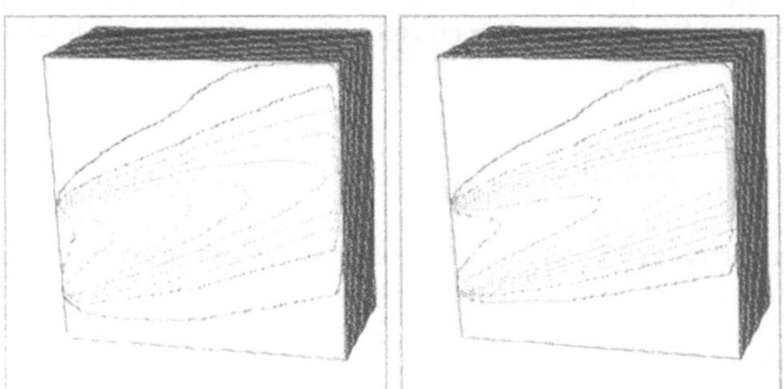

Figure 13: Solution optained by standard (left) and aligned (right) finite volumes

Example 3

We approximate the exact solution with the solution obtained by standard finite volumes and by aligned finite volumes. The exact solution is depending on one parameter specifying the velocity

$$v = (0, \cos\alpha, \sin\alpha)\,,$$
$$\alpha \in \{0.5°, 1°, 1.5°, 2°, \ldots, 179.5°\}$$

and was chosen to be

$$u_\alpha^{exact}(x) = \begin{pmatrix} 0 & \text{if} & x \cdot (n_\alpha - x_{cm}) < 0 \\ 1 & \text{if} & x \cdot (n_\alpha - x_{cm}) \geq 0 \end{pmatrix} \quad \text{for} \quad x \in \partial\Omega$$

with $$n_\alpha = (0, -\sin\alpha, \cos\alpha)\,, \quad x_{cm} = (1/2, 1/2, 1/2)\,.$$

The boundary conditions are the restriction of the exact solution to the boundary

$$u_{0,\alpha}(x) = u_\alpha^{exact}(x) \quad \text{for} \quad x \in \partial\Omega\,.$$

Fig. 14 shows the approximation to the exact solution in the L_2-norm using both discretizations depending on the angle α using a grid with 9x9x9 gridpoints. Fig.15 and 16 show the same results for grids with 17x17x17 resp. 33x33x33 gridpoints. The upper curves in Fig. 14, 15, 16 give the values for the L_2-approximation

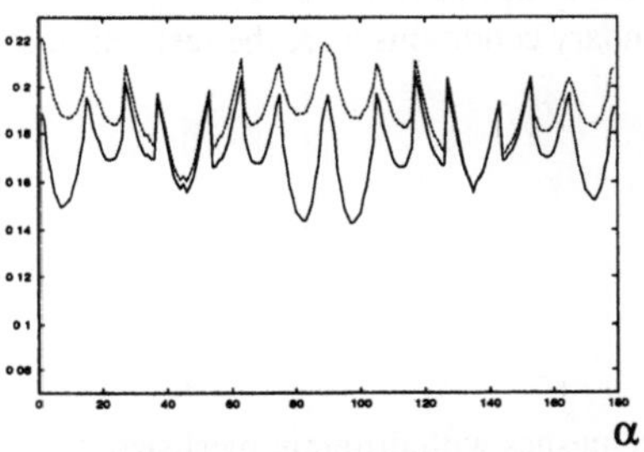

Figure 14: Meshsize: 1/8, 9x9x9 unknowns.

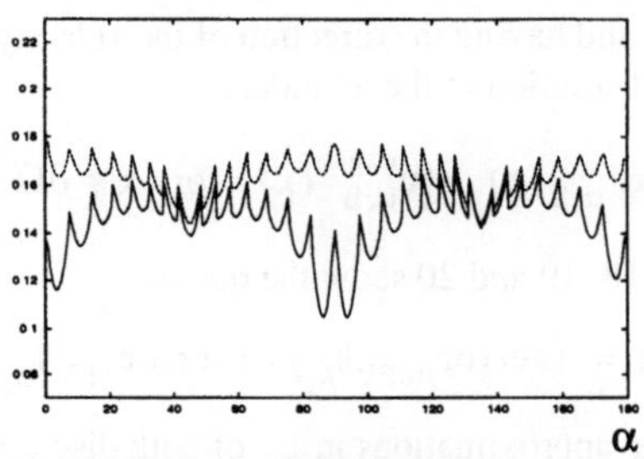

Figure 15: Meshsize 1/16, 17x17x17 unknowns

with standard finite volumes, the lower ones for the L_2-approximation with *AFVs*. Since the plane of discontinuity changes its size depending on the angle, the results have been normalized to a planesize of 1.0.

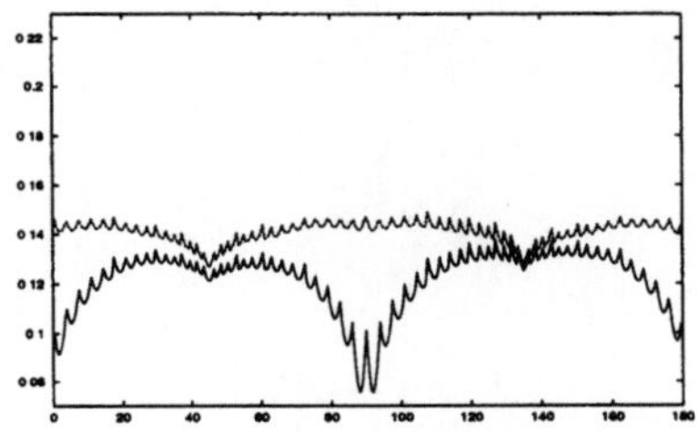

Figure 16: Meshsize 1/32, 33x33x33 unknowns

The interpolation error of a discontinuous function depends not only on the mesh size but also on the position of the discontinuity with respect to the mesh. A simple 1D example shows that the error in L_2 can vary by a factor of two depending on the position of the discontinuity within one grid cell. This effect causes the oscillations in Fig. 14 - 16. Apart from this effect, the best approximation is obtained with AFVs in case of grid-alignment ($\alpha = 90°$). This was expected, since due to the 2-point-star discretization the solution calculated coincides with the interpolated exact solution.

Example 4

We compare the approximation of the calculated solution to an exact solution for both kinds of discretizations. The exact solution depends on two angles specifying the velocity in spherical polar coordinates and was chosen to be

$$u^{exact}_{\alpha,\beta}(x) = \begin{pmatrix} 0 & \text{if} & \|(x - x_{cm}) \times v_{\alpha,\beta}\| > 0.15 \\ 1 & \text{if} & \|(x - x_{cm}) \times v_{\alpha,\beta}\| \leq 0.15 \end{pmatrix} \quad \text{for} \quad x \in \partial\Omega \tag{27}$$

with $\quad x_{cm} = (1/2, 1/2, 1/2)$,

$$v_{\alpha,\beta} = (\sin\alpha\cos\beta, \sin\alpha\sin\beta, \cos\alpha),$$

$$\alpha, \beta \in \{10°, 20°, 30°, \dots, 170°\}$$

which defines the exact solution to be one exactly inside a tube passing through x_{cm} with radius 0.15 and having the direction of the velocity. The boundary conditions were the restriction of the exact solution to the boundary

$$u_{0,\alpha,\beta}(x) = u^{exact}_{\alpha,\beta}(x) \quad \text{for} \quad x \in \partial\Omega.$$

Fig. 18, 19 and 20 show the quotient

$$q = (\|error_{stdFV}\|_{L_2}) / (\|error_{AFV}\|_{L_2})$$

of the approximations in L_2 of both discretizations on meshes with different meshsizes

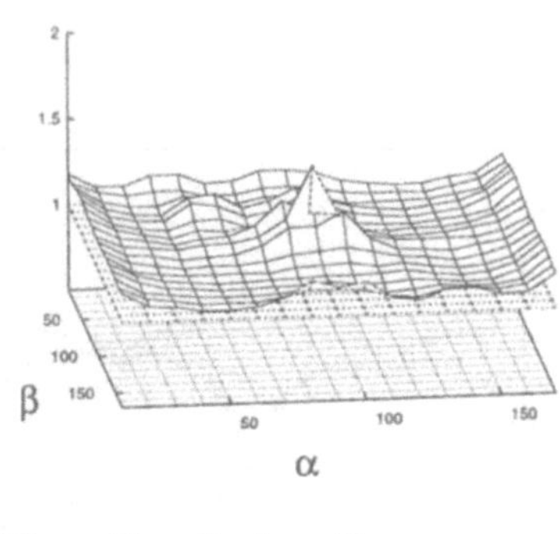

Figure 17: Quotient of L_2-errors of stdFV and AFV, meshsize: 1/8;

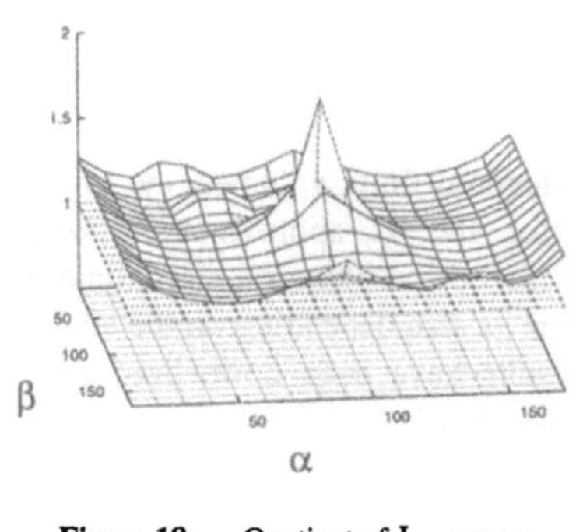

Figure 18: Quotient of L_2-errors of stdFV and AFV, meshsize 1/16;

The effect of grid-alignment can be seen clearly for $\theta = 90°, \varphi = 90°$ in fig. 17-19. Since that peek grows going to finer grids, the order of approximation for the aligned finite volumes is bigger than the one of standard finite volumes. This growth can be observed even in other regions. The surface of discontinuity changes size depending on the angles, therefore the results have been normalized to an area of 1.0.

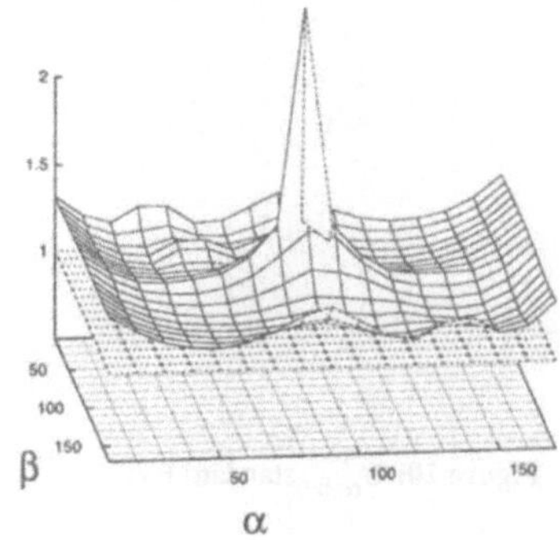

Figure 19: quotient of L_2-errors of stdFV and AFV, meshsize 1/32;

Next we analyse the order of approximation in more detail. The relation of the interpolation errors on two subsequential grids

$$\|(I_{h/2} - 1)\, u_{exact}\|_{L_2} / \|(I_h - 1)\, u_{exact}\|_{L_2}$$

does not neccessarily tend to a fixed number for $h \to 0$. A simple 1D example with a discontinuous function can show that it can vary by a factor of 3. We expect the same behaviour for the errors $\|u_h - u_{exact}\|_{L_2}$. Therefore, we estimate the order of approximation from two subsequent meshes by

$$\sigma^{i,j}_{\alpha,\beta} = \log_2\left(\frac{\left\|e^{h_i}_{\alpha,\beta}\right\|}{\left\|e^{h_j}_{\alpha,\beta}\right\|}\right) - \log_2\left(\frac{\left\|i^{h_i}_{\alpha,\beta}\right\|}{\left\|i^{h_j}_{\alpha,\beta}\right\|}\right) + (j-i)\begin{pmatrix}\text{interpolation}\\ \text{order}\end{pmatrix}$$

with $\quad e^{h}_{\alpha,\beta} := u^{h}_{\alpha,\beta} - u^{exact}_{\alpha,\beta}, \quad i^{h}_{\alpha,\beta} := (I_h - 1)\, u^{exact}_{\alpha,\beta}$

which is more stable. Its limit is the usual order of approximation

$$\sigma_{\alpha,\beta} = \lim_{n \to \infty}\left(\frac{1}{n}\sigma^{0,n}_{\alpha,\beta}\right).$$

We investigated $\sigma^{4,5}_{\alpha,\beta}$ in order to get an idea of the order of approximation. Fig. 20, 21 show the evaluation of $\sigma^{4,5}_{\alpha,\beta}$ for the standard finite volumes and the aligned finite volumes for different parameters α, β on a mesh with $h = 1/32$.

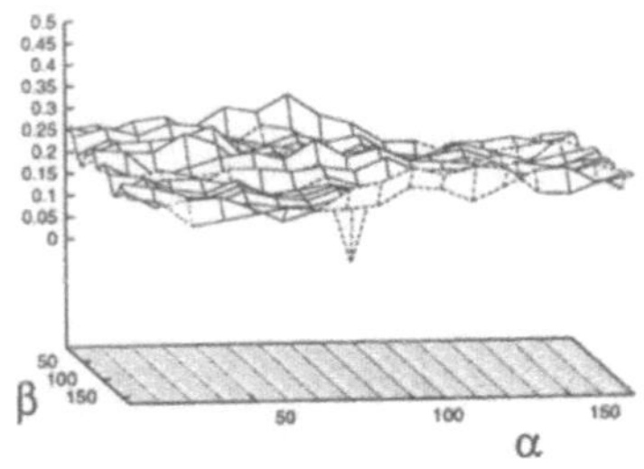

Figure 20: $\sigma_{\alpha,\beta}^{4,5}$, standard FV;

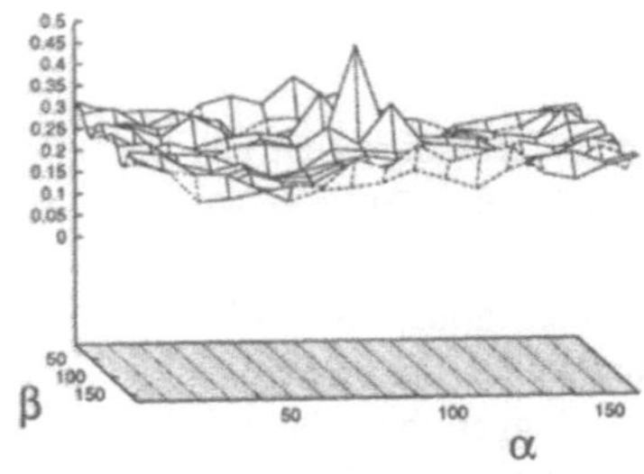

Figure 21: $\sigma_{\alpha,\beta}^{4,5}$, aligned FV;

To have an idea of the order of approximation we approximate

$$\sigma_{4,5}^{m} = \int_0^{2\pi}\int_0^{\pi}\sigma_{\alpha,\beta}^{4,5}\sin\alpha\ d\alpha\ d\beta$$

by a discrete sum over the 17x17 directions (see (27)). We get

$$\sigma_{4,5}^{m} = 0.220 \qquad \text{for standard finite volumes}$$

and

$$\sigma_{4,5}^{m} = 0.263 \qquad \text{for aligned finite volumes.}$$

5. Conclusions

The aligned finite volumes provide a better approximation of convection than the standard finite volumes because of less numerical diffusion. They can be used in combination with standard finite volumes for the diffusion part without loosing the property of local conservation of mass. In the case of grid-alignment one obtains the optimal approximation. Since the method works for unstructured grids, it can be used in combination with alignment strategies for the grid. This would lead to an optimal discretization of equations of convection-diffusion type.

7. References

[1] B. R. Baliga, S. V. Patankar,
A New Finite Element Formulation For Convection-Diffusion Problems, Numer. Heat Transfer **3**, 393-409, (1980)

[2] *P. Bastian*,
private communication.

[3] J. Bey, G. Wittum
Downwind Numbering: A Robust Multigrid Method For Convection Diffusion Problems On Unstructured Grids, ICA Bericht **n95/2**, Universität Stuttgart.

[4] W. Hackbusch,
On First and Second Order Box Schemes, Computing, **41**, p. 277-296, (1989).

[5] G. D. Raithby,
Skew Upstream Differencing Schemes For Problems Involving Fluid Flow, Comput. Methodes Appl. Mech. Eng. **9**, p. 153-164, (1976).

[6] G. E. Schneider, M. J. Raw,
A Skewed Positive Influence Coefficient Upwinding Procedure For Control Volume Based Finite Element Convection Diffusion Computation, Numer. Heat Transfer **9**, p. 1-26, (1986).

[7] Zhiqiang Cai,
On the finite volume element method, Numer. Math. **58**, p 713-735, (1991).

Addresses of the Editors of the Series "Notes on Numerical Fluid Mechanics"

Prof. Dr. Ernst Heinrich Hirschel (General Editor)
Herzog-Heinrich-Weg 6
D-85604 Zorneding
Federal Republic of Germany

Prof. Dr. Kozo Fujii
High-Speed Aerodynamics Div.
The ISAS
Yoshinodai 3-1-1, Sagamihara
Kanagawa 229
Japan

Prof. Dr. Bram van Leer
Department of Aerospace Engineering
The University of Michigan
3025 FXB Building
1320 Beal Avenue
Ann Arbor, Michigan 48109-2118
USA

Prof. Dr. Michael A. Leschziner
UMIST-Department of Mechanical Engineering
P.O. Box 88
Manchester M60 1QD

Prof. Dr. Maurizio Pandolfi
Dipartimento di Ingegneria Aeronautica e Spaziale
Politecnico di Torino
Corso Duca Degli Abruzzi, 24
I-10129 Torino
Italy

Prof. Dr. Arthur Rizzi
Royal Institute of Technology
Dept. of Aeronautics
Aerodynamics Division
S-10044 Stockholm
Sweden

Dr. Bernard Roux
Institut de Mécanique des Fluides
Laboratoire Associé au C.R.N.S. LA 03
1, Rue Honnorat
F-13003 Marseille
France

Brief Instruction for Authors

Manuscripts should have well over 100 pages. As they will be reproduced photomechanically they should be produced with utmost care according to the guidelines, which will be supplied on request.

In print, the size will be reduced linearly to approximately 75 per cent. Figures and diagrams should be lettered accordingly so as to produce letters not smaller than 2 mm in print. The same is valid for handwritten formulae. Manuscripts (in English) or proposals should be sent to the general editor, Prof. Dr. E. H. Hirschel, Herzog-Heinrich-Weg 6, D-85604 Zorneding.

MIX
Papier aus verantwortungsvollen Quellen
Paper from responsible sources
FSC® C105338

If you have any concerns about our products,
you can contact us on
ProductSafety@springernature.com

In case Publisher is established outside the EU,
the EU authorized representative is:
Springer Nature Customer Service Center GmbH
Europaplatz 3, 69115 Heidelberg, Germany

Printed by Libri Plureos GmbH
in Hamburg, Germany